AF352850

# Energy Economics and Policy in Developed Countries

# Energy Economics and Policy in Developed Countries

Editor

**Almas Heshmati**

MDPI • Basel • Beijing • Wuhan • Barcelona • Belgrade • Manchester • Tokyo • Cluj • Tianjin

*Editor*
Almas Heshmati
Professor of Economics,
Jönköping International Business School (JIBS),
Jönköping University
Sweden

*Editorial Office*
MDPI
St. Alban-Anlage 66
4052 Basel, Switzerland

This is a reprint of articles from the Special Issue published online in the open access journal *Energies* (ISSN 1996-1073) (available at: https://www.mdpi.com/journal/energies/special_issues/ Energy_Economics_and_Policy_in_Developed_Countries).

For citation purposes, cite each article independently as indicated on the article page online and as indicated below:

LastName, A.A.; LastName, B.B.; LastName, C.C. Article Title. *Journal Name* **Year**, *Article Number*, Page Range.

ISBN 978-3-03943-246-2 (Hbk)
ISBN 978-3-03943-247-9 (PDF)

# Contents

# About the Editor

**Almas Heshmati** is a Professor of Economics at Jönköping University, Sweden. He held similar positions at the Sogang University, Korea University, Seoul National University, University of Kurdistan Hawler, and MTT Agrifood Research (Finland). He was a Research Fellow at the World Institute for Development Economics Research (WIDER), The United Nations University during 2001–2004. From 1998 until 2001, he was an Associate Professor of Economics at the Stockholm School of Economics. He has a Ph.D. degree from the University of Gothenburg (1994). His research interests include applied microeconomics, globalization, development economics, environmental economics, welfare economics, efficiency, productivity, and growth with applications to manufacturing and services. In addition to more than 150 scientific journal articles, he has published 30 books on the EU Lisbon Process, global inequality, East Asian manufacturing, Chinese economy, technology transfer, information technology, water resources, landmines, power generation, renewable energy, development economics, world values, poverty, well-being, and economic growth.

# Preface to "Energy Economics and Policy in Developed Countries"

Energy use per capita and as a share of GDP in developed countries is high. The high rate is attributed to rapid economic development and welfare that relies heavily on energy-intensive technologies and, in particular, relatively low price and easily accessible fossil fuels. The total energy consumption is continuously increasing but the primary source of electricity generation is increasingly based on renewable and non-fossil fuels. Production and consumption are also increasingly oriented towards the use of more efficient and energy-saving technologies. The global, national, and local environmental degradation and its strong effect on health have promoted the development of institutional capacity, technologies, environmental policies, regulations, and incentive programs to reduce energy use and its negative health and environmental impacts. Efforts are being made to change production and consumption patterns through various command and control policies yet their effects are small and the coordination of costly environmental activities to avert radical climate change is difficult to achieve. The former twin crisis of poverty and environmental degradation is aggravated by the increased gap between north and south and their lack of will and coordination to allocate sufficient investment resources in sustainable development. The world is facing environmental degradation, poverty, inequality, climate change, migration, and frequent natural disasters. This Special Issue invites high-quality research covering a wide range of topics related to energy economics and politics in developed countries. This research is informative on how to encourage sustainable development and facilitate financing technology development, transfer, and applications to mitigate climate change.

**Almas Heshmati**
*Editor*

*Article*

# Development Forecasts for the Zero-Emission Bus Fleet in Servicing Public Transport in Chosen EU Member Countries

Anna Brdulak [1], Grażyna Chaberek [2,*] and Jacek Jagodziński [3]

[1]   Faculty of Computer Science and Management, Wrocław University of Science and Technology, 50-370 Wrocław, Poland; anna.brdulak@pwr.edu.pl
[2]   Faculty of Oceanography and Geography, University of Gdańsk, 81-378 Gdańsk, Poland
[3]   Faculty of Electronics, Wrocław University of Science and Technology, 50-370 Wrocław, Poland; jacek.jagodzinski@pwr.edu.pl
*   Correspondence: grazyna.chaberek@ug.edu.pl

Received: 30 June 2020; Accepted: 12 August 2020; Published: 16 August 2020

**Abstract:** Nearly two-thirds of the emissions that cause smog come from road transport. In April 2019, the European Parliament adopted new regulations on public procurement to encourage investment in clean buses—electric, hydrogen, or gas. Directive 2009/33/EC is to apply from the second half of 2021. The aim of this article is to make an attempt to simulate the number of zero-emission buses (ZEB) in European Union (EU) member countries in two time horizons: 2025 and 2030, and to forecast the number of clean vehicles in the precise time horizons, including before and after 2050. Research questions are as follows: (1) what will be the number of ZEBs in individual EU countries over the next few years; (2) which of the EU countries will reach by 2030 the level of 95% share of ZEBs in all buses, which are a fleet of public transport buses; and (3) in which year will which EU countries reach the level of 95% share of zero-emission buses. The method used is a Bass model. The conducted analyses demonstrate that, by 2050, only four of the EU members will be able to reach 95% level of share of clean buses in the city bus transport fleets. It is likely that other countries may not achieve this even by 2050.

**Keywords:** electric buses; zero-emission buses (ZEB); clean buses; EU policy; zero emission policy; green energy; city management; simulation model; strategy; sustainable development

---

## 1. Introduction

Owing to the significant importance of greenhouse gas emissions for climate change, in particular carbon dioxide, arising during the combustion of solid fuels in transport and the process of electricity, or, heat production, many countries have taken steps to consciously reduce harmful emissions [1]. The European Union (EU) is a particularly active entity in international relations, taking active measures to combat climate change. It aims to create a low-carbon economy in the long term.

The European Commission wants Europe to become climate neutral by 2050. Therefore, the EU has set itself targets for a gradual reduction of greenhouse gas emissions by 2050. The main climate and energy goals have been set out in two documents: the climate and energy package until 2020 [2,3] and under the 2030 climate and energy policy. The assumptions of the climate and energy package were determined by EU leaders in 2007, and in 2009, regulations were adopted in this respect. At the same time, there are the main goals of the Europe 2020 strategy for smart, sustainable, and inclusive growth. The main goals are as follows: a 20% reduction in greenhouse gas emissions (compared with 1990 levels), a 20% share of energy from renewable sources in total energy consumption in the EU, and a 20% increase in energy efficiency [3]. In October 2014, this policy framework was adopted by the Council.

The renewable energy and energy efficiency targets were increased in 2018 [4]. Currently, under the 2030 climate and energy policy, the EU plans to reduce gas emissions by at least 40%. Greenhouse gas emissions (compared with 1990 levels) should increase to at least 32% of the share of energy from renewable sources in total energy consumption. An increase of at least 32.5% in energy efficiency, together with a clause should enable this target to be achieved by 2023. Thus, the original target of at least 27% was corrected in 2018.

According to the management system, Member States are required to adopt integrated national energy and climate plans for 2021–2030 and to develop long-term national strategies, including ensuring coherence between these strategies and their national energy and climate plans. A common approach for the period up to 2030 helps to guarantee regulatory certainty for investors and coordination of the actions of the EU countries. This framework is conducive to changes towards a low-carbon economy and the creation of an energy system.

The upcoming EU Budget and in particular the EU Regional Development Funds spending plans (Operational Programs) for 2021–2027 (to be prepared by the Member States in 2020) also offer a range of opportunities to increase both the climate ambition and implementation of the measures foreseen in the National Energy and Climate Plans (NECPs). Under EU legislation, the EU's current economy broad 40% emission reduction target consists of sector contributions covered by its Emissions Trading System (ETS), mainly the energy and industry sectors. It also consists of the other remaining sectors, such as agriculture, construction, waste, and transport.

Transport is currently responsible for a significant proportion of $CO_2$ emissions. Forecasts assume that, by 2050, carbon dioxide emissions from this sector will increase from 6–7 gigatons to 16–18 gigatons. In addition, around 30% of Europeans live in cities where air pollution exceeds EU quality standards. Conventional fuels burned by buses are one of the largest sources of $CO_2$, nitrogen oxides, and particulate emissions [5].

In this context, the development of a sustainable public transport system is of key importance. The deployment of zero-emission buses to fleets is today a priority for many urban centers around the world. Metropolises see the development of green transport as a basic instrument for combating air pollution. More than 80 cities worldwide have joined the network of C40 Cities Climate Leadership Group. "The cities use to reduce emissions from transportation include switching to effective modes (e.g., public transit or non-motorised transportation) and enhancing the efficiency of fleets via shifting to zero-emission technologies" [6]. According to the Bloomberg New Energy Finance report [7], the total number of buses with electric drive (e-buses) will increase from 386,000 units in 2017 to around 1.2 million in 2025. The share of electrified buses in the global fleet will reach 47% [5]. It is also a solution decided upon by EU member states. The advantages of zero-emission vehicles are being noticed by more and more cities that decide to operate them. Thus, the share of e-buses in urban transport fleets is growing [8,9].

In the short term, the introduction of clean buses can contribute to the implementation of EU 2020 and 2030 targets, as well as national targets and local targets for $CO_2$, air quality, and noise in several ways. On the basis of the '2030 Climate and Energy Policy Framework' [10], at least 80% of the transport work in public collective transport is to be carried out using means of transport that are not powered by conventional fuels. In addition, by 2030, $CO_2$ emissions from the transport sector are expected to be reduced by 40% [5]. The introduction of electric buses to public transport fleets will also allow city authorities to reduce the amount of energy consumed.

In reference to the problems raised, in this article, the authors focused on the forecast of the number of zero-emission buses in individual EU countries by 2025 and 2030, respectively. As mentioned previously, the EU strategies assume two time horizons, 2020 and 2030. Owing to the fact that the most current data, which the authors used to create the simulation, refer to the period 2013–2018, from 2019 and later, a forecast is presented. In order to make it credible and focus on two time horizons that best correspond to the developed EU strategies, the years 2025 and 2030 were taken into consideration. Given the scale of energy consumption by cities in a global perspective, one of the fundamental challenges that the city authorities face is the reduction of energy consumption [11]. The topic taken up by the authors is directly related to the energy consumption market.

It should be emphasized that the vehicles powered by alternatives to the conventional fossil-fuelled engines are a fairly diverse group of vehicles subject to different definitions and classifications. The most promising technologies for use in public bus transport are battery-electric and hydrogen fuel-cells powered engines, which are more energy-efficient and far less pollutant than the conventional diesel engines. Additionally, such vehicles have specific advantages over trolley buses and trams, such as the flexibility of use of road infrastructure without the need for powerlines or rails [12]. In this study, the authors will use the term zero-emission buses (ZEBs), which specifies a group of buses using either of these two fuel technologies, as neither type generates any pollutant emission [12–14].

Such technology applied in public transport is an innovation. Bezruchonak [6] conducted an analysis of the geographical distribution of electric buses in European countries and took into account European cities till 2018. According to him, the increase in European stock suggests that the European market is moving beyond the demonstration phase and into commercial development, and by 2030, the share of battery-electric buses will reach 50%. The United Kingdom, the Netherlands, Germany, Spain, Sweden, Poland, and Lithuania are the major European markets that order and operate fleets of electric buses. However, it is difficult to predict how the new technology will be adopted to the market. The process of adopting an innovation by the market is particularly important from the investor's point of view. In this case, it is extremely important to predict the development of this technology owing to the enormous costs associated with constructing the essential infrastructure and the fact that financing of zero-emission, electric technology in public transport is based almost exclusively on public funds. As investigated by Brozynski and Lejbowicz, predicting the adoption of electric technology in transport is of great importance in investment decisions of policy makers. First of all, it is important when investing public funds. As demonstrated in their research, moving the process forward helps to avoid incurring policy costs repeatedly by lingering in stages affected by the policy [15].

Despite the rapid growth of the number of ZEBs, their share in the entire global bus fleet is still marginal [13]. Referring to the problems raised, this article attempts to create a simulation that shows how quickly EU members will be able to replace traditional buses with zero-emission buses and reach 95% of their share in the public transport bus fleet.

There are several models related to forecasting the development of electric transport technologies in the literature that have been extensively described and analyzed. For example, Meade and Islam present a detailed overview of mathematical (deterministic) models describing the accumulation of adoptions [16]. However, the best known and most frequently quoted model is the one proposed by Bass [17]. The Bass model was chosen, based on the existing data and on the fact that it is a deterministic model that provides precise forecasts [11]. This model is often used alongside the so-called logistic and Gompertz projections, while ordinary predictions based on the Bass model are the most pessimistic [18]. This is an additional argument in favor of this model. An additionally significant aspect is the duality of the model with the Rogers model [19]. The Bass model is the most common model in the literature that discusses forecasting the diffusion of innovation in alternative fuel technology, primarily electric propulsion in transport [20]. Most of the research, however, concerns the diffusion of this innovative technology on the individual automotive market [21], especially e-vehicles [22–24], or on the commercial market of logistics services [25]. It is noteworthy that, despite the popularity of the topic of clean buses and extensive discussions on energy reduction in the scientific literature, the simulation of saturation of ZEBs in public transport bus fleets in the EU member countries has not yet been much presented and described.

Considering the above, the purpose of this article is to make an attempt to simulate the number of ZEBs in EU member countries in two time horizons: 2025 and 2030, and to forecast the number of clean vehicles in the precise time horizons, including before and after 2050. On the basis of the simulation, the year in which the selected countries will reach 95% saturation of their public transport fleets with ZEBs will be indicated.

The research questions posed in the article, to which the authors seek to find answers, are as follows:

Q1: What will be the number of zero-emission buses in individual EU countries over the next few years?

Q2: Which of the EU countries will reach by 2030 the level of 95% share of ZEBs in all buses, which are a fleet of public transport buses?

Q3: In which year which will EU countries reach the level of 95% share of ZEBs in all buses, which are a fleet of public transport buses?

## 2. Materials and Methods

The use of the Bass model to predict the development of new technologies is a common approach. Especially in areas related to new technological solutions in the field of energy. The practical use of diffusion models for prediction has nearly 40 years of history. In 1980, the U.S. Department of Energy used the Bass model to evaluate the adoption of solar batteries and delayed the technology's introduction to the market [26]. In December 2019, this method was also used to evaluate the lighting market (LED and other technology) [27]. The report [28] indicates effective methods of predicting the development of new technologies at various stages of innovation development (introduction, increase acceptance of new technology, mature technology). Diffusion models are the only one effective method at each of these stages. In addition, other diffusion models can be mentioned: the Fourt and Woodlock model, Mansfield model, Blackam Model, Fisher and Pry model, Kalish model, and many others (a list of diffusion models can be found, among others, in [29]; in most cases, these are various extensions of the Bass model).

The model application in practice remains an open issue and different forecasters use different approaches. The Bass model parameters can be obtained on the basis of questionnaire research, historical analogies from similar technologies, and fitting the model to the data. Each of the approaches has its advantages and disadvantages [27]. The article uses the approach of fitting the model to data on the initial development of technology. There are also many approaches for the technical aspects of modeling. For example, the method of estimating parameters based on the data can be performed with one of the following methods: ordinary least squares (OLS), maximum likelihood estimation (MLE), nonlinear least squares estimation (NLS), or algebraic estimation. Mahajan et al. [30] show that the best way to estimate parameters is NLS. However, more recent research shows that the best least squares estimate for the Bass model does not necessarily exist [31]. The main inconvenience with the Bass model, encountered in this article, is a very significant change in the shape of the Bass curve along with the extension of the observations number [32].

In order to better illustrate the proceedings taken by the authors, the sequence of individual stages is presented below (Figure 1).

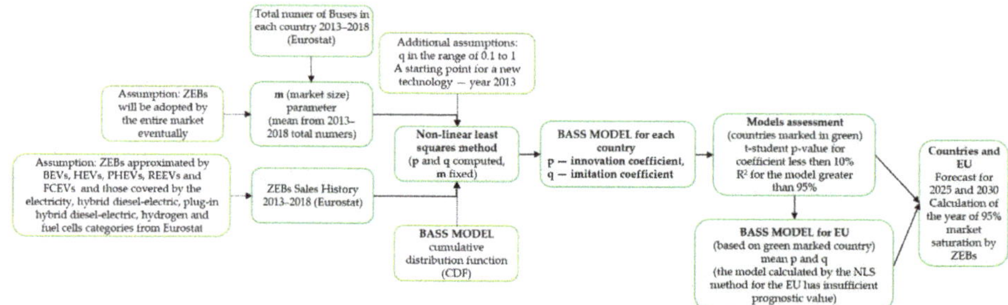


**Figure 1.** Subsequent stages of the proceedings [own study]. ZEB, zero-emission bus; NLS, nonlinear least square; BEV, battery electric vehicle; HEV, hybrid electric vehicle; PHEV, plug-in HEV; REEV, range-extended EV; FCEV, fuel cell EV.

To analyze the development of ZEBs, data from the Eurostat associated with type of motor energy were used. Among the available groups are the following: Petroleum products, Liquefied petroleum gases (LPG), Diesel, Electricity, Alternative Energy, Diesel (excluding hybrids), Hybrid diesel-electric, Plug-in hybrid diesel-electric, Hydrogen and fuel cells, Compressed natural gas (CNG), Liquefied natural gas (LNG), and Other. The method of data collection was established in 2013. Previous statistics only included Petroleum products and Diesel (until 2013). It should be noted that, currently, there are even more vehicle types available in various reports on Eurostat compared with those listed in the statistics. The "Electric vehicles in Europe" report highlights the following [33]:

- Battery electric vehicles (BEVs)—powered by an electric motor and battery with plug-in charging;
- Hybrid electric vehicles (HEVs)—combine a conventional (petrol/diesel) engine and a small electric motor/battery charged via regenerative braking or the engine;
- Plug-in hybrid electric vehicles (PHEVs)—a conventional (petrol/diesel) engine complemented with an electric motor/battery with plug-in charging;
- Range-extended electric vehicles (REEVs)—powered by an electric motor and plug-in battery, with an auxiliary combustion engine used only to supplement battery charging;
- Fuel cell electric vehicles (FCEVs)—use a fuel cell to create on-board electricity, generally using compressed hydrogen and oxygen from the air.

A combination of vehicles from the electricity, hybrid diesel-electric, plug-in hybrid diesel-electric, hydrogen, and fuel cells categories was selected for analysis, which approximate BEVs, HEVs, PHEVs, REEVs, and FCEVs as best as possible, while the REEVs group is not formally indicated.

Unfortunately, Eurostat guidelines on Passenger Mobility Statistics released in 2018 [34] define groups differing from those shown above:

- Petrol;
- Diesel;
- Petrol-electric, covers both non off-vehicle-chargeable hybrid electric vehicle ("Hybrid electric vehicle") and off-vehicle-chargeable hybrid electric vehicle ("Plug-in hybrid electric vehicle");
- Diesel-electric, covers both non off-vehicle-chargeable hybrid electric vehicle ("Hybrid electric vehicle") and off-vehicle-chargeable hybrid electric vehicle ("Plug-in hybrid electric vehicle");
- Electric vehicle (EV), covers pure electric vehicle ("Battery electric vehicle");
- Other, covers bi-fuel petrol/LPG, bi-fuel petrol/CNG, LPG, CNG, flex-fuel, and other fuels than those previously listed.

The categorization is inconsistent with those in the statistics and is not in line with subsequent studies on electric vehicles (for example, the statistics do not include the petrol electric group). In addition, the scope of the alternative energy group, which appears in the statistics, is unfortunately not explained in the document at all. Furthermore, in 2017, the European commission issued the document "Alternative Fuels (Expert group report)" [35], which defines this type of fuel. According to the document, alternative fuels include the following groups: Methane-based fuels (CNG, LNG, bio-methane, E-gas), LPG (propane- and butane-based fuels, BioLPG), Alcohols, Ethers and esters (ethanol, butanol, methanol, MTBE, ETBE, DME, BioDME, FAE), and Synthetic paraffinic and aromatic fuel (GTL, HVO, BTL, SIP, ATJ, CH, SAK). The aforementioned categorization raises doubts about the LPG, LNG, and CNG gas groups included in the statistics, as well as hydrogen cells, which are also sometimes recognized as alternative energy sources.

As part of the data analysis, a number of tests were performed. In most of the countries represented in Eurostat, the sums for individual groups and the total number of buses were not coherent (even after considering that Diesel is available in different variants). Apparently, the numbers distinguished by Eurostat must be included in several groups at the same time.

For the forecast, it was decided to take the number of buses in a given country, not the number of new registrations. Formally, the Bass model in the basic version does not include the replacement of technology. Regular buses (powered conventionally) have a relatively long life cycle. For example, in Poland, there are about 100,000 diesel buses and about 5000 new diesel vehicle registrations per year, which means about 20 years of their life cycle. However, the data show that the life cycle of electric vehicles is extremely short, as for buses (see Table 1). Estonia in 2013 had 91 electric buses, while that number in 2018 is only 1. Similarly, Bulgaria bought 150 buses in 2014, but only 96 came to market, which means that 54 were withdrawn; then, in 2015, 47 were newly registered and 70 were withdrawn. To maintain the number of buses from 2015, one would have to buy as many as 103 buses. This means that the life cycle of these vehicles varies somewhere between 5 and 0 years, or there are other unknown reasons for their withdrawal. Therefore, the data showing the number of buses in a given country are more suitable for estimating the parameters of the Bass model [11,17].

**Table 1.** The comparison of data on new registrations and numbers of electric buses on the example of Bulgaria and Estonia [Source: https://www.eea.europa.eu/data-and-maps/indicators/proportion-of-vehicle-fleet-meeting-4/assessment-4].

| Bulgaria | 2013 | 2014 | 2015 | 2016 | 2017 | 2018 |
|---|---|---|---|---|---|---|
| New registrations | 20 | 150 | 47 | 0 | : | 14 |
| Total number | 467 | 563 | 540 | 437 | 390 | 376 |
| **Estonia** | 2013 | 2014 | 2015 | 2016 | 2017 | 2018 |
| New registrations | 0 | 0 | 0 | 0 | 0 | 0 |
| Total number | 91 | 88 | 75 | 63 | 58 | 1 |

Table 2 summarizes the results for ZEBs based on the data available from Eurostat. The order of the countries with the source data from Eurostat has been preserved. Not all of the presented data are suitable for further analysis, therefore, a preliminary evaluation was carried out. The countries that do not have enough data (data not available) in most data fields (NA marked) are excluded; numbers 1,2,3,4 in Table 2 define which data are missing.

In addition, the trend of collected data was also examined. Linear regression was performed for each country. When the slope was negative, it was assumed that the trend is decreasing; the country was marked with the symbol DT and the data were excluded from further analysis. The reason for that is the assumption of growing sales, which is very important in the Bass model, especially at the beginning of innovation development. In the case of a decreasing trend, the Bass curve fit has very poor estimators. In addition, for these countries, more buses are being decommissioned than registered, which does not indicate the development of technology. Countries where data for analysis were missing or where there was a downward trend were marked in gray in Table 2. Data for 2015, 2016 for Poland and 2015 for Macedonia are gross errors (marked in red); they stand out far above the neighboring trend. It looks like the Hybrid diesel-electric fields were mistakenly copied from the Other field (both values were checked to be identical). Additionally, the values in the Other field are consistent with the others. After taking into account the amendments, the corrected data are placed in brackets; Table 2 (Poland in 2015—504 buses and in 2016—526 buses, Macedonia in 2015—1 bus).

The forecasts presented in the article were obtained using the Bass model, usually defined by the following differential equation [17]:

$$f(t) = \frac{dF(t)}{dt} = \left(p + \frac{q}{m}F(t)\right)(m - F(t)), \tag{1}$$

where

$F(t)$—the total number of new technology users by time $t$ (numbers of ZEBs in the market),
$f(t)$—number of users of new technology that adopt at time $t$,
$m$—the total number of technology users (total number of buses, see Table 1),
$p$—the innovation coefficient,
$q$—the imitation coefficient (for details, see [11]).

**Table 2.** Motor coaches, buses, and trolley buses, by four types of motor energy (electricity, hybrid diesel-electric, plug-in hybrid diesel-electric, and hydrogen and fuel cells) [Source: own study based on data retrieved from Eurostat].

| Country | 2013 | 2014 | 2015 | 2016 | 2017 | 2018 |
|---|---|---|---|---|---|---|
| Belgium | 59 [34] | 63 [34] | 7 [134] | 201 [34] | 213 [34] | 366 [34] |
| Bulgaria[DT] | 467 [234] | 563 [234] | 540 [234] | 437 [234] | 390 [234] | 376 [234] |
| Czech Republic[NA] | 559 [234] | : [1234] | : [1234] | : [1234] | : [1234] | : [1234] |
| Denmark | 0 [2] | 0 [2] | 5 [2] | 7 [2] | 6 [2] | 8 [2] |
| Germany | 99 [234] | 116 [234] | 137 [234] | 168 [234] | 183 [234] | 228 [234] |
| Estonia | 91 [34] | 88 [34] | 99 [34] | 87 [34] | 102 [34] | 45 [34] |
| Ireland[NA] | : [1234] | : [1234] | : [1234] | : [1234] | : [1234] | : [1234] |
| Greece[NA] | : [1234] | : [1234] | : [1234] | : [1234] | : [1234] | : [1234] |
| Spain | : [1234] | 112 | 152 | 274 | 463 | 701 |
| France | 567 [34] | 638 [34] | 1103 [34] | 1682 [34] | 1952 [34] | 2300 [34] |
| Croatia | : [1234] | : [1234] | 2 [234] | 3 [234] | 3 [234] | 3 [234] |
| Italy[DT] | 495 [34] | 488 [34] | 494 [34] | 463 [34] | : [1234] | 488 [34] |
| Cyprus[NA] | 0 | 0 | 1 | 0 | 0 | 0 |
| Latvia | 0 [1] | 257 | 269 | 290 | 255 | 258 |
| Lithuania[DT] | 457 [234] | 434 [234] | 431 [234] | 408 [234] | 424 [234] | 438 [234] |
| Luxembourg | 4 [234] | 2 [234] | 48 [34] | 59 [34] | 7 [234] | 33 [234] |
| Hungary | 2 | 3 | 6 | 25 | 25 | 24 |
| Malta[DT] | 5 [134] | 5 [3] | 5 [3] | 0 | 5 [3] | : [1234] |
| Netherlands[NA] | : [1234] | : [1234] | : [1234] | : [1234] | : [1234] | : [1234] |
| Austria | 143 [34] | 141 [34] | 150 | 154 | 146 | 158 |
| Poland | : [1234] | 458 [23] | 3616 (504) [3] | 3636 (526) [3] | 581 [23] | 803 [3] |
| Portugal | 8 [234] | 15 | 14 | 17 | 19 | 46 |
| Romania | 1 [234] | 2 [234] | 4 [234] | 4 [234] | 4 [234] | 15 [34] |
| Slovenia | : [1234] | 2 [234] | 2 [234] | 4 [234] | 3 [234] | 4 [234] |
| Slovakia[NA] | 250 [234] | : [1234] | : [1234] | : [1234] | : [1234] | : [1234] |
| Finland | 2 [13] | 7 | 8 | 16 | 24 | 24 |
| Sweden | 50 [34] | 56 [34] | 73 | 89 | 108 | 151 |
| United Kingdom | : [1234] | : [1234] | 194 [4] | 261 [4] | 305 [234] | 511 [23] |
| Liechtenstein | 2 | 2 | 2 | 2 | 2 [3] | 2 [3] |
| Norway | : [1234] | 9 [234] | 11 [234] | 37 | 67 | 167 |
| Switzerland | 100 | 100 [3] | 0 [3] | 100 | 100 [3] | 200 [3] |
| Macedonia[6,DT] | 3 [234] | 2 [234] | 2963 (1) [34] | : [1234] | : [1234] | : [1234] |
| Turkey | 0 [234] | 1 [234] | 12 [234] | 24 [234] | 57 | 74 |
| Kosovo[5,NA] | : [1234] | : [1234] | : [1234] | : [1234] | 0 [23] | 0 |

Data with gross errors identified; [1] Electricity data not available; [2] Hybrid diesel-electric data not available; [3] Plug-in hybrid diesel-electric data not available; [4] Hydrogen and fuel cells data not available; [5] Kosovo (under United Nations Security Council Resolution 1244/99); [6] North Macedonia; [DT] Decreasing trend; [NA] Insufficient data to build the model.

The analytical solution of the Bass model (1) is as follows:

$$f(t) = m \frac{(p+q)^2}{p} \frac{e^{-(p+q)t}}{\left(1 + \frac{q}{p} e^{-(p+q)t}\right)^2}. \tag{2}$$

Usually, (2) corresponds to the probability density function (PDF), which represents how many new technology users have arrived in a given time. Thus, in this study, $f(t)$ would correspond to the changing of the number of buses/the changing number in one year to another (the number of new registrations minus the number of withdrawn buses). Because the data are represented differently, for research purposes, the cumulative number of ZEBs was used. The cumulative distribution function (CDF) for $f(t)$ has the following form:

$$F(t) = m \frac{1 - e^{-(p+q)t}}{1 + \frac{q}{p} e^{-(p+q)t}}. \tag{3}$$

Nonlinear least square (NLS) method was used to estimate the parameters $p$ and $q$ of the Bass model for individual countries. Parameter $m$ was fixed arbitrarily on the basis of the average total number of buses in 2013–2018. This is justified because this value remains almost at the same level for each country. This also rests on the assumption that, at some point in time, the entire bus market will be taken by ZEBs. In addition, it was specified that the parameter $q$ should be in the range of 0.1 to 1. In the absence of this limitation, the imitation coefficient for some models was very close to zero. In this case, it would mean that there is no natural diffusion of innovation, which is a requirement of market development. On the basis of previous studies [11,26,36–39], the coefficient $q$ for vehicles using clean energy was usually greater than 0.3. Therefore, setting the limit at 0.1 does not constitute a significant interference in the parameterization of the model. The estimation results of the Bass model parameters are summarized in Table 3. Table 3 includes estimates, standard errors, $t$-test statistics, and $p$-values for each parameter, as well as the coefficient of determination $R^2$ for each country model.

**Table 3.** Estimation of Bass model parameters for EU countries [Source: own study based on data retrieved from Eurostat].

| Country | Parameter | Estimate | Standard Error | T-Statistic | p-Value | R$^2$ |
|---|---|---|---|---|---|---|
| Belgium | $p$ | 0.000885394 | 0.000565631 | 1.565322522 | 0.192561721 | 0948650663 |
|  | $q$ | 0.410295825 | 0.172394185 | 2.379986445 | 0.075990387 |  |
| Denmark | $p$ | 0.0000771299 | 0.000045924 | 1.679509668 | 0.168351712 | 0.926736146 |
|  | $q$ | 0.100003852 | 0.212953914 | 0.469603257 | 0.663094394 |  |
| Germany | $p$ | 0.000396186 | 0.000204596 | 1.936428221 | 0.124888934 | 0.943931675 |
|  | $q$ | 0.1 | 0.184882586 | 0.540883823 | 0.617301126 |  |
| Estonia | $p$ | 0.002887886 | 0.004669849 | 0.618411127 | 0.56977991 | 0.633401476 |
|  | $q$ | 0.1 | 0.583379797 | 0.171414918 | 0.872219773 |  |
| Spain | $p$ | 0.000457022 | 0.0000644608 | 7.089928856 | 0.002089651 | 0.99731596 |
|  | $q$ | 0.401918954 | 0.038185946 | 10.52531079 | 0.000460806 |  |
| France | $p$ | 0.003082591 | 0.000640235 | 4.814783299 | 0.008555448 | 0.990588119 |
|  | $q$ | 0.1 | 0.074974179 | 1.333792525 | 0.253150892 |  |
| Croatia | $p$ | 0.0000807419 | 0.0000502966 | 1.605314643 | 0.183695642 | 0.920360181 |
|  | $q$ | 0.100009858 | 0.222797329 | 0.448882661 | 0.676761267 |  |
| Latvia | $p$ | 0.009131224 | 0.008486617 | 1.075955683 | 0.34251624 | 0.842631191 |
|  | $q$ | 0.1 | 0.34176469 | 0.292598981 | 0.78437893 |  |
| Luxembourg | $p$ | 0.002770994 | 0.004986776 | 0.555668447 | 0.608049994 | 0.582779443 |
|  | $q$ | 0.100000027 | 0.649018282 | 0.154078906 | 0.885008825 |  |
| Hungary | $p$ | 0.000181564 | 0.000113826 | 1.595099572 | 0.185919392 | 0.919813533 |
|  | $q$ | 0.10466556 | 0.223167127 | 0.469000794 | 0.663489577 |  |
| Austria | $p$ | 0.002718866 | 0.002978315 | 0.91288739 | 0.412946675 | 0.790341288 |
|  | $q$ | 0.1 | 0.394989813 | 0.253171086 | 0.812615238 |  |
| Poland | $p$ | 0.00091753 | 0.000469269 | 1.955231138 | 0.122221307 | 0.945055774 |
|  | $q$ | 0.1 | 0.183399162 | 0.545258763 | 0.614554536 |  |
| Portugal | $p$ | 0.000171208 | 0.000111008 | 1.542300231 | 0.197864381 | 0.933889305 |
|  | $q$ | 0.293996079 | 0.191424236 | 1.535835201 | 0.199380212 |  |
| Romania | $p$ | 0.00000303032 | 0.00000320946 | 0.944183717 | 0.398537742 | 0.934794737 |
|  | $q$ | 0.709422062 | 0.238694759 | 2.972088974 | 0.041059215 |  |
| Slovenia | $p$ | 0.000206534 | 0.000116033 | 1.779960219 | 0.14969122 | 0.93427419 |
|  | $q$ | 0.100016751 | 0.201013473 | 0.497562423 | 0.644902165 |  |
| Finland | $p$ | 0.00017463 | 0.000050292 | 3.47231985 | 0.02552966 | 0.982203041 |
|  | $q$ | 0.118605229 | 0.100991957 | 1.174402713 | 0.305378156 |  |
| Sweden | $p$ | 0.001315502 | 0.000452593 | 2.906593192 | 0.043827409 | 0.974409779 |
|  | $q$ | 0.1 | 0.123521821 | 0.809573557 | 0.463593199 |  |
| United Kingdom | $p$ | 0.000169943 | 0.0000722559 | 2.351959527 | 0.078347481 | 0.972186036 |
|  | $q$ | 0.326811667 | 0.122009779 | 2.678569456 | 0.055310761 |  |
| Liechtenstein | $p$ | 0.003518262 | 0.004142058 | 0.849399656 | 0.44350691 | 0.765882685 |
|  | $q$ | 0.100000655 | 0.425556385 | 0.234988027 | 0.825757503 |  |
| Norway | $p$ | 0.0000590045 | 0.0000141281 | 4.176397302 | 0.013958849 | 0.99743895 |
|  | $q$ | 0.829605613 | 0.051582606 | 16.0830496 | 0.0000874111 |  |
| Switzerland | $p$ | 0.001742495 | 0.001962474 | 0.887907319 | 0.424757571 | 0.796458861 |
|  | $q$ | 0.177589975 | 0.373654193 | 0.475278957 | 0.659377964 |  |
| Turkey | $p$ | 0.00000330122 | 0.00000142233 | 2.320995094 | 0.081048171 | 0.979453761 |
|  | $q$ | 0.480497008 | 0.109960075 | 4.36974062 | 0.011970507 |  |

Countries marked in gray have very bad parameter estimators. It has been assumed that the criterion for such countries would be the coefficient of determination $R^2$ below 0.9 (determining the quality of model fit). Bad fit of the model to the data can also be recognized. The parameter q set itself at the boundary of the range; that is, it adopted the lowest possible value of 0.1, which is accompanied by a high standard error value for this parameter. The green color indicates countries with a relatively high quality matching of $p$ and $q$ parameters ($p$-value less than 10%). Spain, the United Kingdom, Norway, and Turkey, are the countries for which forecasts are most likely, and detailed results are presented in the Results chapter.

The forecast for the entire EU was made on the basis of the analysis of the $p$ and $q$ coefficient and its arbitrary selection based on the average. This approach was dictated by the poor fit of the model to the data for the entire EU (see Table 4). Two models were presented where the market size $m$ was calculated automatically. Unfortunately, the values are very low in relation to $m = 1.6 \times 10^6$ for the entire Union. In addition, the best $p$-value for the imitation coefficient that was obtained is at the level of 63%, which practically cancels any possibility of forecast based on those models.

**Table 4.** Estimation of Bass model parameters for EU [Source: own study based on data retrieved from Eurostat].

| Model Type | Parameter | Estimate | Standard Error | $t$-Statistic | $p$-Value | $R^2$ |
|---|---|---|---|---|---|---|
| | $m$ | 7508.84 | 6356.48 | 1.18129 | 0.322606 | |
| EU | $p$ | 0.338689 | 0.19937 | 1.69879 | 0.187921 | 0.980254 |
| | $q$ | 0.0001 | 1.44641 | 0.000069137 | 0.999949 | |
| EU fixed $m$ | $p$ | 0.000614179 | 0.000333394 | 1.8422 | 0.139242 | 0.938461 |
| | $q$ | 0.1 | 0.19447 | 0.514219 | 0.634204 | |
| EU fixed $p$ | $m$ | 611100 | 260598 | 2.34499 | 0.0789462 | 0.958096 |
| | $q$ | 0.0001 | 0.172262 | 0.000580512 | 0.999565 | |

## 3. Results

Resulting from the analysis, the following answers to research questions were proposed:

A1 (Q1): According to the methodology, the analysis is feasible in four countries, that is, Spain, the United Kingdom, Norway, and Turkey. In 2025, there will be 10,761 ZEBs in Spain, 5530 in the United Kingdom, 12,658 in Norway, and 2373 in Turkey. However, in 2030, the number of zero-emission buses in the same countries will increase significantly and will amount to 37,854 for Spain, 25,056 for the United Kingdom, 16,267 for Norway, and 25,372 for Turkey.

A2 (Q2): With this predicted number of clean buses, it seems that only Norway will be able to reach 95% level of ZEBs share in all buses possessed by this country.

A3 (Q3): On the basis of analyses conducted, the majority of EU members will have a 95% share of ZEBs in a fleet consisting of all types of buses after 2050. Detailed data are presented in the Table 5 with the forecast of the ZEBs' share in the market in the EU countries in 2025 and 2030. Figure 2 presents the geographical distribution of the results.

Table 5 provides a forecast of the share of ZEBs in the market in 2025 and 2030 for all European countries for which data were available. In addition, it presents in which year the number of buses in this category will constitute 95% of all buses ($m$). Out of the concern about the quality of the model, the results are marked with different colors. Countries with poor fit parameters are marked in gray, while those in which the quality of the model fits the data very well are shown in green (details described in the Materials and Methods chapter). The presented results are negatively affected by the following factors: short reporting period, data quality, and the issue of technology definitions that are inconsistently interpreted by different countries.

**Table 5.** Development forecast for the electricity, hybrid diesel-electric, plug-in hybrid diesel-electric, and hydrogen and fuel cells bus market [own study based on data retrieved from Eurostat].

| Country | E-Bus Market | | | | Year 95% of $m$ | Number of Buses |
|---|---|---|---|---|---|---|
| | 2025 Total | 2025 % | 2030 Total | 2030 % | | |
| Belgium | 4965 | 30.99% | 12,479 | 77.90% | 2034 | 16,019 |
| Denmark | 27 | 0.20% | 51 | 0.38% | 2113 | 13,353 |
| Germany | 824 | 1.05% | 1548 | 1.97% | 2096 | 78,591 |
| Estonia | 354 | 7.30% | 635 | 13.10% | 2075 | 4847 |
| Spain | 10,761 | 17.44% | 37,854 | 61.34% | 2036 | 61,712 |
| France | 7674 | 7.77% | 13,711 | 13.89% | 2074 | 98,701 |
| Croatia | 11 | 0.21% | 21 | 0.39% | 2112 | 5365 |
| Latvia | 1046 | 20.75% | 1709 | 33.90% | 2061 | 5041 |
| Luxembourg | 132 | 7.01% | 237 | 12.59% | 2075 | 1883 |
| Hungary | 91 | 0.50% | 176 | 0.96% | 2100 | 18,324 |
| Austria | 674 | 6.89% | 1213 | 12.41% | 2076 | 9776 |
| Poland | 2724 | 2.41% | 5062 | 4.47% | 2087 | 113,175 |
| Portugal | 369 | 2.53% | 1506 | 10.35% | 2047 | 14,557 |
| Romania | 1971 | 4.14% | 28,545 | 60.02% | 2033 | 47,563 |
| Slovenia | 14 | 0.53% | 27 | 1.02% | 2103 | 2658 |
| Finland | 92 | 0.54% | 185 | 1.08% | 2091 | 17,087 |
| Sweden | 484 | 3.43% | 892 | 6.31% | 2083 | 14,126 |
| United Kingdom | 5530 | 3.47% | 25,056 | 15.72% | 2044 | 159,404 |
| Liechtenstein | 8 | 8.00% | 15 | 15.00% | 2073 | 100 |
| Norway | 12,658 | 77.46% | 16,267 | 99.54% | 2027 | 16,342 |
| Switzerland | 788 | 8.28% | 1813 | 19.05% | 2054 | 9516 |
| Turkey | 2373 | 0.35% | 25,372 | 3.77% | 2042 | 672,885 |

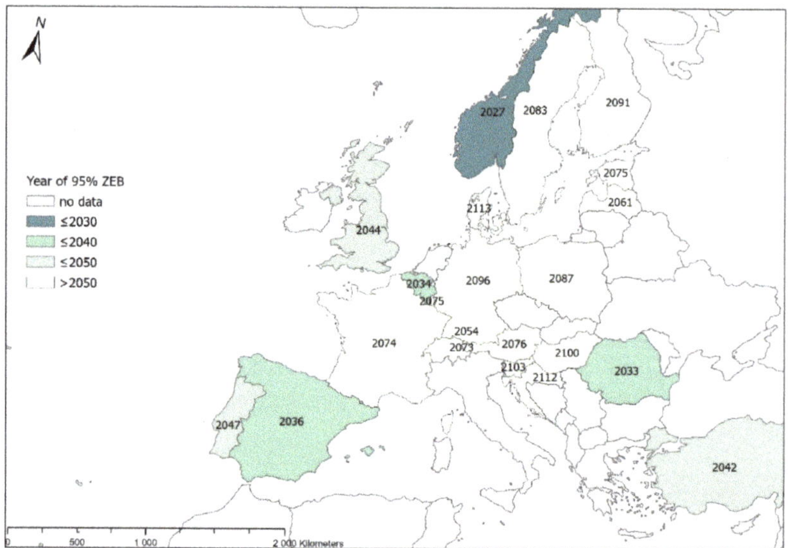


**Figure 2.** Geographical distribution of the year when countries reach the 95% of ZEBs in their fleets [own study].

The analysis traced the situation of 22 EU countries (Table 5). The detailed analysis of the data showed that only four countries out of all the countries considered show activity related to the replacement of their bus fleet with electric ones. The reasons for this endeavor can be twofold. Either it results from a high level of environmental awareness of the mentioned countries, such as Norway, or it testifies to the countries' high commitment and efficiency in obtaining EU subsidies. The lack of reliable data, including consistent historical sequences for 2013–2018, in the case of the remaining countries may indicate a low level of their activity in this area.

Table 5 contains columns presenting the market share for 2025 and 2030 calculated on the basis of the Bass model. The percentage values are related to the total market share (*m*). The number of buses in individual countries was estimated based on the average number of all buses in a given country for 2013–2018. Additionally, the year in which market penetration by new bus generations will reach the level of 95% is indicated. The 95% level was chosen arbitrarily and results from the slow growth of the S-shaped Bass curve at the end of a given technology development.

Figure 3 presents a histogram, based on the calculations. It indicates the years when traditional buses should be replaced by buses using electricity, hybrid diesel-electric, plug-in hybrid diesel-electric, and hydrogen or fuel cells.

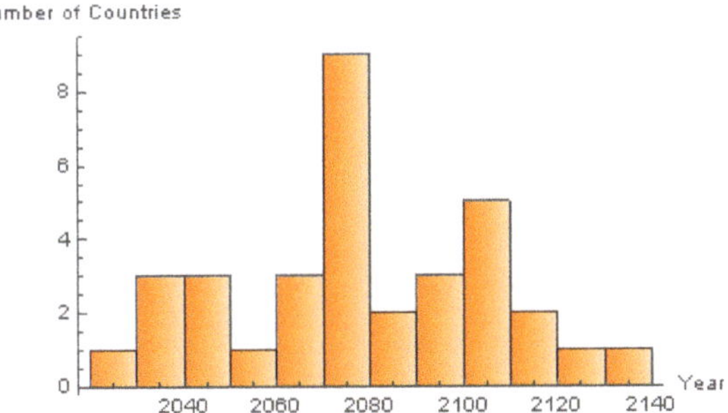


**Figure 3.** Histogram of years of 95% market adaptation by ZEBs (all EU countries included) [own study].

It should be noted that all European countries were considered in Figure 3 regardless of the quality of the model. The chart shows that the average adaptation to the market should take place around 2077 (average), with a standard deviation of about 28 years. In addition, the above histogram shows compliance with the normal distribution based on the Kolomogorov–Smirnov test, with $p$-value $2.23745 \times 10^{-7}$ and statistic 0.473591; however, the Shapiro–Wilk test gives statistic 0.973548 and $p$-value 0.56575.

In the case of countries for which the coefficient of determination $R^2$ was higher than 0.9 (see Table 3), the histogram is presented in Figure 4.

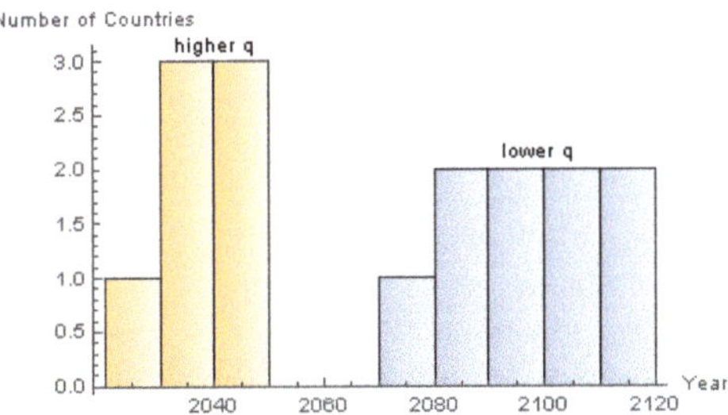


**Figure 4.** Histogram of years of 95% market adaptation by ZEBs (country with $R^2 > 0.9$ included) [own study].

The countries were divided into two groups with a higher and lower imitation coefficient. A higher imitation factor $q > 0.3$ means that countries are adopting the new technology relatively quickly. Countries classified in this area include Belgium, Spain, Portugal, Romania, the United Kingdom, Norway, and Turkey, with the average of full adaptation in 2038 and a standard deviation of about 7 years. In contrast, the second group with a lower q usually close to 0.1 constitutes the following countries: Denmark, Germany, France, Croatia, Hungary, Poland, Slovenia, Finland, and Sweden, with the average of around 2095 and a 13-year standard deviation. However, statistics for parameters in these countries give a low level of confidence for calculations of the Bass model variables. The forecast was based on available data. The country in which the forecast indicates a very distant time of market acceptance has considerable uncertainty in estimating this year of adjustment. This is partly owing to the fact that countries have not shown significant activities in this area. It should be noted that, if a given country has already begun investment in a given technology and the process of diffusion of innovation, then adjustment could take place quite quickly. With the data we have, there is no basis to assess what will happen in a given country if it changes its policy and makes significant investments in ZEBs (for a more detailed comment, see the Materials and Methods section).

The most reliable results were obtained for countries marked in green (Table 5). They have the best parameter estimators and a very good model fit factor. In Spain, the United Kingdom, Norway, and Turkey, the average saturation of the market with zero-emission buses should occur around 2037. The process of technology adaptation calculated from the Bass model is presented in Figure 5. The vertical axis presents market adoption expressed as a percentage and the horizontal axis represents time in years. The cumulated number of buses for selected countries allows for the assessment of innovation diffusion. For example, on the basis of Figure 4, in 2030 in Norway, the percentage saturation of ZEBs in the total bus transport fleet will reach around 50%.

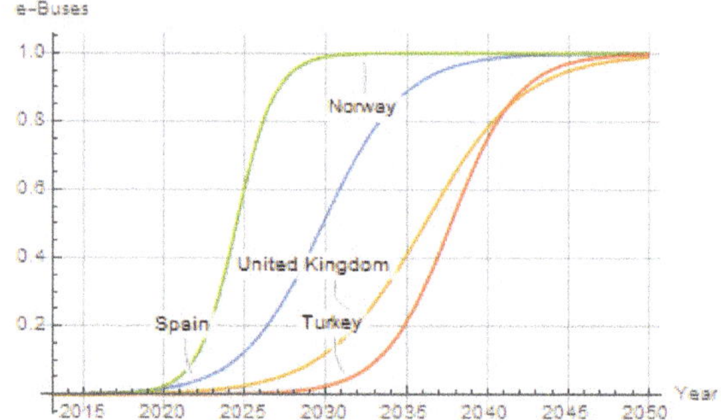


**Figure 5.** The cumulative Bass curves for Spain, the United Kingdom, Norway, and Turkey [own study].

The development forecast for the entire European Union was made in two versions (Figure 6). The first optimistic variant assumes a high imitation factor. Parameters of the Bass model were calculated on the basis of the average for the best four models, that is, Spain, Norway, the United Kingdom, and Turkey ($p = 0.000173$, $q = 0.5195$). The pessimistic variant was created on the basis of average parameter values for all countries from Table 5 ($p = 0.001407$, $q = 0.2178$). The above approach results from the fact that an attempt to compile data for all EU countries gave a model with unsatisfactory estimators (please see details in Materials and Methods).

It has to be stated that the predicted number of buses could be estimated only for the chosen EU members. This is owing to the lack of a uniform definition of zero-emission vehicles. An additional factor causing calculation difficulties was errors in Eurostat statistics. In addition, there were gaps in the data collected for individual countries. Thus, the authors could only conduct the correct simulation for four countries: Spain, Great Britain, Norway, and Turkey, as shown in Table 5.

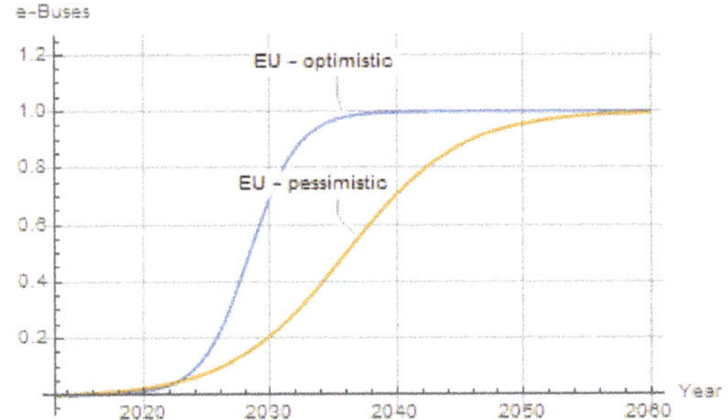


**Figure 6.** The cumulative Bass curves for Spain, the United Kingdom, Norway, and Turkey [own study].

## 4. Discussion

A discussion should first refer to the existing research on similar issues, using the same methodology. Although an in-depth review of the literature in the Scopus, Web of Science, and EBSCO databases has not brought the expected results, a number of similar topics were raised within scientific publications. However, it is worth emphasizing that no description of studies alike using the Bass model to describe the diffusion of innovation understood as an increase in the number of ZEBs in the total bus fleet of a given city has been identified. Therefore, the research was conducted in a similar thematic scope, although with the use of different tool or with the same tool, but referring to other issues.

In the study from 2018 conducted by Ma and Zhang [40], the Bass model was used to optimize and predict the number of charging stations for electric vehicles. In order to solve this issue, the researchers used the exhaustion method, regarding minimum cost as the objective function. To finish up their work, they tested the given model using data from a particular Chinese city.

Akbari, Brenna, and Longo [41] adopted similar assumptions when using the Bass model. The authors focused on Milan, Italy. The main purpose of using the model was to calculate how many electric vehicles (EV) will be in 2024, and thus will need charging stations. In further analyses, the aim was to indicate the optimal location of the stations so as to meet the demand generated by customers on the one hand, and on the other hand, to minimize the costs of vehicle charging and management.

What appears to be complementary in the presented approach is the reference of the optimal number of charging stations for electric vehicles to the public transport infrastructure. Thus, the subject of city management in the context of creating optimal urban spaces and the optimal use of ZEBs would find a wider application. A similar objection regarding the narrow approach to the subject of electric vehicles can be formulated against the authors of this article. Taking into account the holistic approach to electric vehicles in cities, it seems to be an interesting research direction.

Rogge, van der Hurk, Larsen, and Sauer [42] also looked at the problem of electric vehicles in an interesting way. Similarly to the authors of this article, they analysed public transport in the context of developing the most optimal saturation of the city transport fleet with electric buses. What was different about their approach was that they did not assume that the fleet should be fully electrified, but rather that the fleet should have different proportions of both electric and conventionally powered vehicles. Undoubtedly, this is a beneficial direction for further research with high potential.

The results of the 2018 research by Mohamed, Ferguson, and Kanaroglou [43] may provide some kind of valuable inspiration for the authors to carry out in-depth EU country-by-country analyses. This study identified factors that hinder the implementation of the electric bus in the public transit context as seen from the perspective of Canadian-based service providers.

With regard to modelling the market share of a specific group of electric vehicles, such as the zero-emission buses described in the article, it should be emphasized that there is a clear inconsistency between the actual market share of ZEBs and the feasibility models. This inconsistency shows a clear gap between the theoretical evidence for the positive environmental impact of ZEBs as well as the benefits from electric buses in the public transport fleet, and its practical application. While some argue that a lack of political support, technological immaturity, and inertia to change are key factors contributing to low participation, others attribute this situation to the sensitivity of technical-economic models related to the operational context. This sensitivity is common ground for all evaluation models of electric buses [44–46]. As a result, this sensitivity increases the uncertainty about the operational benefits of the electric bus in the context of the network, thus limiting market share.

The adoption of the electric bus globally is geographically uneven and limited in scale [47]. Predictions for the development of the ZEB fleet are quite difficult, as replacing the fleet with an electric one faces many obstacles. These obstacles can primarily be divided into two groups dependent on each other. It is a matter of available technologies that affect bus electricity demand and both initial and operational costs that affect the economic efficiency of investment in these solutions. Polish geographers identified the main factors and mechanisms behind the development of low-emission public transport vehicles in Polish cities. They included energy challenges, environmental requirements, governance strategies, and manufacturing capacities [48].

One issue is the high upfront cost of zero-emission buses. The huge costs associated with the investment, both the purchase of a new fleet and the appropriate infrastructure, cause a number of considerations about this technology, as well as its ecological, economic, and organizational effectiveness [49]. A few studies have analyzed the contracting and financing mechanisms that can help accelerate electric bus adoption [47]. The justification for using public funds when purchasing battery electric vehicles is the anticipated reduction of $CO_2$ emissions [50,51]. Using the Bass model, Brito and others investigated how governmental incentives can influence the diffusion of low emission technology in individual transport decisions. They were able to demonstrate how, for example, tax regulations can affect the increase in the adoption of zero-emission (to be precise, electric) technologies by the individual customer market [21]. There are also several other studies of the impact of the economic and social policy on the development of electric technologies in transport based on technology diffusion models [52].

Therefore, one of the most important discussions in the literature on clean buses is the issue of their real impact on reducing greenhouse gases. The environmental benefits of ZEB in cities should be calculated and assessed from two points of view: emissivity and operational harmfulness as well as emissivity, harmful to the environment at the time of energy production. Ultimately, however, the environmental benefits of ZEB will really depend on what sources electricity is obtained from in the country or city. Some research papers propose the application of a life cycle assessment (LCA) [46,53,54], or through a combination of LCA with an economic analysis [55], or through a cost–benefit analysis [56,57]. The methods adopted to evaluate the transport impact of $CO_2$ emissions are rather heterogenic, including different phases of the fuel production and the emission phase. Cavallaro et al. [50] propose a well-to-wheel analysis, including the well-to-tank and tank-to-wheel phases, while Topal and Nakir [58] propose a total cost of ownership calculation model. Total cost of ownership from well-to-wheel has been proposed for the three groups of transportation, namely diesel, CNG (compressed natural gas), and electric buses. The analysis showed that the total costs of ownership for electric buses is greater than for those with diesel and hybrid engines. Nonetheless, the authors conclude that high initial costs and amortization points in electric buses can be caught because of low operating costs. Total costs of ownership are directly influenced by the costs of purchasing technology (buses and necessary infrastructure, that is, charging stations) as well as operating costs.

The total operation costs also depend on several factors. The buses' electricity demand depends on the operating time on one battery charge. Electric buses have a limited driving range and need to be charged during the day [13,59–62], which turns them off for some time. That creates the need for more buses to fulfill the transportation demand than the diesel ones. This goes for battery electric buses and, to a lesser extent, for hydrogen fuel cell buses. Regular diesel buses can drive all day without refueling, while battery electric buses need to recharge after about 200–250 km, depending on circumstances such as the climate and road conditions, and hydrogen fuel cell buses need to refuel after about 200–400 km [63]. This also affects the problem of planning courses. The electric bus scheduling problem requires not only satisfying timetable constraints, but also considering battery range limitation and vehicles recharging plans, including available charging infrastructure [64]. Moreover, for example, May [65] proposed planning bus lines served by ZEB based on local environmental impact assessment. The interdisciplinary approach was applied to evaluate the local environmental relief potential of electric buses in comparison with diesel buses. The issue of timetable planning is connected with the problem of charging station location and charging method solutions [62,66–68].

Bus demand for electricity depends on battery type, driving cycle and style, number of stops, traffic level, elevation profile [69,70], and weather conditions including temperature and humidity [8]. In order to improve the driving style, and thus reduce the bus's energy demand, various technological solutions are proposed, for example, the robotized manual gearbox [56,71,72]. Cost effectiveness also varies depending on the energy storage systems in electric buses [71,73] and the choice of charging technology [74]. Operating costs also depend on the price of the battery and the battery life [75]. Ufert and Bäker [9] propose a model for predicting battery life. According to them, the ability to predict battery life can reduce total operation costs by up to 17%.

In turn, Bakker and Konings [13] argue that the technological barriers for replacing the diesel bus fleet with ZEB are not that great compared with institutional barriers in individual countries and cities. Veeneman [76] also draws attention to the tender processes that must take place when purchasing a bus fleet owing to the fact that the funds involved in the purchase are public. What is important here is the low quality of tenders, which are based on the lowest price, which ultimately leads to the purchase of low quality products and services. There are also some studies addressing the public's willingness to pay for environmental-friendly buses [12,77,78].

Knowing the multidimensionality of issues related to the ecological and economic efficiency of electric buses, it is difficult to predict to what extent the forecast presented in this article will be implemented. This will probably depend on both, the direction of technology development, which will allow reduction of electricity demand and greater operational efficiency and on the funds available for cities in the future. The latest reports from the technology market inform about a new type of battery. Catl, a Chinese car battery-maker, says it is ready to manufacture a product capable of powering a vehicle for 1.2 million miles (two million kilometers) across the course of a 16-year lifespan [79]. For now, this technology is to be used in cars, but it is probably only a matter of time before it will also be used in larger vehicles with a higher energy demand.

The current global coronavirus pandemic problem should also be considered. It seems that the impact of Coronavirus disease 2019 (COVID-19) on financing clean buses in the near future cannot be predicted. According to international experts, despite many barriers and the crisis in the automotive industry, electromobility, including the ZEB sector, is developing dynamically and will continue to develop. This is owing to the fact that its expansion is based on very solid foundations, such as EU, national, and regional legal standards or multi-billion investments by automotive concerns [80].

The obligations arising from the Act on electromobility and alternative fuels in the field of zero-emission collective transport remain unchanged, new tenders are announced, subsequent deliveries are carried out, and leading manufacturers such as Solaris record a dynamic increase in orders (506 contracted ZEBs in 2020 compared with 162 in 2019) [80].

However, the survival of many enterprises often operating in key sectors to promote zero-emission transport, which is directly connected with the ZEB sector's development, depends on the government's rapid and decisive action on both local and international levels.

## 5. Conclusions

It should be stressed that it is likely that most European countries will not be able to replace the traditional fleet of diesel buses with ZEBs by 2050. The analysis shows that most countries will replace the fleet after 2050, around 2077.

As a result of the study based on available European data, the authors isolated two groups of EU countries. The first group consists of four countries for which the fit of the model is appropriate and these countries seem able to achieve the saturation of their fleet by 95% by 2040. For the remaining countries, owing to insufficient data, the model fit is low and these countries do not seem to be able to replace their bus fleet before 2050.

The fact is that, today, the number of ZEBs in EU urban spaces is rising, despite many technological, organizational, and financial barriers. The future of the bus fleet will depend on which way the technology develops and how it can be financed. The economic situation of countries will also be of great importance, especially in the face of COVID-19, to which extent ecological priorities can be further financed in the face of the need to save national economies.

**Author Contributions:** Conceptualization, G.C., A.B. and J.J; data curation, G.C., A.B. and J.J.; formal analysis, G.C., A.B. and J.J.; funding acquisition, G.C., A.B. and J.J.; investigation, G.C., A.B. and J.J.; methodology, G.C., A.B. and J.J.; project administration, G.C., A.B. and J.J.; resources, G.C., A.B. and J.J.; software, G.C., A.B. and J.J.; supervision, G.C.; validation, G.C., A.B. and J.J.; visualization, G.C., A.B. and J.J.; writing—original draft, G.C., A.B. and J.J.; writing—review & editing, G.C., A.B. and J.J. All authors have read and agreed to the published version of the manuscript.

**Funding:** This work was partially supported by the Ministry of Science and Higher Education (MNiSW, Poland) core funding for statutory R&D activities.

**Conflicts of Interest:** The authors declare no conflict of interest.

## References

1. Wojtkowska-Łodej, G. Wyzwania klimatyczne i energetyczne a polityka Unii Europejskiej. *Polit. Energ. Energy Policy J.* **2014**, *17*, 39–52.
2. European Commission. Citizens' Summary: Analysis of Options for Reducing the EU's Greenhouse Gas Emissions by 30 % by 2020. 2009. Available online: https://ec.europa.eu/clima/sites/clima/files/summary/docs/greenhouse_gas_2020_en.pdf (accessed on 2 June 2020).
3. European Union. Citizens' Summary EU Climate and Energy Package 2020. 2012. Available online: https://ec.europa.eu/clima/sites/clima/files/strategies/2020/docs/climate_package_en.pdf (accessed on 2 June 2020).
4. Committee of the Regions of European Union. Opinia Komitetu Regionów—Ramy Polityczne Na Okres 2020–2030 Dotyczące Klimatu i Energii. 2014. Available online: https://eur-lex.europa.eu/legal-content/PL/TXT/PDF/?uri=CELEX:52014IR2691&from=HR (accessed on 31 May 2020).
5. Polskie Stowarzyszenie Paliw Alternatywnych; Izba Gospodarcza Komunikacji Miejskiej Take E-Bus! Elektromobilność i Zrównoważony Rozwój Publicznego Transportu Zbiorowego w Miastach. 2019. Available online: https://pspa.com.pl/assets/uploads/2019/09/take_e-bus_raport_S.pdf (accessed on 25 May 2020).
6. Bezruchonak, A. Geographic Features of Zero-Emissions Urban Mobility: The Case of Electric Buses in Europe and Belarus. *Eur. Spat. Res. Policy* **2019**, *26*, 81–99. [CrossRef]
7. Blomberg. New Energy Outlook 2019. 2019. Available online: https://about.bnef.com/new-energy-outlook/ (accessed on 13 June 2020).
8. Todoruț, A.; Cordoș, N.; Iclodean, C. Replacing Diesel Buses with Electric Buses for Sustainable Public Transportation and Reduction of $CO_2$ Emissions. *Pol. J. Environ. Stud.* **2020**, *29*, 3339–3351. [CrossRef]
9. Ufert, M.; Bäker, B. Battery Ageing as Part of the System Design of Battery Electric Urban Bus Fleets. *Sci. Tech.* **2020**, *19*, 12–19. [CrossRef]
10. Barroso, J.M. Climate and Energy Priorities for Europe: The Way Forward—A Presentation of J.M. Barroso. 2014. Available online: https://ec.europa.eu/clima/sites/clima/files/strategies/2030/docs/climate_energy_priorities_en.pdf (accessed on 17 June 2020).
11. Brdulak, A.; Chaberek, G.; Jagodziński, J. Determination of electricity demand by personal light electric vehicles (PLEVs): An example of e-motor scooters in the context of large city management in Poland. *Energies* **2020**, *13*, 194. [CrossRef]
12. Pedrosa, G.; Leontyeva, Y.; Mayburov, I. Promoting zero emissions buses programs: A study of ekaterinburg residents' willingness to pay. *Int. J. Energy Prod. Manag.* **2018**, *3*, 253–265. [CrossRef]
13. Bakker, S.; Konings, R. The transition to zero-emission buses in public transport–The need for institutional innovation. *Transp. Res. Part D Transp. Environ.* **2018**, *64*, 204–215. [CrossRef]

14. Alfonsin, V.; Suarez, A.; Maceiras, R.; Sanchez, A. Modeling and simulation of a zero emission urban bus with battery and fuel cell energy systems under real conditions. *Environ. Prog. Sustain. Energy* **2018**, *37*, 832–838. [CrossRef]

15. Brozynski, M.T.; Leibowicz, B.D. Markov models of policy support for technology transitions. *Eur. J. Oper. Res.* **2020**, *286*, 1052–1069. [CrossRef]

16. Meade, N.; Islam, T. Modelling and forecasting the diffusion of innovation—A 25-year review. *Int. J.* **2006**, *22*, 519–545. [CrossRef]

17. Bass, F.M. A new product growth for model consumer durables. *Manag. Sci.* **2004**, *15*, 215–227. [CrossRef]

18. Bauckhage, C.; Kersting, K. Strong Regularities in Growth and Decline of Popularity of Social Media Services. *arXiv* **2014**, arXiv:1406.6529.

19. Rogers, E.M.; Singhal, A.; Quinlan, M.M. Diffusion of innovations. In *An Integrated Approach to Communication Theory and Research*, 3rd ed.; Stacks, D.W., Salwen, M.B., Eichhorn, K.C., Eds.; Routledge: New York, NY, USA, 2019; ISBN 9781351358712.

20. Bernards, R.; Morren, J.; Slootweg, H. Development and Implementation of Statistical Models for Estimating Diversified Adoption of Energy Transition Technologies. *IEEE Trans. Sustain. Energy* **2018**, *9*, 1540–1554. [CrossRef]

21. Brito, T.L.F.; Islam, T.; Stettler, M.; Mouette, D.; Meade, N.; Moutinho dos Santos, E. Transitions between technological generations of alternative fuel vehicles in Brazil. *Energy Policy* **2019**, *134*, 110915. [CrossRef]

22. Kong, D.Y.; Bi, X.H. Impact of social network and business model on innovation diffusion of electric vehicles in China. *Math. Probl. Eng.* **2014**, *2014*, 1–7. [CrossRef]

23. Li, S.; Chen, H.; Zhang, G. Comparison of the short-term forecasting accuracy on battery electric vehicle between modified bass and Lotka-Volterra model: A case study of China. *J. Adv. Transp.* **2017**, *2017*, 1–6. [CrossRef]

24. Ayyadi, S.; Maaroufi, M. Diffusion Models for Predicting Electric Vehicles Market in Morocco. In Proceedings of the EPE 2018—Proceedings of the 2018 10th International Conference and Expositions on Electrical and Power Engineering, Iasi, Romania, 17–18 October 2018.

25. Cagliano, A.C.; Carlin, A.; Mangano, G.; Rafele, C. Analyzing the diffusion of eco-friendly vans for urban freight distribution. *Int. J. Logist. Manag.* **2017**, *28*, 1218–1242. [CrossRef]

26. Van Den Bulte, C. Want to know how diffusion speed varies across countries and products? Try using a Bass model. *Pdma Vis.* **2002**, *XXVI*, 12–15.

27. Navigant Consulting, I. *Energy Savings Forecast of Solid-State Lighting in General Illumination Applications*; Navigant Consulting Inc.: Washington, DC, USA, 2016; pp. 1–119.

28. Packey, D.J. Market Penetration of New Energy Technologies. *Energy Policy* **1993**, *34*, 3317–3326.

29. Jeyaraj, A.; Sabherwal, R. The Bass Model of Diffusion: Recommendations for Use in Information Systems Research and Practice. *J. Inf. Technol. Theory Appl.* **2014**, *15*, 5–30.

30. Mahajan, V.; Mason, C.H.; Srinivasan, V. An Evaluation of Estimation Procedures for New Product Diffusion Models. *New Prod. Diffus.* **1986**, *851*, 203–232.

31. Marković, D.; Jukić, D. On parameter estimation in the bass model by nonlinear least squares fitting the adoption curve. *Int. J. Appl. Math. Comput. Sci.* **2013**, *23*, 145–155. [CrossRef]

32. Van Den Bulte, C.; Lilien, G.L. Bias and systematic change in the parameter estimates of macro-level diffusion models. *Mark. Sci.* **1997**, *16*, 338–353. [CrossRef]

33. European Environment Agency. *Electric Vehicles in Europe*; European Office of the European Union: Copenhagen, Denmark, 2016.

34. Eurostat EU Transport Statistics. *Eurostat Guidelines on Passenger Mobility Statistics*; Eurostat EU Transport Statistics: Brussels, Belgium, 2018.

35. European Union. *Alternative Fuels. Expert Group Report*; European Union: Luxembourg, 2017.

36. Sultan, F.; Farley, J.U.; Lehmann, D.R. A Meta-Analysis of Applications of Diffusion Models. *J. Mark. Res.* **1990**, *27*, 70–77. [CrossRef]

37. Wong, D.; Yap, K.; Turner, B.; Rexha, N. Predicting the Diffusion Pattern of Internet-Based Communication Applications Using Bass Model Parameter Estimates for Email. *J. Internet Bus.* **2011**, *9*, 1–25.

38. Turk, T.; Trkman, P. Bass model estimates for broadband diffusion in European countries. *Technol. Soc. Chang.* **2012**, *79*, 85–96. [CrossRef]

39. Massiani, J.; Gohs, A. The choice of Bass model coefficients to forecast diffusion for innovative products: An empirical investigation for new automotive technologies. *Res. Transp. Econ.* **2015**, *50*, 17–28. [CrossRef]
40. Ma, J.; Zhang, L. A deploying method for predicting the size and optimizing the location of an electric vehicle charging stations. *Information* **2018**, *9*, 170. [CrossRef]
41. Akbari, M.; Brenna, M.; Longo, M. Optimal locating of electric vehicle charging stations by application of Genetic Algorithm. *Sustaiability* **2018**, *10*, 1076. [CrossRef]
42. Rogge, M.; van der Hurk, E.; Larsen, A.; Sauer, D.U. Electric bus fleet size and mix problem with optimization of charging infrastructure. *Appl. Energy* **2018**, *211*, 282–295. [CrossRef]
43. Mohamed, M.; Ferguson, M.; Kanaroglou, P. What hinders adoption of the electric bus in Canadian transit? Perspectives of transit providers. *Transp. Res. Part D Transp. Environ.* **2018**, *64*, 134–149. [CrossRef]
44. Nurhadi, L.; Borén, S.; Ny, H. A sensitivity analysis of total cost of ownership for electric public bus transport systems in swedish medium sized cities. *Transp. Res. Procedia* **2014**, *3*, 818–827. [CrossRef]
45. Xu, Y.; Gbologah, F.E.; Liu, H.; Rodgers, M.O.; Guensler, R.L. Corrigendum to Assessment of alternative fuel and powertrain transit bus options using real-world operations data: Life-cycle fuel and emissions modeling. *Appl. Energy* **2015**, *154*, 143–159. [CrossRef]
46. Zhou, B.; Wu, Y.; Zhou, B.; Wang, R.; Ke, W.; Zhang, S.; Hao, J. Real-world performance of battery electric buses and their life-cycle benefits with respect to energy consumption and carbon dioxide emissions. *Energy* **2016**, *96*, 603–613. [CrossRef]
47. Li, X.; Castellanos, S.; Maassen, A. Emerging trends and innovations for electric bus adoption—A comparative case study of contracting and financing of 22 cities in the Americas, Asia-Pacific, and Europe. *Res. Transp. Econ.* **2018**, *69*, 470–481. [CrossRef]
48. Taczanowski, J.; Kołoś, A.; Gwosdz, K.; Domański, B.; Guzik, R. The development of low-emission public urban transport in Poland. *Bull. Geogr.* **2018**, *41*, 79–92. [CrossRef]
49. Pelletier, S.; Jabali, O.; Mendoza, J.E.; Laporte, G. The electric bus fleet transition problem. *Transp. Res. Part C Emerg. Technol.* **2019**, *109*, 174–193. [CrossRef]
50. Cavallaro, F.; Danielis, R.; Nocera, S.; Rotaris, L. Should BEVs be subsidized or taxed? A European perspective based on the economic value of CO2 emissions. *Transp. Res. Part D Transp. Environ.* **2018**, *64*, 70–89. [CrossRef]
51. Pedrosa, G.; Leontyeva, Y.; Mayburov, I. Financing of buses with zero emissions: The willingness of consumers to pay for public marketing. In Proceedings of the the 32nd International Business Information Management Association Conference, IBIMA 2018—Vision 2020: Sustainable Economic Development and Application of Innovation Management from Regional expansion to Global Growth, Seville, Spain, 15–16 November 2018.
52. Plötz, P.; Schneider, U.; Globisch, J.; Dütschke, E. Who will buy electric vehicles? Identifying early adopters in Germany. *Transp. Res. Part A Policy Pract.* **2014**, *67*, 96–109. [CrossRef]
53. Cooney, G.; Hawkins, T.R.; Marriott, J. Life cycle assessment of diesel and electric public transportation buses. *J. Ind. Ecol.* **2013**, *17*, 689–699. [CrossRef]
54. Kliucininkas, L.; Matulevicius, J.; Martuzevicius, D. The life cycle assessment of alternative fuel chains for urban buses and trolleybuses. *J. Environ. Manag.* **2012**, *99*, 98–103. [CrossRef] [PubMed]
55. Nurhadi, L.; Borén, S.; Ny, H. Advancing from Efficiency to Sustainability in Swedish Medium-sized Cities: An Approach for Recommending Powertrains and Energy Carriers for Public Bus Transport Systems. *Procedia Soc. Behav. Sci.* **2014**, *111*, 586–595. [CrossRef]
56. Lajunen, A. Energy consumption and cost-benefit analysis of hybrid and electric city buses. *Transp. Res. Part C Emerg. Technol.* **2014**, *38*, 1–15. [CrossRef]
57. Noel, L.; McCormack, R. A cost benefit analysis of a V2G-capable electric school bus compared to a traditional diesel school bus. *Appl. Energy* **2014**, *126*, 246–255. [CrossRef]
58. Topal, O.; Nakir, İ. Total cost of ownership based economic analysis of diesel, CNG and electric bus concepts for the public transport in Istanbul City. *Energies* **2018**, *11*, 2369. [CrossRef]
59. Brecher, A. Transit Bus Applications of Lithium-Ion Batteries. Progress and Prospects. In *Lithium-Ion Batteries: Advances and Applications*; Pistoia, G., Ed.; Elsevier: Amsterdam, The Netherlands, 2014; ISBN 9780444595133.
60. Li, J.Q. Battery-electric transit bus developments and operations: A review. *Int. J. Sustain. Transp.* **2016**, *10*, 157–169. [CrossRef]
61. Zivanovic, Z.; Nikolic, Z. The Application of Electric Drive Technologies in City Buses. In *New Generation of Electric Vehicles*; Stevic, Z., Ed.; Intech Open: London, UK, 2012.

62. An, K. Battery electric bus infrastructure planning under demand uncertainty. *Transp. Res. Part C Emerg. Technol.* **2020**, *111*, 572–587. [CrossRef]
63. CIVITAS. Smat choices for cities: Clean buses for your city. *Police Note* **2013**.
64. Tang, X.; Lin, X.; He, F. Robust scheduling strategies of electric buses under stochastic traffic conditions. *Transp. Res. Part C Emerg. Technol.* **2019**, *105*, 163–182. [CrossRef]
65. May, N. Local environmental impact assessment as decision support for the introduction of electromobility in urban public transport systems. *Transp. Res. Part D Transp. Environ.* **2018**, *64*, 192–203. [CrossRef]
66. Lin, Y.; Zhang, K.; Shen, Z.J.M.; Ye, B.; Miao, L. Multistage large-scale charging station planning for electric buses considering transportation network and power grid. *Transp. Res. Part C Emerg. Technol.* **2019**, *107*, 423–443. [CrossRef]
67. Jing, W.; An, K.; Ramezani, M.; Kim, I. Location Design of Electric Vehicle Charging Facilities: A Path-Distance Constrained Stochastic User Equilibrium Approach. *J. Adv. Transp.* **2017**, *2017*, 1–15. [CrossRef]
68. Liu, H.; Wang, D.Z.W. Locating multiple types of charging facilities for battery electric vehicles. *Transp. Res. Part B Methodol.* **2017**, *103*, 30–55. [CrossRef]
69. Vepsäläinen, J.; Kivekäs, K.; Otto, K.; Lajunen, A.; Tammi, K. Development and validation of energy demand uncertainty model for electric city buses. *Transp. Res. Part D Transp. Environ.* **2018**, *63*, 347–361. [CrossRef]
70. Grijalva, E.R.; López Martínez, J.M. Analysis of the Reduction of CO2 Emissions in Urban Environments by Replacing Conventional City Buses by Electric Bus Fleets: Spain Case Study. *Energies* **2019**, *12*, 525. [CrossRef]
71. Wu, X.; Wang, T. Optimization of battery capacity decay for semi-active hybrid energy storage system equipped on electric city bus. *Energies* **2017**, *10*, 792. [CrossRef]
72. Gokce, K. Performance evaluation of a newly designed robotized gearbox for electric city buses. *Mechanika* **2017**, *23*, 639–645. [CrossRef]
73. Wieczorek, M.; Lewandowski, M.; Jefimowski, W. Cost comparison of different configurations of a hybrid energy storage system with battery-only and supercapacitor-only storage in an electric city bus. *Bull. Pol. Acad. Sci. Tech. Sci.* **2019**, *67*, 1095–1106. [CrossRef]
74. Xylia, M.; Silveira, S. The role of charging technologies in upscaling the use of electric buses in public transport: Experiences from demonstration projects. *Transp. Res. Part A Policy Pract.* **2018**, *118*, 399–415. [CrossRef]
75. Chiodo, E.; Lauria, D.; Andrenacci, N.; Pede, G. Accelerated life tests of complete lithium-ion battery systems for battery life statistics assessment. In Proceedings of the 2016 International Symposium on Power Electronics, Electrical Drives, Automation and Motion, SPEEDAM 2016, Pisa, Italy, 22–24 June 2016.
76. Veeneman, W. Developments in public transport governance in the Netherlands; the maturing of tendering. *Res. Transp. Econ.* **2018**, *69*, 227–234. [CrossRef]
77. Lin, B.; Tan, R. Are people willing to pay more for new energy bus fares? *Energy* **2017**, *130*, 365–372. [CrossRef]
78. Heo, J.Y.; Yoo, S.H. The public's value of hydrogen fuel cell buses: A contingent valuation study. *Int. J. Hydrog. Energy* **2013**, *38*, 4232–4240. [CrossRef]
79. Tesla battery supplier Catl says new design has one million-mile lifespan. *BBC News*, 8 June 2020.
80. Frost & Sullivan. *Global Electric Vehicle Market Outlook*; Frost & Sullivan: San Antonio, TX, USA, 2020.

*Article*

# The Balanced Energy Mix for Achieving Environmental and Economic Goals in the Long Run

**Anh Hoang To and Duc Hong Vo ***

Business and Economics Research Group, Ho Chi Minh City Open University, 97 Vo Van Tan Street, District 3, Ho Chi Minh City 7000, Vietnam; anh.th@vnp.edu.vn
* Correspondence: duc.vhong@ou.edu.vn

Received: 4 June 2020; Accepted: 22 July 2020; Published: 28 July 2020

**Abstract:** In this paper, we seek to find a balanced structure of energy sources that can simultaneously achieve two essential goals: (i) the environmental (degradation) goal and (ii) the economic (growth) goal. This study combines quantitative and qualitative methods to estimate and then rank each of the energy sources (including coal, gas, oil, hydropower, and renewable energy) to achieve the above two goals. This paper uses the weighted scoring method, the most popular method in multi-criteria decision-making techniques, to combine the rankings using five energy sources and two goals from panel data of 28 countries from Organization for Economic Co-operation and Development (OECD) countries for the period 1980–2017. Techniques for estimating the mean group long-run effect, including fully modified ordinary least squares (FMOLS) and dynamic ordinary least squares (DOLS), are used. The empirical findings of this paper reveal that, in the long term, in achieving both environmental goals and economic goals, the OECD countries should consider adopting a balanced energy mix in which the following structure is preferred: (i) hydropower, (ii) renewables and (iii) fossil fuels (oil, gas, coal).

**Keywords:** environmental degradation goal; economic growth goal; mean group analysis; weighted scoring method

---

## 1. Introduction

Energy is of widespread concern because of its effects on life, development, and the existence of current as well as future generations. Early in human history, fire was the primary energy source. Since then, we have exploited energy from various sources, such as coal, oil, hydropower, wind, solar, geothermal, and nuclear. Each source of energy has different advantages and disadvantages. In the past, fossil fuels were cheaper than renewables and had stable production. However, they caused pollution, whereas renewables were clean but limited in production. However, the selection of a source of energy depends on governmental direction, without deep and overall analysis of the economy and environment simultaneously.

Following the general trend of sustainable economic growth and development, which is generally known as green growth, the sustainable aspect of economic growth focuses on policies that can achieve economic growth not only for this generation, but also for many generations to come. The OECD countries have been formulating and implementing energy policies that are based on limiting $CO_2$ emissions by cutting and moving towards zero oil and coal use. The energy use of the United Kingdom has been transferred dramatically from fossil to clean energies, which accounted for 52 percent of the total energy consumption in 2017 [1]. In the US, the current government is still interested in fossil fuels. However, the government is planning to switch to solar and wind because of its low cost and environmentally friendly attributes. However, the US economy is still heavily dependent on fossil energy, particularly coal and oil. Limiting the use of fossil fuels will reduce the amount of $CO_2$ released into the environment, but at the same time, slow down economic growth [2].

Some countries take advantage of their natural advantages to develop and exploit renewable energy. In Sweden, renewable energy (mainly wind, nuclear and hydro) accounts for a share of more than half of the domestic demand. This current level is expected to increase further by 100 percent by 2040 [1]. The most significant problem for energy policies with a focus on renewable energy is the guarantee of energy security for the nation.

On the other hand, the problem with oil-use countries is the fluctuations in oil prices. We note that oil prices are unpredictable and uncontrollable, especially in the event of unexpected events like the COVID-19 pandemic. The impact of oil prices on the consumer price index (CPI) in these countries is very significant, requiring quick actions from the government in seeking alternative energy sources or supporting the economy with stimulus packages. The need to balance sustainable economic growth and development and environmental protection should be a top priority in energy policy.

In general, energy policies are based on various factors, including internal and external factors, such as price stabilization, reducing $CO_2$ emissions and ensuring energy security, affordability, and suitability to the economy. However, decisions are mainly based on specific information for each factor, without considering the balance of economics and the environment simultaneously. As such, we consider that this paper will provide an additional piece of empirical evidence for governments to consider when they formulate and implement energy mix policies in their countries.

The empirical papers on energy economics to date appear to focus on the investigation of a relationship among variables of interest. For example, empirical studies on the environmental Kuznets curve hypothesis (EKC), a highly cited concept in energy economics studies, generally focus on the three main streams of analyses. The first stream empirically examines the change in the traditional EKC theory. The second stream investigates the nexus between environmental quality and total energy use. The last stream of research investigates the inter-relationship between trade openness, proxied by foreign investment flows, and environmental quality.

This paper is unique and different from other empirical papers in the area of energy economics and policy implications. In this paper, we seek to find a balanced structure of energy sources that can simultaneously achieve important goals in both domains: (i) the environment and (ii) the economy. Advanced countries such as Japan, the United Kingdom and France (and many others) have been advancing towards the use of cleaner energy sources to minimize the negative impacts of energy consumption on environmental degradation. The governments of developing and emerging countries appear to prioritize economic growth and development. The debate about striking the right balance between what we call the environmental goal and the economic goal appears to have been ignored in the current literature.

The structure of this paper is as follows: Section 2 discusses selected empirical studies on energy economics to date, with a focus on the three strands of research in response to the environmental Kuznets curve hypothesis. Section 3 presents the research methodology and data. The empirical findings of this paper, including the sensitivity analyses, are included in Section 4, followed by the conclusions in Section 5 of the paper.

## 2. Literature Review

This paper is based on two traditional theories/hypotheses, including the Kuznets environment curve from environmental studies and economic growth theory. For the first hypothesis, in the 1950s, Simon Kuznets examined the relationship between economic growth and initial inequality. The Kuznets curve hypothesis states that when a nation follows industrialization, especially in agricultural mechanization, the economic center of a nation will move gradually towards the urban zones. The consequence of this development is that farmers and unskilled laborers from rural areas have to change their workplace by moving to large cities in order to earn more. This movement causes a substantial gap in earnings between people living in rural areas and downtown areas. Business owners earn profits. Workers in these industries receive an increase in income at a slower rate. However, the incomes of farmers fall because the population in rural areas declines, while the urban population increases. Nevertheless, inequality then declines as economic growth reaches the highest level of average income. An increase in Gross Domestic Product (GDP) per capita follows after the country reaches the optimal level of industrialization. Kuznets states that this inequality tends to resemble a U-shaped curve, where it increases first and then decreases with the increase in GDP per capita.

After decades of hypothesizing, Kruger and Grossman [3] apply the concept to their research in the field of the environment. Kuznets' curve is used to illustrate the relationship between economic growth and environmental degradation. An inverted U-shaped correlation is found. These results indicate that the nation's early stage of economic growth can be associated with the sacrifice of environmental quality. This view supports observations from developing and emerging countries. However, when a relatively high level of economic growth and development is achieved, the concerns for environmental quality emerge and increase. As a result, the inverted U-shape of the EKC curve has been supported by various empirical studies, including Shafik [4] and Omotor and Orubu [5]. This relationship has been considered a standard feature in engineering for formulating and implementing environmental policies. Onafowora and Owoye [6] indicate the long-term relationship between economic development and $CO_2$ emissions. Al-Mulali and Oxturk [7] also confirm the U-shaped relationship between GDP and $CO_2$ emissions.

Given the importance of the concept, may empirical studies have been conducted to examine the validity of the EKC hypothesis. In the beginning, time series analyses with many different techniques are employed such as Auto Regressive Distributed Lag (ARDL), Vector Auto Regression (VAR), the Vector Error Correction Model (VECM), and Granger causality to investigate the nexus of economic growth and environmental quality for specific countries or groups of countries. Empirical papers to date appear to focus on the investigation of a relationship among variables of interest using panel data. Mixed results on this relationship are reported, including studies from Ang [8], Hossain [9], Lean and Smyth [10], Magazzino [11], and Magazzino [12]. In particular, Rahman and Velayutham [13] confirm no causal relationship or unidirectional causality in the short and long run. In contrast, Zhang [14], Shahbaz et al. [15], Salahuddin et al. [16] report on the bidirectional relationship between economic growth and environmental quality. Niu et al. [17] indicate the unidirectional causality findings in the short run, and the directional relationship is observed in the long run.

In addition to testing the validity of traditional EKC theory (the inverted U-shape curve), a new stream of research examines whether a so-called N-shaped relationship between income and $CO_2$ emissions does exist. This stream of research has raised many empirical research questions, which have been explored by various scholars such as Rahman and Velayutha [13], To et al. [18], Sarkodie and Strezov [19], Churchill et al. [20], Zhou et al. [21], Magazzino [22], and Magazzino [14]. Churchill et al. [20] test the N-shape relationship for the OECD countries in the period 1870–2014 using mean group estimators (Mean Group (MG), Pooled Mean Group (PMG), Augmented Mean Group (AMG), and Common Correlated Effects Mean Group (CCEMG)). They find two turning points in terms of GDP per capita, i.e., the relationship exhibits an N-shape in some countries such as Australia, Canada, and Japan, but not in others, such as Spain and the UK. Moreover, Sarkodie and Strezov [19] also test this N-shape in the top five developing countries that emit a significant level of greenhouse gases, including China, Iran, Indonesia, India, and South Africa, using panel quantile regression with data from 1982 to 2016. The findings of this study confirm the N-shaped relationship between per capita income and $CO_2$ emissions in selected countries, leading to support for the validity of the EKC hypothesis.

In addition, unlike other papers, the second background theory utilized in this paper is based on economic growth theory. Economic growth has been considered an important and interesting topic for many economists. Barro [23], who supplements the classical and neoclassical growth models, studies the growth model, which considers additional variables of energy and other macro variables, including the impact of government on economic growth in the long run. This model simultaneously tests the validity of Keynes' theory and provides evidence on an unclear relationship between economic growth and environmental quality. By assuming government spending is complementary to private-sector production, Barro's model points to the non-monotonous relationship between government spending and economic growth. Hence, the neoclassical growth model with the participation of government such as the model of Barro [23] is often used in the research to test the factors affecting economic growth [24].

Empirical research based on growth theory and the growth model of Barro [23] conducts two main empirical tests. The first empirical test examines the impact of the government's role on economic development. The second test examines the link between economic growth and energy consumption, a stream that normally runs alongside EKC research.

The mechanism and level of the economic impact of public spending remain controversial and are explained by different theories. In Keynesian economics, economic output is determined by aggregate demand. Meanwhile, along with factors such as consumption, income, and net exports, public spending is seen as an important derivative of aggregate demand [25]. Consequently, therefore, Keynes argued that government involvement in the economy is necessary. When the economy is in recession, the government needs to maintain demand for investment to stimulate private investment with large public investment programs, also known as the "crowding-in effects" hypothesis of public spending with private investment [26,27].

In contrast, neoclassical growth models argue for the "crowding-out effects" of public spending on private investment [28–32]. Government spending can directly substitute private investment, thereby slowing future growth [31]. Furthermore, government demand for goods and services may cause interest rates to rise. As a result, capital becomes more expensive, negatively affecting access to private sector capital. By raising taxes or borrowing to finance public spending, public spending also makes it difficult for the private sector to access scarce financial resources [30,31].

Many economists, such as Devarajan et al. [33], Chen [34], and Ghosh and Gregoriou [35], extended Barro's model to examine the impact of different components of government spending on economic growth. By assigning different elasticity coefficients to different sectors of government expenditure, their models can determine the optimal scale and structure of the public sector for economic growth.

The empirical findings of other papers confirm the negative effect of public spending on economic growth [36,37]. In contrast, a positive contribution of public spending to economic growth has also been found in other studies [38]. Meanwhile, a few studies have found public spending to have non-linear effects on economic growth [39]. Interpreting the results of a mixed test, Gemmell et al. [24] point out the role of budget constraints in the relationship between public spending and economic growth. Nevertheless, empirical studies examining the role of budget constraints in the relationship between public spending and economic growth are quite limited [24,40].

In other words, empirical studies that test the relationship between energy consumption and economic growth face problems with inconsistency and conflicting results among researchers. In the beginning, researchers found a one-directional effect of this nexus; however, the direction between them is actually the opposite. For instance, Soytas and Sari [41] find no bidirectional nexus between the two, while Lee [42] confirms a causal relationship in which energy consumption affects economic growth and vice versa.

Huang et al. [43]; To et al. [18]; Vo et al. [44] state the reasons for their inconsistent results: (1) the difference in the period of the time series; (2) the use of time series techniques without analyzing or controlling for structural change (change in the short run) and the business cycle; (3) the sample period not being long enough to analyze the long-run effects. Thus, to address these issues, especially the disadvantages of time series data, To et al. [18] used macro panel data (panel data with a large time dimension) on 25 emerging and developing countries to determine the causality nexus between energy consumption, foreign direct investment (FDI), $CO_2$ emissions, and GDP. They found an inverted N-shaped relationship between GDP and environmental degradation. An inverted U-shaped nexus between FDI and $CO_2$ emissions is also found in these emerging and developing countries, which implies a trade-off between economic growth and the quality of the environment, in which environmental standards are relaxed to attract more foreign investment. Moreover, they also stated the positive impact of energy consumption on $CO_2$ emissions. This finding is consistent with the results of Chandran and Tang [45] and Acaravci and Ozturk [46].

In this paper, the authors simultaneously use the EKC hypothesis on the environment and Barro's economic growth model. Unlike previous papers, our study does not utilize energy consumption as the total amount of energy. In contrast, our study breaks down energy consumption into various energy sources. We then carefully analyze the impact of each energy source on the environment and economic growth. This analysis is done together with the use of macroeconomic panel data to estimate the long-run effect for each energy source. Based on the estimated coefficients, all energy sources are ranked in order of those that are the least harmful to the environment and that provide the most significant contribution to economic growth. We believe that the approach which was taken for solving our research objective is new. Multi-criteria decision making (MCDM) analyses, including five energy sources and two criteria, are considered to combine these two rankings from two criteria (environmental degradation and economic growth). This study uses the weighted scoring method (WSM), the most popular method for MCDM [47–49], to score all rankings from five sources and two criteria. This method chooses a set of several alternatives (energy sources), which depend on the score for each alternative and the weighting for each criterion. The final optimal structure of energy sources is a set of five sources that satisfy the two most important criteria, including: (i) the most positive and significant effect on economic growth, and (ii) the least harmful effect on the environment by reducing $CO_2$ emissions.

## 3. Methodology and Data

### 3.1. Models Representing for the Environmental Goal and the Economic Goal

To determine an energy structure that simultaneously achieves both the environmental degradation goal and the economic growth goal, we construct a model including two distinct parts for achieving these two goals simultaneously: (i) the environmental degradation goal and (ii) the economic growth goal.

- The first part, with a focus on the environmental goal, ranks five energy sources based on the level of $CO_2$ emissions.
- The second part, with a focus on the economic goal, ranks five energy sources (coal, gas, oil, hydropower, and renewable energy) based on their impact on economic growth.

To combine the rankings from these two parts, we develop a multi-criterion decision-making technique (MCDM) using five energy sources (including coal, gas, oil, hydropower, and renewable energy) and two criteria (environmental goal and economic goal). The weighted score method (WSM), the most popular method in MCDM [47–49], is used to score all the ranking results. The final structure for the energy mix demonstrates a source of energy (coal, gas, oil, hydropower, and renewable energy) in the order of preferences that satisfy the following two conditions: (i) a particular source of energy does the least harm to the environment or has the lowest $CO_2$ emissions and (ii) a particular source of energy boosts economic growth the most.

#### 3.1.1. The Environmental Goal

The rankings related to the environmental goal are employed following an examination of the validity of the traditional EKC hypothesis [50]. The non-linear relationships between environmental quality and income are reported. More specifically, the relationship has an inverted U-shape, which means that, after a threshold level of income, an increase in income will reduce the negative effect on environmental quality. On this basis, the model takes the following form [18]:

$$EQ = f(GDP, GDP^2, EC)$$

where EQ stands for the environment quality, which can be proxied by the level of emissions, such as $CO_2$. In order to raise the reliability of the analysis and estimation, the proxy variable should have a long time period. As such, we consider that employing $CO_2$ emissions as the proxy is appropriate. Income and square income are widely used in empirical analyses testing a non-linear relationship between economic growth and environmental degradation. EC stands for energy consumption, which comprises the consumption of five energy sources: oil, gas, coal, hydropower, and renewable energy. Employing various energy sources allows us to separate the contribution of each energy source to environmental quality. With these variables, we construct a regression model, model 1, as follows:

$$CO_{2\,it} = \pi_0 + \pi_1 GDP_{it} + \pi_2 GDP^2_{it} + \pi_3 Coa_{it} + \pi_4 Gas_{it} + \pi_5 Oil_{it} + \pi_6 Hyd_{it} + \pi_7 Ren_{it} + \varepsilon_{it} \tag{1}$$

where carbon emissions ($CO_2$) are used as the dependent variable.

Various independent variables are used, including per capita real GDP, per capita real GDP squared and cubed, and per capita consumption of coal, gas, oil, hydropower, and renewable energy. All variables are transformed into their logarithmic form.

### 3.1.2. The Economic Growth Goal

According to growth theories, empirical studies on testing and estimating the effect of growth factors are commonly based on the production function, especially the Cobb-Douglas [51] function, divided into four main factors: technology, capital, human resources, and natural resources. These empirical studies normally transform the model into a logarithmic form to facilitate analysis. The output growth model of Barro [52] is basically presented as follows:

$$\Delta Y = F(Y, Y^*)$$

where $\Delta Y$ is the growth rate of income/output, $Y$ is per capita income/output, and $Y^*$ is the long-run level of income/output or potential income/output of an economy. The value of $Y^*$ is based on government policies such as investment in education, research activities, and increases in capital. $\Delta Y$ is positively related to $Y^*$ and negatively related to $Y$. The Barro model, which includes control variables, is as follows:

$$\Delta Y_{it} = \pi_0 + \pi_1 Y_{o\,i} + \pi_2 \Delta EC_{it} + \pi_3 X_{it} + \varepsilon_{it}$$

where $\Delta Y_{it}$ is the economic growth rate of country i at year t, $Y_{o\,I}$ stands for the logarithm of initial per capita GDP of country i, and $EC_{it}$ denotes the log of energy consumption of country i at year t. Barro [52] and Huang et al. [43] used control variables ($X_{it}$), including inflation, capital stock, government spending, growth of labor, and degree of international openness. Model 2 is written as follows:

$$\Delta lnY_{it} = \pi_0 + \pi_1 lnY_{o\,i} + \pi_2 \Delta lnCoa_{it} + \pi_3 \Delta lnGas_{it} + \pi_4 \Delta lnOil_{it} + \pi_5 \Delta lnHyd_{it} + \pi_6 \Delta lnRen_{it} +$$
$$\pi_7 INF_{it} + \pi_8 CAP_{it} + \pi_9 GEX_{it} + \pi_{10} \Delta lnLF_{it} + \pi_{11} TRADE_{it} + \varepsilon_{it} \tag{2}$$

where:

$\Delta lnY_{it}$: the first difference in the logarithm of per capita income for the country i at year t;

$lnY_{o\,i}$: the log of initial per capita income of country i;

$\Delta lnCoa/Gas/Oil/Hyd/Ren_{it}$: the first difference in the logarithm of coal, natural gas, oil, hydropower, and renewable energy consumption for country i at year t;

$INF_{it}$: inflation rate of country i at year t;

$CAP_{it}$: gross fixed capital formation for country i at year t (%GDP);

$GEX_{it}$: general government final consumption expenditure for country i at year t (%GDP);

$\Delta lnLF_{it}$: the first difference in the logarithm of the labor force for country i at year t;

$TRADE_{it}$: total export and import as a share of GDP for country i at year t (%GDP).

### 3.2. Econometric Techniques

Two models, model 1 representing the environmental goal and model 2 representing the economic goal, are analyzed using the same econometric techniques. Various econometric analyses, including the cross-sectional test, the stationary test, and panel cointegration test, are conducted. Nguyen and Vo [53] and To et al. [18] state that this procedure can be estimated by three steps. First, macro panel data have a long time dimension; thus, the first step for the macro panel data is the same as with the time series data. Second, the order of integration for each variable of the macro panel data needs to be tested and determined. Third, a prerequisite for the existence of a long-run relationship is the presence of cointegration between variables. Once cointegration is confirmed, the long-run relationship between the group of integrated variables can be investigated using the Vector Error Correction Model (VECM) (see [54,55]). These tests are discussed in detail by To et al. [18].

If these tests are verified, and the results show that there is at least one long-run nexus between explanatory variables and the dependent variable, then the long-run estimators are employed. The most popular models for estimating the mean group long-run effect, which can treat endogeneity problems and serial correlation in macro panel data, are fully modified ordinary least squares (FMOLS) and dynamic ordinary least squares (DOLS). These techniques, FMOLS and DOLS, are used in this paper. The purpose of this study is to find a balanced structure of energy sources (coal, gas, oil, hydropower, and renewable energy). As such, after conducting regressions, these five energy sources are ranked based on their regressors.

- For the first criterion (with a focus on the environmental degradation goal, using the model of EKC as presented in Equation (1)), the first priority is the source of energy that has the smallest effect on environmental degradation or the smallest regression coefficient obtained in model 1. As such, these five sources of energy are ranked based on the magnitude of their respective regressors, in the following order: (i) negative impact on $CO_2$ emissions or negative coefficients (statistically significant); (ii) no impact (zero coefficients with statistically significant) or the coefficients are statistically insignificant; (iii) positive coefficient (statistically significant).
- For the second criterion (with a focus on the economic growth goal, using the growth model as presented in Equation (2)), the first priority is the source of energy that has the largest contribution to economic growth. As such, the following order is used to rank energy sources: (i) statistically significant with a positive sign, (ii) no impact (zero coefficient with insignificant coefficient), and (iii) negative contribution (statistically significant).

### 3.3. Weighted Scoring Method

The weighted scoring method (WSM) is then used in the next step to combine the two sets of rankings (one set for the environmental goal and the other set for the economic goal) based on the score for each of the five energy sources (coal, gas, oil, hydropower, and renewable energy). The following equation is used:

$$S(A_i) = \sum W_j \times S_{ij}.$$

The multi-criteria decision-making techniques (MCDM) in this paper now consist of two criteria $\{C_1, C_2\}$ (being the environmental criterion and the economic criterion) and five energy sources $\{A_1, A_2, A_3, A_4, A_5\}$ (being coal, gas, oil, hydropower, and renewable energy) in a decision matrix of all choices $\{S_{ij}\}$, where $\{S_{ij}\}$ is the score after calculating and evaluating the performance of choices using criterion $\{C_j\}$. The weights $\{W_1, W_2\}$ indicate the importance or the role of a specific criterion. A sensitivity analyses using different sets of weights are also conducted to ensure the robustness of the findings.

### 3.4. Data

Data are used for both model 1 (on the environmental goal) as presented in Equation (1) and model 2 (on the environmental goal) as presented in Equation (2). Data are collected for the period from 1980 to 2017 for 28 developed countries in the Organization for Economic Co-operation and Development (OECD), which include the following countries: Australia, Austria, Belgium, Canada, Chile, Czech Republic, Denmark, Estonia, Finland, France, Germany, Greece, Hungary, Iceland, Ireland, Israel, Italy, Japan, Korea, Rep., Latvia, Lithuania, Luxembourg, Mexico, Netherlands, New Zealand, Norway, Poland, Portugal, Slovak Republic, Slovenia, Spain, Sweden, Switzerland, Turkey, United Kingdom, and the United States.

The description of all variables in both model 1 and model 2 are presented in Table 1.

**Table 1.** Definitions of the variables.

| Variable | Measurement | Definition | Source |
|---|---|---|---|
| Carbon emissions ($CO_2$) | Metric tons per capita | $CO_2$ emissions are generated by burning fossil fuels and by consumption of solid, liquid and gas fuel, and gas. | International Energy Agency * |
| Per capita income (GDP) | GDP per capita (current US$) | GDP per capita is the ratio between the gross domestic product and the midyear population | WDI, World Bank |
| Oil consumption (Oil) | Tons of per capita oil equivalent) | Crude oil is an unrefined oil, and it is classified as fossil fuels. It includes hydrocarbon residues and other organic materials. It can be refined to produce usable products such as gasoline, diesel, petrochemicals (such as plastics), fertilizers, and even drugs. | BP Statistical Review of World Energy ** |
| Gas consumption (Gas) | Consumption of natural gas (tons of oil equivalent per capita) | Natural gas is a fossil fuel that is a mixture of combustible gases, including most of the hydrocarbons. | BP Statistical Review |
| Coal consumption (Coa) | Consumption of coal (tons of oil equivalent per capita) | Coal consumption is commercial coal, which is primarily used as a solid fuel for electricity generation and combustion. | BP Statistical Review |
| Hydropower consumption (Hyd) | Consumption of hydropower (tons of oil equivalent per capita) | Hydropower consumption, based on total primary hydropower output, does not account for transboundary electricity supply. Consumption is converted from energy generation data, with an assumption of efficiency of 38% based on data from modern thermal power plants | BP Statistical Review |
| Renewable energy consumption (Ren) | Consumption of Renewable energy (tons of oil equivalent per capita) | Renewable energy consumption, based on the total output from renewable sources, including wind, geothermal, solar, biomass, and waste, and does not account for cross-border power supplies | BP Statistical Review |
| Inflation rate (INF) | Based on consumer prices (annual %) | Inflation represents an increase in the general price of goods and services over time in the economy. | World Bank |

**Table 1.** *Cont.*

| Variable | Measurement | Definition | Source |
|---|---|---|---|
| Gross fixed capital formation (CAP) | Measured by dividing investment by income (%) | Total fixed capital includes improvements in land, factories, machines, vehicles, weapons, intellectual property, rare assets (gold, silver and others), underground assets (oil, coal and others), and other natural assets. | World Bank |
| Government expenditure (GEX) | Measured by dividing public spending by income (annual %) | General government expenditure includes all current government spending on the purchase of goods and services (excluding military spending) | World Bank |
| Labor force (LF) | Total labor force | The labor force includes all people who are of working age who have a job or are looking for work. | World Bank |
| International openness (TRADE) | Measured as % of GDP | It is calculated by dividing the sum of exports and imports by GDP. | World Bank |

* $CO_2$ emissions are affected by burning data from fossil fuels, soil, and cement equipment, collected by the Carbon Dioxide Information Analysis Center (CD CDIAC). The center collected global carbon dioxide emissions between 1950 and 1982, estimated by Marland and Rotty [56] from fuel production data from the UN's Energy Statistics Yearbook [57]. We consider that the main reason for the use of fuel production data is due to a higher level of reliability in comparison with fuel consumption data at the global level. This choice of using fuel production data is widely utilized in empirical analyses. Moreover, doing so will also avoid creating an accounting identity in Equation (1). We consider that when energy consumption data is used, the total of estimated coefficients for GDP, $GDP^2$ in Equation (1) is equal to zero. ** Collected from government sources and published data, including data from the Energy Research of the Institute of Geosciences and Natural Resources which is available in BP Statistical Review [58].

Table 2 presents the descriptive statistics of the variables utilized in our two models.

**Table 2.** Descriptive statistics.

| Variable * | Mean | Standard Error | Skewness | Kurtosis | Min | Max | Obs. |
|---|---|---|---|---|---|---|---|
| $CO_2$ (log) | 2.11 | 0.53 | −0.36 | 3.34 | 0.50 | 3.44 | 1064 |
| GDP (log) | 9.91 | 0.86 | −0.80 | 3.52 | 7.13 | 11.69 | 1064 |
| $GDP^2$ (log) | 98.95 | 16.55 | −0.55 | 3.08 | 50.81 | 136.63 | 1064 |
| Coal consumption (log) | 5.56 | 1.88 | −2.03 | 6.97 | −1.38 | 8.52 | 1064 |
| Natural gas consumption (log) | 5.56 | 2.18 | −1.67 | 4.84 | −0.89 | 7.91 | 991 |
| Oil consumption (log) | 0.54 | 0.49 | −0.70 | 4.39 | −1.07 | 1.93 | 1064 |
| Hydropower consumption (log) | 4.33 | 2.75 | −0.53 | 2.63 | −2.90 | 9.15 | 1054 |
| Renewable energy consumption ** (log) | −11.92 | 7.60 | 0.06 | 2.80 | −34.79 | 9.51 | 986 |
| Inflation | 7.67 | 20.91 | 9.73 | 137.60 | −4.48 | 373.22 | 1064 |
| Capital formation | 22.81 | 3.92 | 0.67 | 4.33 | 11.54 | 39.40 | 1064 |
| Government expenditure | 18.55 | 4.61 | −0.02 | 3.12 | 7.52 | 38.24 | 1064 |
| Labor force (dlog) | 0.01 | 0.02 | 4.33 | 53.78 | −0.05 | 0.25 | 961 |
| Trade openness | 73.56 | 49.04 | 3.11 | 16.95 | 16.01 | 416.39 | 1064 |

* The unit production data is a million tonnes of oil equivalent (Mtoe). ** Renewable energy includes biomass, geothermal, solar, wind, and other renewable sources.

## 4. Empirical Results

### 4.1. Test of Presence of Cross-Sectional Dependence

The first step in the regression technique is to test for the presence of cross-sectional dependence, and the results from this test affect all the techniques used in the subsequent steps. As a result, to obtain strong test results, we use three simultaneous tests: Pesaran [59], Friedman [60], and Frees [61]. Although the three tests have their advantages and disadvantages, they also provide an overview of the robustness of the results. Moreover, two of the specifications (fixed effect and random effect) are employed in the three tests in both models to reveal the change in the test results.

If the null hypothesis is accepted, there is no cross-sectional dependence, and the appropriate unit root test for all data is the Pesaran test [62]—the second generation of panel unit root tests, and the long-run estimator methods are pooled using FMOLS and DOLS [18]. In contrast, all the results in Table 3, including six tests for each model, strongly reject the null hypothesis at the one percent and five percent significance levels. That means that all data samples have cross-sectional dependence or are sample country specific. This leads to a change in the test used in the following steps. In this case, the unit root test used should be the one by Im, Pesaran, and Shin (IPS test; [63]), which is expanded in the Choi test for cross-sectional dependence. Furthermore, mean group regressions, such as main mean group analysis, including fully modified ordinary least squares (FMOLS), dynamic ordinary least squares (DOLS) and the other mean group analysis methods, including Mean Group (MG), Common Correlated Effects Mean Group (CCEMG) and Augmented Mean Group (AMG), should be used to determine the long-run effects [64].

**Table 3.** Sectional independence tests.

| | Pesaran | | Friedman | | Frees | |
|---|---|---|---|---|---|---|
| | **CD Test** | *p*-**Value** | **CD** | *p*-**Value** | **CD (Q)** | *p*-**Value** |
| | Model 1: Environment | | | | | |
| FE model | 2.089 ** | 0.0367 | 42.632 ** | 0.0211 | 3.949 *** | 0.000 |
| RE model | 2.173 ** | 0.0298 | 46.507 *** | 0.0080 | 4.130 *** | 0.000 |
| | Model 2: Economic | | | | | |
| FE model | 50.617 *** | 0.000 | 230.101 *** | 0.000 | 8.563 *** | 0.000 |
| RE model | 47.956 *** | 0.000 | 228.886 *** | 0.000 | 8.499 *** | 0.000 |

Notes: Fixed effects (FE) and random effects (RE) models. *** and ** indicate statistical significance at the one and five percent level, respectively.

### 4.2. Panel Unit Root Tests

A unit root test is conducted to determine the stationarity and the integration of the same order for variables used in the paper [18]. This test is required before the cointegration tests are conducted to examine the long-run nexus between $CO_2$ emissions and each source of energy (model 1) and economic growth and each source of energy (model 2). The results of the unit root tests and robustness checks are presented in Table 4 below. The robustness checks from all four tests are presented for all variables, with the constant and the trend and constant shown in both the level and first difference forms.

**Table 4.** Unit root test using Pesaran test.

| Variables | Level | | First Difference | | Order of Integration |
| --- | --- | --- | --- | --- | --- |
| | **Constant** | **Constant and Trend** | **Constant** | **Constant and Trend** | |
| $lnCO_2$ | 0.969 (0.834) | −0.672 (0.251) | −16.158 (0.000) | −15.002 (0.000) | I(1) |
| lnGDP | −0.11 (0.456) | −0.103 (0.459) | −13.964 (0.000) | −11.918 (0.000) | I(1) |
| $lnGDP^2$ | 0.034 (0.514) | 0.014 (0.506) | −13.602 (0.000) | −11.442 (0.000) | I(1) |
| lncoa | 2.032 (0.979) | 2.407 (0.992) | −13.365 (0.000) | −11.547 (0.000) | I(1) |
| lngas | 0.661 (0.746) | 3.816 (1.000) | −12.924 (0.000) | −12.06 (0.000) | I(1) |
| lnoil | 3.495 (1.000) | −1.203 (0.114) | −14.358 (0.000) | −12.208 (0.000) | I(1) |
| lnhyd | −0.651 (0.258) | −1.037 (0.150) | −19.948 (0.000) | −18.359 (0.000) | I(1) |
| lnren | −0.237 (0.406) | −0.403 (0.344) | −12.499 (0.000) | −10.724 (0.000) | I(1) |
| INFCPI | −1.911 (0.028) | −0.171 (0.432) | −14.968 (0.000) | −13.106 (0.000) | I(1) |
| CAP | −0.281 (0.389) | 1.351 (0.912) | −13.002 (0.000) | −10.442 (0.000) | I(1) |
| GEX | −0.863 (0.194) | 1.154 (0.876) | −9.131 (0.000) | −7.272 (0.000) | I(1) |
| lnLF | 0.922 (0.822) | 0.260 (0.603) | −6.546 (0.000) | −5.563 (0.000) | I(1) |
| TRADE | −1.171 (0.121) | 0.190 (0.575) | −12.965 (0.000) | −10.474 (0.000) | I(1) |

Standard errors in parentheses.

### 4.3. Panel Cointegration Test Results

Cointegration tests, including those by Kao [65], Pedroni [66], and Westerlund [67], are employed after confirming the stationarity at the same order I(1) in Table 4. This step helps to avoid spurious results [18], and we conduct these three tests at the same time to obtain robust results. All the results are in Tables 5 and 6.

**Table 5.** Cointegration tests for model 1 (Equation (1)).

| | **Test Statistic** | **$p$-Value** |
| --- | --- | --- |
| **Kao test for cointegration** | | |
| Modified Dickey–Fuller t | −2.6220 *** | 0.0044 |
| Dickey–Fuller t | −1.1550 | 0.1239 |
| Augmented Dickey–Fuller t | −0.3082 | 0.3790 |
| Unadjusted modified Dickey–Fuller t | −3.1510 *** | 0.0008 |
| Unadjusted Dickey–Fuller t | −1.4200 * | 0.0778 |
| **Pedroni test for cointegration** | | |
| Modified Phillips–Perron t | 2.3250 *** | 0.0100 |
| Phillips–Perron t | −4.1450 *** | 0.0000 |
| Augmented Dickey–Fuller t | −4.0770 *** | 0.0000 |
| **Westerlund test for cointegration** | | |
| Variance ratio | −2.3404 ** | 0.0333 |

Notes: ***, **, and * show the rejection of the null hypothesis of no cointegration is statistically significant at the 1, 5, and 10 percent levels, respectively.

**Table 6.** Cointegration tests for model 2 (Equation (2)).

|  | **Test Statistic** | ***p*-Value** |
|---|---|---|
| *Kao test for cointegration* | | |
| Modified Dickey–Fuller t | −28.5952 *** | 0.0000 |
| Dickey–Fuller t | −18.8550 *** | 0.0000 |
| Augmented Dickey–Fuller t | −16.9937 *** | 0.0000 |
| Unadjusted modified Dickey–Fuller t | −34.1630 *** | 0.0000 |
| Unadjusted Dickey–Fuller t | −19.3050 *** | 0.0000 |
| *Pedroni test for cointegration* | | |
| Modified Phillips–Perron t | −2.2520 ** | 0.0122 |
| Phillips–Perron t | −8.6760 *** | 0.0000 |
| Augmented Dickey–Fuller t | −8.4680 *** | 0.0000 |
| *Westerlund test for cointegration* | | |
| Variance ratio | −2.0190 ** | 0.0218 |

Notes: *** and ** show the rejection of the null hypothesis of no cointegration is statistically significant at the 1, 5, percent levels, respectively.

In Table 5, the result is highly statistically significant at one percent in both the Kao and Pedroni tests and at five percent in the Westerlund test using model 1, a model of the environment. We conclude that some or all panels show cointegration between variables. In other words, there may be at least one long-run nexus between the variables in model 1. Therefore, the use of Panel Vector Auto Regression (P-VAR), which is used for evaluating the short-run effect, is not considered in this research.

Similarly, the growth model for the environment has the same results in the cointegration test. All the results, which are in Table 6, reject the null hypothesis at a highly significant level (one percent). This critical step ensures that at least one variable in model 2 has a long-run relationship with the dependent variable (economic growth rate). Overall methods, both models 1 and 2 show robustness in their test results, which predict high reliability in conclusion to this study.

*4.4. Regression and Ranking Results*

We consider that ordinary least squares (OLS) regression is inappropriate, leading to a biasedness in estimating the long-run equilibrium relationship. In this paper, we apply fully modified ordinary least squares (FMOLS) in order to take the endogeneity problems, as well as the serial correlation issues, into account [68,69]. In addition, dynamic ordinary least squares (DOLS) is also employed, as this DOLS technique can also eliminate endogeneity problems and serial correlation issues using contemporaneous values, leads, and lags in the first difference. Due to the greater use of assumptions and the reduction in the degrees of freedom by using leads and lags [70,71], FMOLS is the preferred model in this study. However, we use the result of the DOLS model to confirm the direction of the estimated coefficients.

### 4.4.1. Model 1: Environmental Goal

We employ the traditional model of the environment (Kuznet [50]) with the control variables (income and square of income) and the proxy variable for energy consumption, which is divided into five main sources of energy (coal, gas, oil, hydropower, and renewable energy) that have enough data for analysis. Furthermore, in this model, the multicollinearity problem between GDP and sources of energy is very clear in the variables in model 2 (the impact of energy use on the growth rate). To analyze the effect of the multicollinearity problem on the coefficient of the variables of concern, we use both models (FMOLS and DOLS) and other mean group models with and without control variables (GDP and $GDP^2$). The regression results are shown in Tables 7 and 8.

**Table 7.** Regression results for model 1 (dependent variable: $lnCO_2$).

| Variable | Rank | Main Mean Group Models | | Other Mean Group Methods | | |
|---|---|---|---|---|---|---|
| | | **FMOLS** | **DOLS** | **MG** | **CCEMG** | **AMG** |
| Coal | 4 | 0.273 *** | 0.273 *** | 0.225 *** | 0.245 *** | 0.232 *** |
| | | (0.007) | (0.004) | (0.025) | (0.025) | (0.021) |
| Gas | 3 | 0.166 *** | 0.190 *** | 0.136 *** | 0.135 *** | 0.123 *** |
| | | (0.012) | (0.008) | (0.021) | (0.021) | (0.021) |
| Oil | 5 | 0.533 *** | 0.522 *** | 0.585 *** | 0.560 *** | 0.591 *** |
| | | (0.013) | (0.007) | (0.036) | (0.036) | (0.031) |
| Hydropower | 1 | −0.0073 * | −0.0134 *** | −0.0186 *** | −0.012 ** | −0.0151 ** |
| | | (0.004) | (0.002) | (0.007) | (0.006) | (0.008) |
| Renewable | 2 | −0.0001 | −0.0001 *** | −0.0000 | −0.0005 | 0.0001 |
| | | (0.000) | (0.000) | (0.000) | (0.001) | (0.001) |
| GDP | | 0.0013 | −0.0785 | −0.0454 | −0.086 | 0.112 |
| | | (0.107) | (0.079) | (0.268) | (0.248) | (0.260) |
| $GDP^2$ | | −0.0011 | 0.0026 | 0.0024 | 0.0061 | −0.0055 |
| | | (0.005) | (0.004) | (0.013) | (0.012) | (0.013) |
| Constant | | 2.191 *** | 2.596 *** | 2.136 | 3.146 *** | 1.3900 |
| | | (0.540) | (0.399) | (1.390) | (1.196) | (1.340) |
| Observation | | 912 | 912 | 912 | 912 | 912 |

Standard errors in parentheses, *** $p < 0.01$, ** $p < 0.05$, * $p < 0.1$.

**Table 8.** Robustness check for model 1 by excluding the income variables.

| Variable | Rank | Main Mean Group Models | | Other Mean Group Methods | | |
|---|---|---|---|---|---|---|
| | | **FMOLS** | **DOLS** | **MG** | **CCEMG** | **AMG** |
| Coal | 4 | 0.255 *** | 0.253 *** | 0.229 *** | 0.242 *** | 0.254 *** |
| | | (0.009) | (0.014) | (0.023) | (0.024) | (0.027) |
| Gas | 3 | 0.152 *** | 0.174 *** | 0.140 *** | 0.113 *** | 0.120 *** |
| | | (0.020) | (0.035) | (0.024) | (0.021) | (0.020) |
| Oil | 5 | 0.595 *** | 0.586 *** | 0.558 *** | 0.547 *** | 0.552 *** |
| | | (0.024) | (0.035) | (0.040) | (0.033) | (0.038) |
| Hydropower | 1 | 0.0275 *** | 0.0204 ** | −0.0177 *** | −0.0147 ** | −0.0095 |
| | | (0.005) | (0.010) | (0.006) | (0.007) | (0.007) |
| Renewable | 2 | −0.0003 * | −0.0007 | −0.0002 | −0.0002 | 0.0002 |
| | | (0.000) | (0.000) | (0.000) | (0.000) | (0.001) |
| Constant | | 2.078 *** | 2.058 *** | 1.937 *** | 0.765 ** | 1.987 *** |
| | | (0.030) | (0.050) | (0.068) | (0.380) | (0.062) |
| Observation | | 912 | 912 | 912 | 912 | 912 |

Standard errors in parentheses, *** $p < 0.01$, ** $p < 0.05$, * $p < 0.1$. All variables are transferred into logarithmic form.
Note: A rank of one denotes the least environmental harm, and a rank of five denotes the most environmental harm.

This step aims to determine the impact of energy sources on environmental degradation before we do any further rankings. The main concern is how the coefficient or final rankings of these sources change across multicollinearity problems and multiple specifications. If there is no difference or the deviation in regressors between specifications is small, or the statistical significance is still high, the final rank based on these coefficients is the most reliable.

With these concerns in mind, the first considerations in the results in both Tables 7 and 8 are the sign and magnitude of the estimated coefficients for all five energies (coal, gas, oil, hydropower, and renewable energy). In the main models, FMOLS and DOLS with and without a multicollinearity check, the coefficient of coal consumption is from +0.253 to +0.273. This deviation is quite low and highly significant (one percent). This coefficient means that when coal consumption increases/decreases by one percent, on average, $CO_2$ emissions increase/decrease by 0.273 percent (FMOLS model), ceteris paribus. The coefficient for gas consumption is from +0.152 to +0.19, and all regressors are also highly statistically significant (at one percent). The economic meaning is similar to that for coal consumption, in that a change in the use of gas of one percent leads to a 0.17 percent change in $CO_2$ emissions (in the same direction). Oil and hydropower consumption have high statistical significance as well, but oil use has the biggest effect on $CO_2$ emissions (a one percent increase/reduce in oil use, on average, leads to a 0.53 percent increase/decrease in $CO_2$ emissions). In contrast, the effect of renewable energy is unclear and has a weak significance. These results indicate no evidence that the use of renewables leads to environmental degradation.

Based on the magnitude and signs of the regressors, the following steps are used to rank the five energy sources: (1) a negative impact on $CO_2$ emissions or a negative coefficient (statistically significant); (2) no impact (zero coefficient with statistically significant) or the coefficient is statistically insignificant; (3) positive coefficient (statistically significant). The final rank is as follows: hydropower, renewable energy, gas, coal and oil in the order of the least environmental harm to the most environmental harm, as indicated in Table 8.

### 4.4.2. Model 2: Economic Goal

This step considers the impact of each energy source on economic growth in the long run by applying Barro's growth model. In accordance with the approach adopted for the environmental model (Equation (1)), this analysis for the economic growth model (Equation (2)) also uses the five mean group methods, FMOLS, DOLS, MG, CCEMG and AMG, to regress the effects. The results are presented in Table 9.

**Table 9.** Regression and ranking results.

| $\Delta$lnGDP | Rank | Main Mean Group Models | | | Other Mean Group Methods | |
|---|---|---|---|---|---|---|
| | | FMOLS | DOLS | MG | CCEMG | AMG |
| $\text{lnGDP}_0$ | | −0.0132 *** | | 0.000 | 0.000 | 0.000 |
| | | (0.005) | | 0.000 | 0.000 | 0.000 |
| $\Delta$lncoa | 5 | −0.0483 * | −0.0219 | 0.0017 | −0.0537 | 0.0114 |
| | | (0.029) | (0.014) | (0.050) | (0.064) | (0.039) |
| $\Delta$lngas | 2 | 0.112 *** | 0.125 *** | −0.0661 | −0.0442 | 0.001 |
| | | (0.036) | (0.021) | (0.075) | (0.081) | (0.053) |
| $\Delta$lnoil | 1 | 0.548 *** | 0.176 *** | 0.577 *** | 0.0153 | 0.419 *** |
| | | (0.064) | (0.049) | (0.119) | (0.125) | (0.098) |
| $\Delta$lnhyd | 3 | 0.0158 * | 0.0184 *** | 0.0152 | 0.0819 * | 0.0265 |
| | | (0.009) | (0.004) | (0.024) | (0.044) | (0.017) |
| $\Delta$lnren | 4 | −0.0014 *** | −0.0009 *** | −0.0016 | −0.0116 | −0.0044 |
| | | (0.000) | (0.000) | (0.004) | (0.009) | (0.004) |
| INFCPI | | 0.0093 *** | 0.0043 *** | 0.0054 * | −0.0015 | −0.0047 * |
| | | (0.001) | (0.000) | (0.003) | (0.005) | (0.003) |
| CAP | | 0.0047 *** | 0.0101 *** | −0.0079 ** | −0.0044 | −0.004 |
| | | (0.001) | (0.001) | (0.004) | (0.005) | (0.004) |
| GEX | | 0.0014 | −0.0003 | −0.0313 *** | −0.0621 *** | −0.0530 *** |
| | | (0.002) | (0.001) | (0.007) | (0.017) | (0.006) |
| dlnLF | | 0.332 | −1.437 *** | 0.671 * | −1.187 | 0.152 |
| | | (0.306) | (0.202) | (0.385) | (0.863) | (0.292) |
| TRADE | | 0.0007 * | −0.0016 *** | −0.0016 ** | 0.0000 | −0.004 *** |
| | | (0.000) | (0.000) | (0.001) | (0.002) | (0.001) |
| Constant | | 0.0000 | −0.130 *** | 0.900 *** | −18.03 | 1.508 *** |
| | | 0.000 | (0.013) | (0.191) | (14.380) | (0.186) |
| $R^2$ | | 0.739 | 0.945 | | | |

Standard errors in parentheses, *** $p < 0.01$, ** $p < 0.05$, * $p < 0.1$. Note: A rank of one denotes the most contribution to economic growth, and a rank of five denotes the least contribution to economic growth.

*4.5. A Combination of These Two Criteria Using the Weighted Scoring Method*

We then use the weighted scoring method (WSM) to obtain a score for the five sources, in which the order of these scores shows their final contribution to the balanced structure of energy sources (including coal, gas, oil, hydropower, and renewable energy). The most preferred source of energy is the energy source with the highest score. In this method, the weight of each criterion plays an important role in the score and directly affects the final ranking. Thus, to obtain an overview of the final rank and its role in achieving both environmental and economic goals, we analyze changes in the final rank across the two scenarios step by step with sensitivity analyses. Using the weights of each criterion range from 70% to 80%, Table 10 shows how the WSM method works and the influence of this weight on the final structure.

**Table 10.** Final ranking for each of the energy sources.

| Source * | Model 1 (Environment Goal) | Model 2 (Economic Goal) | Weighting Range ** | | Final Ranking *** | |
|---|---|---|---|---|---|---|
| | | | Environment Scenario | Economic Scenario | Environment Scenario | Economic Scenario |
| Coal | 4 | 5 | | | 5 | 5 |
| Gas | 3 | 2 | | | 3 | 2 |
| Oil | 5 | 1 | 70% to 80% | 70% to 80% | 4 | 1 |
| Hydro | 1 | 3 | | | 1 | 3 |
| Renew | 2 | 4 | | | 2 | 4 |

* Hydro stands for hydropower, and Renew stands for other types of renewable energy, including wind, geothermal, solar, biomass, and waste. ** With the environmental scenario, the weighting of the environmental goal ranges from 70% to 80% and the remaining proportion of 20–30% applies to the economic goal. The same approach applies to the economic scenario, with a weight of 70–80%, leaving 20–30% for the environmental goal. *** The final ranking takes into account the rankings from estimated coefficients from models 1 and 2, together with the assumed weighting range. Robustness analyses are presented in the Appendix A of this paper. Note: A rank of one denotes the most contribution to economic growth, and a rank of five denotes the least contribution to economic growth. A rank of one denotes the least environmental harm, and a rank of five denotes the most environmental harm.

In the environmental goal scenario, we assume that a government prioritizes the environmental goal rather than the economic goal. We assign scores for each energy source using the WSM, which increases the weight of the environmental goal from 70 percent to 80 percent, and the remaining proportion for the economic goal is from 30 percent to 20 percent. This selected range demonstrates the overwhelming priority of one goal, being the environmental goal, which in this case might create a "crowding-out effect" on the other goal, being the economic goal. We also conduct sensitivity analyses, which are included in Appendix A, Table A1.

Table 10 shows that, with the priority of the environmental goal, the top ranking belongs to clean energy such as hydropower (ranked 1) and renewable (ranked 2) and fossil sources including gas, oil and coal. These findings have important implications for countries who make the environmental goal a policy priority. These countries should focus on policies that can encourage the use of clean energies. This scenario may be relevant for developed countries who have achieved a certain level of economic development, as these countries are not completely reliant on fossil fuels. For instance, many countries such as Germany, France, and Britain have set targets to ban the sale of petrol and diesel vehicles in the future [72,73]. Similarly, many cities around the world have started to convert public transportation to electric vehicles, and have banned or put taxes on diesel vehicles coming into their cities, such as Paris, Athens, Mexico City, Madrid and London [74].

The economic scenario uses the same method to analyze the role of the economic goal in the final ranking. We assign a weight from 70 percent to 80 percent to prioritize the economic goal. We also conduct sensitivity analyses using various weights for the economic scenario, which are included in Appendix A, Table A2.

When the priority is on the economic goal, a major change in ranking is observed from the first scenario (with a focus on the environmental goal) to the second scenario (with a focus on the economic goal). Table 10 shows that fossil fuels are ranked first (oil is ranked first, and gas is ranked second), and then clean energy follows (hydro and renewable energy). Making the economic goal a priority may be relevant in practice for underdeveloped countries and some developing countries. Currently, at a low economic growth rate, these countries are willing to trade off environmental degradation to attract more foreign investment in order to boost the economy [18]. It is argued that the widespread use of fossil fuel-based energy in these countries, such as oil and gas, in the process of industrialization and modernization, will lead to significant economic growth.

## 5. Conclusions and Policy Implications

This paper aimed to determine a balanced energy structure, in the long run, using data from the OECD countries for the period from 1980 to 2017 [75]. In this paper, five energy sources were considered including coal, gas, oil, hydropower, and renewable energy. The proposed optimal energy mix was developed with the view of achieving two fundamental goals at the same time: (i) to minimize environmental degradation; and (ii) to support economic growth.

In this paper, the weighted scoring method (WSM), the most popular method of the multi-criteria decision-making (MCDM) techniques, was used to combine the rankings using five energy sources and two goals. Various tests, including the cross-sectional test, the stationarity test, and panel cointegration test, were conducted in this paper. Furthermore, this paper employed mean group regressions to consider the long-run effect of the estimates. These mean group techniques included two groups: (i) the main mean group analysis, including fully modified ordinary least squares (FMOLS) and dynamic ordinary least squares (DOLS); the other mean group analysis, including Mean Group (MG), Common Correlated Effects Mean Group (CCEMG) and Augmented Mean Group (AMG). These techniques were used to determine the long-run effects between the variables utilized in the paper. Sensitivity analyses were also conducted to ensure the robustness of the findings.

Our empirical findings indicate that, in the long term, in achieving both the environmental goal and economic goals, the OECD countries may consider adopting a balanced energy mix in which the following structure, associated with preferences for each source of energy, is considered: (i) hydropower, (ii) renewables, and (iii) fossil fuels (oil, then gas, and then coal). However, we are aware that determining an optimal energy structure is not a solid scientific process because the decision on optimal energy mix heavily depends on various factors, including internal and external factors. Some of these factors may be well beyond the control of the governments of the OECD countries. For example, in designing an optimal energy structure, affordability is very important. Affordability represents the financial capacity the general public can pay to use energy. An energy structure is not optimal if the general public is unable to pay for its energy consumption. In addition, security is also a very important aspect of any optimal energy mix because the economy and society cannot be without energy. Last but not least, sustainability in economic growth and development, together with sustainability in energy consumption, are equally important compared to any other aspects. Designing an optimal energy structure is not only for current generations, but also for the many generations to come. As a consequence, we are aware of and agree with the view that designing and implementing an optimal energy structure is an extremely complicated issue. In addition, there may not be a one-size-fits-all approach because each country will face different challenges in the process of designing an optimal energy policy. The members of the OECD are mainly advanced countries, and they may share similarities in terms of their economic growth and development progress, social inclusion and culture. However, this does not mean that one policy for an optimal energy structure can be developed and applied to all members. We also consider that there may not be an optimal energy structure for any nation because energy policy has been moving and changing very quickly, particularly due to the current progress of technology. An optimal energy structure for a country today may no longer be optimal in the very near future as technology can change at the pace of days or months.

Based on the above observations, we consider that the findings of this paper should be considered as an additional piece of empirical evidence for the governments of the OECD countries to take into account, alongside all other pieces of evidence currently available within their constraints and contexts. As a result, based on the findings of this paper, the policy implications can be summarized as follows. When the environmental goal is prioritized, the optimal energy structure will start with clean energy sources, including hydropower and renewable energy. Fossil fuel energy will follow, including oil, gas and then coal. This scenario appears to be relatively consistent with the current environment for most of the developed countries in the OECD. On the other hand, in our economic scenario, in which the economic growth goal is prioritized, the important role of fossil fuel in boosting the economy is observed. This scenario confirms the view that it is difficult to replace fossil fuels with cleaner sources of energy when the first priority is to achieve economic goals. This scenario reflects the reality of the developing and emerging markets in the process of industrialization and modernization.

**Author Contributions:** Conceptualization, A.H.T. and D.H.V.; methodology, D.H.V.; software, A.H.T.; validation, A.H.T. and D.H.V.; formal analysis, A.H.T.; investigation, A.H.T.; resources, D.H.V.; data curation, A.H.T.; writing—original draft preparation, A.H.T. and D.H.V.; writing—review and editing, A.H.T. and D.H.V.; visualization, A.H.T.; supervision, D.H.V.; project administration, A.H.T. and D.H.V.; funding acquisition, D.H.V. All authors have read and agreed to the published version of the manuscript.

**Funding:** This study was funded by the Ministry of Education and Training of Vietnam under grant B2020-MBS-03.

**Acknowledgments:** The authors acknowledge constructive comments from the Editor of the journal and three reviewers. With their expert views on the issue, three reviewers provided us with very helpful, insightful and practical comments on this important topic. We also greatly appreciate the comments and suggestions from participants at Vietnam's Business and Economics Research Conference (VBER2019), July 2019, Ho Chi Minh City, Vietnam.

**Conflicts of Interest:** The authors declare no conflict of interest. The funders had no role in the design of the study; in the collection, analyses, or interpretation of data; in the writing of the manuscript, and in the decision to publish the results.

## Appendix A

**Table A1.** Sensitivity analysis for the environmental goal.

| The Weighting of Environment Goal | Score Results | | | | | Ranking | | | | |
|---|---|---|---|---|---|---|---|---|---|---|
| | Coal | Gas | Oil | Hydro | Renew | Coal | Gas | Oil | Hydro | Renew |
| 70% | 4.30 | 2.70 | 3.80 | 1.60 | 2.60 | 5 | 3 | 4 | 1 | 2 |
| 71% | 4.29 | 2.71 | 3.84 | 1.58 | 2.58 | 5 | 3 | 4 | 1 | 2 |
| 72% | 4.28 | 2.72 | 3.88 | 1.56 | 2.56 | 5 | 3 | 4 | 1 | 2 |
| 73% | 4.27 | 2.73 | 3.92 | 1.54 | 2.54 | 5 | 3 | 4 | 1 | 2 |
| 74% | 4.26 | 2.74 | 3.96 | 1.52 | 2.52 | 5 | 3 | 4 | 1 | 2 |
| 75% | 4.25 | 2.75 | 4.00 | 1.50 | 2.50 | 5 | 3 | 4 | 1 | 2 |
| 76% | 4.24 | 2.76 | 4.04 | 1.48 | 2.48 | 5 | 3 | 4 | 1 | 2 |
| 77% | 4.23 | 2.77 | 4.08 | 1.46 | 2.46 | 5 | 3 | 4 | 1 | 2 |
| 78% | 4.22 | 2.78 | 4.12 | 1.44 | 2.44 | 5 | 3 | 4 | 1 | 2 |
| 79% | 4.21 | 2.79 | 4.16 | 1.42 | 2.42 | 5 | 3 | 4 | 1 | 2 |
| 80% | 4.20 | 2.80 | 4.20 | 1.40 | 2.40 | 5 | 3 | 4 | 1 | 2 |

Note: A rank of one denotes the least environmental harm, and a rank of five denotes the most environmental harm.

**Table A2.** Sensitivity analysis for the economic goal.

| The Weighting of Economic Goal | Score Results | | | | | Ranking | | | | |
|---|---|---|---|---|---|---|---|---|---|---|
| | Coal | Gas | Oil | Hydro | Renew | Coal | Gas | Oil | Hydro | Renew |
| 70% | 4.80 | 2.20 | 1.80 | 2.60 | 3.60 | 5 | 2 | 1 | 3 | 4 |
| 71% | 4.79 | 2.21 | 1.84 | 2.58 | 3.58 | 5 | 2 | 1 | 3 | 4 |
| 72% | 4.78 | 2.22 | 1.88 | 2.56 | 3.56 | 5 | 2 | 1 | 3 | 4 |
| 73% | 4.77 | 2.23 | 1.92 | 2.54 | 3.54 | 5 | 2 | 1 | 3 | 4 |
| 74% | 4.76 | 2.24 | 1.96 | 2.52 | 3.52 | 5 | 2 | 1 | 3 | 4 |
| 75% | 4.75 | 2.25 | 2.00 | 2.50 | 3.50 | 5 | 2 | 1 | 3 | 4 |
| 76% | 4.74 | 2.26 | 2.04 | 2.48 | 3.48 | 5 | 2 | 1 | 3 | 4 |
| 77% | 4.73 | 2.27 | 2.08 | 2.46 | 3.46 | 5 | 2 | 1 | 3 | 4 |
| 78% | 4.72 | 2.28 | 2.12 | 2.44 | 3.44 | 5 | 2 | 1 | 3 | 4 |
| 79% | 4.71 | 2.29 | 2.16 | 2.42 | 3.42 | 5 | 2 | 1 | 3 | 4 |
| 80% | 4.70 | 2.30 | 2.20 | 2.40 | 3.40 | 5 | 2 | 1 | 3 | 4 |

Note: A rank of one denotes the most contribution to economic growth and a rank of five denotes the least contribution to economic growth.

## References


1. International Energy Agency. 2019. Available online: https://webstore.iea.org/download/summary/2784 (accessed on 1 July 2020).
2. U.S. Energy Information Administration. 2019. Available online: https://www.eia.gov/beta/international/data/browser (accessed on 20 April 2019).
3. Grossman, G.M.; Krueger, A.B. *Environmental Impacts of a North American Free Trade Agreement*; Working Paper No. w3914; National Bureau of Economic Research: Cambridge, MA, USA, 1991.
4. Shafik, N. Economic development and environmental quality: An econometric analysis. *Oxf. Econ. Pap.* **1994**, *46*, 757–774. [CrossRef]
5. Orubu, C.O.; Omotor, D.G. Environmental quality and economic growth: Searching for environmental Kuznets curves for air and water pollutants in Africa. *Energy Policy* **2011**, *39*, 4178–4188. [CrossRef]
6. Onafowora, O.A.; Owoye, O. Bounds testing approach to analysis of the environment Kuznets curve hypothesis. *Energy Econ.* **2014**, *44*, 47–62. [CrossRef]
7. Al-Mulali, U.; Ozturk, I. The investigation of environmental Kuznets curve hypothesis in the advanced economies: The role of energy prices. *Renew. Sustain. Energy Rev.* **2016**, *54*, 1622–1631. [CrossRef]
8. Ang, J.B. Economic development, pollutant emissions and energy consumption in Malaysia. *J. Policy Model.* **2008**, *30*, 271–278. [CrossRef]
9. Hossain, M.S. Panel estimation for $CO_2$ emissions, energy consumption, economic growth, trade openness and urbanization of newly industrialized countries. *Energy Policy* **2011**, *39*, 6991–6999. [CrossRef]
10. Lean, H.H.; Smyth, R. $CO_2$ emissions, electricity consumption and output in ASEAN. *Appl. Energy* **2010**, *87*, 1858–1864. [CrossRef]
11. Magazzino, C. Economic growth, $CO_2$ emissions and energy use in Israel. *Int. J. Sustain. Dev. World Ecol.* **2015**, *22*, 89–97.
12. Magazzino, C. The relationship between $CO_2$ emissions, energy consumption and economic growth in Italy. *Int. J. Sustain. Energy* **2016**, *35*, 844–857. [CrossRef]
13. Rahman, M.M.; Velayutham, E. Renewable and non-renewable energy consumption-economic growth nexus: New evidence from South Asia. *Renew. Energy* **2020**, *147*, 399–408. [CrossRef]
14. Zhang, M.; Li, H.; Zhou, M.; Mu, H. Decomposition analysis of energy consumption in Chinese transportation sector. *Appl. Energy* **2011**, *88*, 2279–2285. [CrossRef]
15. Shahbaz, M.; Mahalik, M.K.; Shah, S.H.; Sato, J.R. Time-varying analysis of $CO_2$ emissions, energy consumption, and economic growth nexus: Statistical experience in next 11 countries. *Energy Policy* **2016**, *98*, 33–48. [CrossRef]
16. Salahuddin, M.; Alam, K.; Ozturk, I.; Sohag, K. The effects of electricity consumption, economic growth, financial development and foreign direct investment on $CO_2$ emissions in Kuwait. *Renew. Sustain. Energy Rev.* **2018**, *81*, 2002–2010. [CrossRef]

17. Niu, S.; Ding, Y.; Niu, Y.; Li, Y.; Luo, G. Economic growth, energy conservation and emissions reduction: A comparative analysis based on panel data for 8 Asian-Pacific countries. *Energy Policy* **2011**, *39*, 2121–2131. [CrossRef]

18. To, A.H.; Ha, D.T.T.; Nguyen, H.M.; Vo, D.H. The impact of foreign direct investment on environment degradation: Evidence from emerging markets in Asia. *Int. J. Environ. Res. Public Health* **2019**, *16*, 1636. [CrossRef] [PubMed]

19. Sarkodie, S.A.; Strezov, V. Effect of foreign direct investments, economic development and energy consumption on greenhouse gas emissions in developing countries. *Sci. Total Environ.* **2019**, *646*, 862–871. [CrossRef]

20. Churchill, S.A.; Inekwe, J.; Ivanovski, K.; Smyth, R. The environmental Kuznets curve in the OECD: 1870–2014. *Energy Econ.* **2018**, *75*, 389–399. [CrossRef]

21. Zhou, Y.; Fu, J.; Kong, Y.; Wu, R. How foreign direct investment influences carbon emissions, based on the empirical analysis of Chinese urban data. *Sustainability* **2018**, *10*, 2163. [CrossRef]

22. Magazzino, C. $CO_2$ emissions, economic growth, and energy use in the Middle East countries: A panel VAR approach. *Energy Sources Part B Econ. Plan. Policy* **2016**, *11*, 960–968. [CrossRef]

23. Barro, R.J. Economic growth in a cross section of countries. *Q. J. Econ.* **1991**, *106*, 407–443. [CrossRef]

24. Gemmell, N.; Misch, F.; Moreno-Dodson, B. Public spending and long-run growth in practice: Concepts, tools, and evidence. In *Is Fiscal Policy the Answer? A Developing Country Perspective*; Moreno–Dodson, B., Ed.; World Bank: Washington, DC, USA, 2012; pp. 69–107.

25. Coddington, A. Keynesian economics: The search for first principles. *J. Econ. Lit.* **1976**, *14*, 1258–1273.

26. Coenen, G.; Straub, R. Does government spending crowd in private consumption? Theory and empirical evidence for the euro area. *Int. Financ.* **2005**, *8*, 435–470. [CrossRef]

27. Jahan, S.; Mahmud, A.S.; Papageorgiou, C. What is Keynesian economics? *Financ. Dev.* **2014**, *51*, 53–54.

28. Aschauer, D.A. Is government spending productive? *J. Monet. Econ.* **1989**, *23*, 177–200. [CrossRef]

29. Carlson, K.M.; Spencer, R.W. *Crowding Out and Its Critics*; Federal Reserve Bank of St. Louis Review: St. Louis, MO, USA, 1975.

30. Greene, J.; Villanueva, D. Private investment in developing countries: An empirical analysis. *Staff Pap.* **1991**, *38*, 33–58. [CrossRef]

31. Ramirez, M.D. The impact of public investment on private investment spending in Latin America: 1980–95. *Atl. Econ. J.* **2000**, *28*, 210–225. [CrossRef]

32. Spencer, R.W.; Yohe, W.P. The "crowding out" of private expenditures by fiscal policy actions. *Fed. Reserve Bank St. Louis Rev.* **1970**, *10*, 12–24. [CrossRef]

33. Devarajan, S.; Swaroop, V.; Zou, H.-F. The composition of public expenditure and economic growth. *J. Monet. Econ.* **1996**, *37*, 313–344. [CrossRef]

34. Chen, B.L. Economic growth with an optimal public spending composition. *Oxf. Econ. Pap.* **2006**, *58*, 123–136. [CrossRef]

35. Ghosh, S.; Gregoriou, A. The composition of government spending and growth: Is current or capital spending better? *Oxf. Econ. Pap.* **2008**, *60*, 484–516. [CrossRef]

36. Dar, A.A.; Amir Khalkhali, S. Government size, factor accumulation, and economic growth: Evidence from OECD countries. *J. Policy Model.* **2002**, *24*, 679–692. [CrossRef]

37. Schaltegger, C.A.; Torgler, B. Growth effects of public expenditure on the state and local level: Evidence from a sample of rich governments. *Appl. Econ.* **2006**, *38*, 1181–1192. [CrossRef]

38. Karras, G. Employment and output effects of government spending: Is government size important? *Econ. Inq.* **1993**, *31*, 354–369. [CrossRef]

39. Herath, S. Size of government and economic growth: A nonlinear analysis. *Econ. Ann.* **2012**, *57*, 7–30. [CrossRef]

40. Teles, V.K.; Mussolini, C.C. Public debt and the limits of fiscal policy to increase economic growth. *Eur. Econ. Rev.* **2014**, *66*, 1–15. [CrossRef]

41. Soytas, U.; Sari, R. Energy consumption and GDP: Causality relationship in G-7 countries and emerging markets. *Energy Econ.* **2003**, *25*, 33–37. [CrossRef]

42. Lee, C.C. Energy consumption and GDP in developing countries: A cointegrated panel analysis. *Energy Econ.* **2005**, *27*, 415–427. [CrossRef]

43. Huang, B.N.; Hwang, M.J.; Yang, C.W. Does more energy consumption bolster economic growth? An application of the nonlinear threshold regression model. *Energy Policy* **2008**, *36*, 755–767. [CrossRef]

44. Vo, D.H.; Vo, T.A.; Ho, M.C.; Nguyen, M.H. The Role of Renewable Energy, Alternative and Nuclear Energy in Mitigating Carbon Emissions in the CPTPP Countries Renewable Energy. *Renew. Energy* **2020**. forthcoming.

45. Chandran, V.G.R.; Tang, C.F. The impacts of transport energy consumption, foreign direct investment and income on $CO_2$ emissions in ASEAN-5 economies. *Renew. Sustain. Energy Rev.* **2013**, *24*, 445–453. [CrossRef]

46. Acaravcı, A.; Ozturk, I. On the relationship between energy consumption, $CO_2$ emissions and economic growth in Europe. *Energy* **2010**, *35*, 5412–5420. [CrossRef]

47. Jadhav, A.; Sonar, R. Analytic hierarchy process (AHP), weighted scoring method (WSM), and hybrid knowledge-based system (HKBS) for software selection: A comparative study. In Proceedings of the 2009 Second International Conference on Emerging Trends in Engineering & Technology, Nagpur, India, 16–18 December 2009; pp. 991–997.

48. Fishburn, P.C. Methods of estimating additive utilities. *Manag. Sci.* **1967**, *13*, 435–453. [CrossRef]

49. Mendoza, G.A.; Prabhu, R. Multiple criteria decision-making approaches to assessing forest sustainability using criteria and indicators: A case study. *For. Ecol. Manag.* **2000**, *131*, 107–126. [CrossRef]

50. Kuznets, S. Economic growth and income inequality. *Am. Econ. Rev.* **1955**, *45*, 1–28.

51. Douglas, P.H. The Cobb-Douglas production function once again: Its history, its testing, and some new empirical values. *J. Polit. Econ.* **1976**, *84*, 903–915. [CrossRef]

52. Barro, R.J. Human capital and growth. *Am. Econ. Rev.* **2001**, *91*, 12–17. [CrossRef]

53. Nguyen, V.P.; Vo, H.D. Macroeconomics Determinants of Exchange Rate Pass-Through: New Evidence from the Asia-Pacific Region. *Emerg. Mark. Financ. Trade* **2019**, 1–16. [CrossRef]

54. Vo, D.H.; Nguyen, V.P.; Nguyen, M.H.; Vo, T.A.; Nguyen, C.T. Derivatives market and economic growth nexus: Policy implications for emerging markets. *N. Am. J. Econ. Financ.* **2018**, 100866. [CrossRef]

55. Vo, T.A.; Vo, H.D.; Le, T.T.Q. $CO_2$ Emissions, Energy Consumption, and Economic Growth: New Evidence in the ASEAN Countries. *J. Risk Financ. Manag.* **2019**, *12*, 145. [CrossRef]

56. Marland, G.; Rotty, R.M. Carbon dioxide emissions from fossil fuels: A procedure for estimation and results for 1950–1982. *Tellus B Chem. Phys. Meteorol.* **1984**, *36*, 232–261. [CrossRef]

57. United Nations Statistical Yearbook 1983–1984. Available online: https://www.un-ilibrary.org/economic-and-social-development/statistical-yearbook-1983-1984-thirty-fourth-issue_0d8efb97-en-fr (accessed on 1 July 2020).

58. BP Statistical Review of World Energy. Available online: http://www.bp.com/statisticalreview (accessed on 20 June 2019).

59. Pesaran, M.H. General Diagnostic Tests for Cross Section Dependence in Panels. 2004. Available online: http://ftp.iza.org/dp1240.pdf (accessed on 25 July 2019).

60. Friedman, M. The use of ranks to avoid the assumption of normality implicit in the analysis of variance. *J. Am. Stat. Assoc.* **1937**, *32*, 675–701. [CrossRef]

61. Frees, E.W. Assessing cross-sectional correlation in panel data. *J. Econom.* **1995**, *69*, 393–414. [CrossRef]

62. Pesaran, M.H. A simple panel unit root test in the presence of cross-section dependence. *J. Appl. Econom.* **2007**, *22*, 265–312. [CrossRef]

63. Im, K.S.; Pesaran, M.H.; Shin, Y. Testing for unit roots in heterogeneous panels. *J. Econom.* **2003**, *115*, 53–74.

64. Eberhardt, M. Estimating panel time-series models with heterogeneous slopes. *Stata J.* **2012**, *12*, 61–71. [CrossRef]

65. Kao, C. Spurious regression and residual-based tests for cointegration in panel data. *J. Econom.* **1999**, *90*, 1–44. [CrossRef]

66. Pedroni, P. Panel cointegration: Asymptotic and finite sample properties of pooled time series tests with an application to the PPP hypothesis. *Econom. Theory* **2004**, *20*, 597–625. [CrossRef]

67. Westerlund, J. Testing for error correction in panel data. *Oxf. Bull. Econ. Stat.* **2007**, *69*, 709–748. [CrossRef]

68. Pedroni, P. Purchasing power parity tests in cointegrated panels. *Rev. Econ. Stat.* **2001**, *83*, 727–731. [CrossRef]

69. Vo, T.A.; Ho, M.C.; Vo, H.D. Understanding the exchange rate pass-through to consumer prices in Vietnam: The SVAR approach. *Int. J. Emerg. Mark.* **2019**, *15*, 971–989. [CrossRef]

70. Ouedraogo, N.S. Energy consumption and human development: Evidence from a panel cointegration and error correction model. *Energy* **2013**, *63*, 28–41. [CrossRef]

71. Huynh, V.S.; Vo, H.D.; Vo, T.A.; Ha, T.T.D. The Importance of the Financial Derivatives Markets to Economic Development in the World's Four Major Economies. *J. Risk Financ. Manag.* **2019**, *12*, 35. [CrossRef]

72. Riley, C. Britain Bans Gasoline and Diesel Cars Starting in 2040. 2017. Available online: https://money.cnn.com/2017/07/26/news/uk-bans-gasoline-diesel-engines-2040/index.html (accessed on 12 October 2019).
73. Petroff, A. These Countries Want to Ditch Gas and Diesel Cars. 2017. Available online: https://money.cnn.com/2017/07/26/autos/countries-that-are-banning-gas-cars-for-electric/index.html (accessed on 12 October 2019).
74. Forrest, A. The Death of Diesel: Has the One-Time Wonder Fuel Become the New Asbestos? 2017. Available online: https://www.theguardian.com/cities/2017/apr/13/death-of-diesel-wonder-fuel-new-asbestos (accessed on 12 October 2019).
75. World Bank. World Development Indicators. 2019. Available online: https://data.worldbank.org/indicator (accessed on 18 June 2019).

*Article*

# Forecasting Hierarchical Time Series in Power Generation

**Tiago Silveira Gontijo [1,*] and Marcelo Azevedo Costa [1,2]**

[1]   Graduate Program in Industrial Engineering, Universidade Federal de Minas Gerais, Av.
     Antônio Carlos 6627, Belo Horizonte 31270-901, MG, Brazil; azevedo@est.ufmg.br
[2]   Department of Industrial Engineering, Universidade Federal de Minas Gerais, Av. Antônio Carlos 6627,
     Belo Horizonte 31270-901, MG, Brazil
*   Correspondence: tsgontijo@hotmail.com

Received: 26 June 2020; Accepted: 15 July 2020; Published: 20 July 2020

**Abstract:** Academic attention is being paid to the study of hierarchical time series. Especially in the electrical sector, there are several applications in which information can be organized into a hierarchical structure. The present study analyzed hourly power generation in Brazil (2018–2020), grouped according to each of the electrical subsystems and their respective sources of generating energy. The objective was to calculate the accuracy of the main measures of aggregating and disaggregating the forecasts of the Autoregressive Integrated Moving Average (ARIMA) and Error, Trend, Seasonal (ETS) models. Specifically, the following hierarchical approaches were analyzed: (i) bottom-up (BU), (ii) top-down (TD), and (iii) optimal reconciliation. The optimal reconciliation models showed the best mean performance, considering the primary predictive windows. It was also found that energy forecasts in the South subsystem presented greater inaccuracy compared to the others, which signals the need for individualized models for this subsystem.

**Keywords:** power generation; electrical subsystems; time series

---

## 1. Introduction

The advent of Industry 4.0 revolutionized factories worldwide, since it allowed the connectivity between measuring machines and the automation of companies, distributing the capacity to collect massive volumes of data [1]. In high-level data analysis, forecasting models allow the extraction of behavior patterns, as well as the prediction of future values for the collected data set [2].

In the above-mentioned scenario, the construction of predictive models is gaining prominence in the literature [3–5], since economic agents deal with uncertainty in multiple spheres and aim to achieve the best results using available resources [6]. Developing acceptably accurate models presents a meaningful challenge, as prediction is a technique that deals with risk and there will always be a fundamental error associated with it. The best model is the one that most adequately represents the phenomenon of interest.

In relation to the object of our study, power generation, there are several forecasting applications: (i) classical time series models like the autoregressive moving average, autoregressive integrated moving average, and generalized autoregressive conditional heteroscedastic among others [7,8]; (ii) pre-processing techniques like spectrum analysis, wavelets, and Fourier analysis [9]; and, (iii) machine learning approaches such as neural networks, fuzzy systems, and support vector machine [10]. Alternatively, hybrid models aim to combine machine learning representations with different methods. These methods include focused time-delay neural networks [11], wavelet neuro-fuzzy systems [12], finite-impulse response neural networks [13], local feedback dynamic fuzzy neural networks [14], type recurrent fuzzy networks [15], and neuro-fuzzy inference systems [16] among others.

Additionally, an alternative class known as hierarchical forecasting [17–19] deals with organized time series that can be aggregated at different levels into groups based on geography, sources of energy, or other, specific features. Despite this being a recent topic, there is already research that has addressed the use of hierarchical forecasting models in the energy sector. Examples of hierarchical forecasting include electrical grids [20], solar power generation [21], energy transport [22], short-term load forecasting [23], long-term load forecasting [24], energy consumption [25], and air pollution [26] among others.

The papers identified above have calibrated the forecasts using only the bottom-up, top-down or Ordinary least squares (OLS) assumptions [19]. Thus, the following research question is formulated: how is it possible to make hierarchical predictions using advanced linear regression models with regularization? In this way, it is expected to obtain more reliable forecasts by rewriting the hierarchical problem in terms of finding a set of unbiased, minimum variance measures of projected values across the whole array of data. It is possible to minimize the sum of variances of the reconciled estimate errors under the property of unbiasedness, using the procedure called MinT (minimum trace) reconciliation [27].

The present paper presents a case study using a power generation data set from Brazil (2018–2020) organized by electrical subsystems and different generating sources. Specifically, the main approaches used to aggregate and disaggregate predictions made for grouped time series are examined, namely: (i) bottom-up, (ii) top-down and (iii) optimal reconciliation models (OLS, WLS and MinT). The ARIMA and ETS predictive models were used to test the performance of these reconciliation methods, since these are the default models available in the R-package HTS. Further descriptions can be found in the materials and methods section.

The remainder of the present paper is organized as follows. Section 2 defines the study methodology, describing the data set, hierarchical procedures, and forecasting models employed. Section 3 presents the results and discussions of the techniques, in addition to the limitations of this paper. Finally, Section 4 presents the conclusions and guidelines for future work.

## 2. Materials and Methods

The secondary data used in this study correspond to the amounts of power generated by each of the Brazilian electrical subsystems (North, Northeast, Southeast/Midwest, and South). We separated these data according to the source of energy (wind, hydroelectric, thermal, solar, and nuclear). Data were obtained from the National Electric System Operator [28], due to their reliability. The observations of hourly power generation (GWh) were made during the period from January 2018 to January 2020, making a total of 17,521 h.

Based on Hyndman et al. [19], we present a schematic representation of the Brazilian energy generation system, comprising a three-level hierarchical structure (Figure 1). Level 0 represents the total energy generated in Brazil (completely aggregated series). Level 1 denotes each of Brazil's electrical subsystems (first level of disaggregation). The last level, Level 2, represents each of the energy generating sources (Level $k$). According to this framework, it is possible to identify the most disaggregated time series (in this case $k = 2$).

Table 1 shows the amounts of power generation in Brazil (GWh), according to generating sources and electrical subsystems. There is a predominance of hydroelectric generation (73%), making the Brazilian electrical matrix one of the cleanest in the world. At the same time, the Southeast/Midwest subsystem accounts for more than half (56%) of all energy generated in the country.

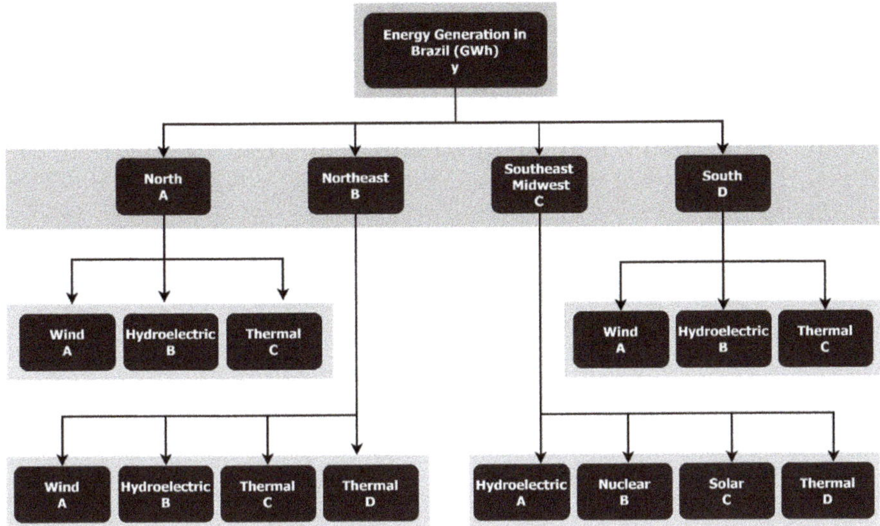

**Figure 1.** Hierarchical aggregation structure for the energy generation in Brazil.

**Table 1.** Amounts of power generation in Brazil (GWh).

| Subsystem/Source | | Wind | Hydro | Thermal | Solar | Nuclear | Total (GWh—Subsystem) | % |
|---|---|---|---|---|---|---|---|---|
| North | (A) | 2688 | 125,182 | 31,489 | 0 | 0 | 159,359 | 14.3% |
| Northeast | (B) | 85,377 | 37,705 | 36,699 | 4626 | 0 | 164,407 | 14.7% |
| Southeast/Midwest | (C) | 0 | 518,714 | 73,555 | 2437 | 31,805 | 626,511 | 56.1% |
| South | (D) | 11,326 | 135,914 | 19,472 | 0 | 0 | 166,712 | 14.9% |
| **Total (GWh—Source)** | | 99,391 | 817,516 | 161,215 | 7063 | 31,805 | 1,116,989 | **100%** |
| % | | **8.9%** | **73.2%** | **14.4%** | **0.6%** | **2.8%** | 100% | - |

Routines were implemented using the $R^{®}$ programming language [29]. The R-package HTS was used to calculate the bottom-up, top-down, optimal combination reconciliation and trace minimization reconciliation. HTS is available at: https://cran.r-project.org/web/packages/hts/index.html. Although HTS includes functions for creating, plotting and forecasting hierarchical time series, it has some limitations. Those limitations include the fact that it has only three built-in forecasting options: ARIMA, ETS, and random walks [19]. This paper will use the ARIMA and the ETS models since they have automatic adjustment and allow consideration of factors such as the trend and seasonality of the data set. The computer used to execute the algorithms had CPU Intel Core i5-7200 2.70 GHz, RAM of 16 GB, and operating system Windows 10 x64. In the next subsection, we present the hierarchical reconciliation models used in the present paper, as well as the forecasting models.

## 2.1. The Bottom-Up (BU) Approach

The BU procedure requires first providing forecasts for every series at the bottom-level, and then summing these to generate forecasts for all the levels of the hierarchical structure [30]. In its simplicity, this approach neglects the relations between time series and works, mainly unsuccessfully, on highly disaggregated data. These data tend to have a low signal-to-noise ratio [27]. According to the hierarchy (Figure 1), we first make h-step-ahead forecasts for all the bottom-level time series ($n = 14$):

$$\hat{y}_{AA,t}, \; \hat{y}_{AB,t}, \; \hat{y}_{AC,t}, \; \hat{y}_{BA,t}, \; \hat{y}_{BB,t}, \; \hat{y}_{BC,t}, \; \hat{y}_{BD,t}, \; \hat{y}_{CA,t}, \; \hat{y}_{CB,t}, \; \hat{y}_{CC,t}, \; \hat{y}_{CD,t}, \; \hat{y}_{DA,t}, \; \hat{y}_{DB,t}, \; \hat{y}_{DC,t}. \tag{1}$$

Summing these, we obtain h-step-ahead forecasts for the rest of the series:

$$\widetilde{y}_t = \hat{y}_{AA,t} + \hat{y}_{AB,t} + \hat{y}_{AC,t} + \hat{y}_{BA,t} + \hat{y}_{BB,t} + \hat{y}_{BC,t} + \hat{y}_{BD,t} + \hat{y}_{CA,t} + \hat{y}_{CB,t}$$
$$+\hat{y}_{CC,t} + \hat{y}_{CD,t} + \hat{y}_{DA,t} + \hat{y}_{DB,t} + \hat{y}_{DC,t}.$$
$$\widetilde{y}_{A,t} = \hat{y}_{AA,t} + \hat{y}_{AB,t} + \hat{y}_{AC,t}.$$
$$\widetilde{y}_{B,t} = \hat{y}_{BA,t} + \hat{y}_{BB,t} + \hat{y}_{BC,t} + \hat{y}_{BD,t}. \tag{2}$$
$$\widetilde{y}_{C,t} = \hat{y}_{CA,t} + \hat{y}_{CB,t} + \hat{y}_{CC,t} + \hat{y}_{CD,t}.$$
$$\widetilde{y}_{D,t} = \hat{y}_{DA,t} + \hat{y}_{DB,t} + \hat{y}_{DC,t}.$$

According to [19], it is possible to arrange the equations expressed in (2) into an algebra notation. Below is a complete notation for this problem:

$$
\begin{bmatrix}
\widetilde{y}_t \\
\widetilde{y}_{A,t} \\
\widetilde{y}_{B,t} \\
\widetilde{y}_{C,t} \\
\widetilde{y}_{D,t} \\
\widetilde{y}_{AA,t} \\
\widetilde{y}_{AB,t} \\
\widetilde{y}_{AC,t} \\
\widetilde{y}_{BA,t} \\
\widetilde{y}_{BB,t} \\
\widetilde{y}_{BC,t} \\
\widetilde{y}_{BD,t} \\
\widetilde{y}_{CA,t} \\
\widetilde{y}_{CB,t} \\
\widetilde{y}_{CC,t} \\
\widetilde{y}_{CD,t} \\
\widetilde{y}_{DA,t} \\
\widetilde{y}_{DB,t} \\
\widetilde{y}_{DC,t}
\end{bmatrix}
=
\begin{bmatrix}
1 & 1 & 1 & 1 & 1 & 1 & 1 & 1 & 1 & 1 & 1 & 1 & 1 & 1 \\
1 & 1 & 1 & 0 & 0 & 0 & 0 & 0 & 0 & 0 & 0 & 0 & 0 & 0 \\
0 & 0 & 0 & 1 & 1 & 1 & 1 & 0 & 0 & 0 & 0 & 0 & 0 & 0 \\
0 & 0 & 0 & 0 & 0 & 0 & 0 & 1 & 1 & 1 & 1 & 0 & 0 & 0 \\
0 & 0 & 0 & 0 & 0 & 0 & 0 & 0 & 0 & 0 & 0 & 1 & 1 & 1 \\
1 & 0 & 0 & 0 & 0 & 0 & 0 & 0 & 0 & 0 & 0 & 0 & 0 & 0 \\
0 & 1 & 0 & 0 & 0 & 0 & 0 & 0 & 0 & 0 & 0 & 0 & 0 & 0 \\
0 & 0 & 1 & 0 & 0 & 0 & 0 & 0 & 0 & 0 & 0 & 0 & 0 & 0 \\
0 & 0 & 0 & 1 & 0 & 0 & 0 & 0 & 0 & 0 & 0 & 0 & 0 & 0 \\
0 & 0 & 0 & 0 & 1 & 0 & 0 & 0 & 0 & 0 & 0 & 0 & 0 & 0 \\
0 & 0 & 0 & 0 & 0 & 1 & 0 & 0 & 0 & 0 & 0 & 0 & 0 & 0 \\
0 & 0 & 0 & 0 & 0 & 0 & 1 & 0 & 0 & 0 & 0 & 0 & 0 & 0 \\
0 & 0 & 0 & 0 & 0 & 0 & 0 & 1 & 0 & 0 & 0 & 0 & 0 & 0 \\
0 & 0 & 0 & 0 & 0 & 0 & 0 & 0 & 1 & 0 & 0 & 0 & 0 & 0 \\
0 & 0 & 0 & 0 & 0 & 0 & 0 & 0 & 0 & 1 & 0 & 0 & 0 & 0 \\
0 & 0 & 0 & 0 & 0 & 0 & 0 & 0 & 0 & 0 & 1 & 0 & 0 & 0 \\
0 & 0 & 0 & 0 & 0 & 0 & 0 & 0 & 0 & 0 & 0 & 1 & 0 & 0 \\
0 & 0 & 0 & 0 & 0 & 0 & 0 & 0 & 0 & 0 & 0 & 0 & 1 & 0 \\
0 & 0 & 0 & 0 & 0 & 0 & 0 & 0 & 0 & 0 & 0 & 0 & 0 & 1
\end{bmatrix}
\begin{bmatrix}
\hat{y}_{AA,t} \\
\hat{y}_{AB,t} \\
\hat{y}_{AC,t} \\
\hat{y}_{BA,t} \\
\hat{y}_{BB,t} \\
\hat{y}_{BC,t} \\
\hat{y}_{BD,t} \\
\hat{y}_{CA,t} \\
\hat{y}_{CB,t} \\
\hat{y}_{CC,t} \\
\hat{y}_{CD,t} \\
\hat{y}_{DA,t} \\
\hat{y}_{DB,t} \\
\hat{y}_{DC,t}
\end{bmatrix}
\tag{3}
$$

Alternatively, the notation presented in (3) can be reformulated in a compact way by applying the summing matrix. Thus, the bottom-up approach can be represented as:

$$\widetilde{y}_t = S\hat{b}_t, \tag{4}$$

where $\widetilde{y}_t$ is an $n$-dimensional vector of $h$-step-ahead forecasts for the total energy, $S$ is the summing matrix, and $\hat{b}_t$ is an $m$-dimensional vector of $h$-step-ahead forecasts for each of the sources of energy at bottom-level. An advantage of this procedure is that we are forecasting at the bottom-level of a hierarchy. Consequently, no information is missed due to aggregation [17].

### 2.2. The Top-Down (TD) Approach

Top-down methods operate with strictly hierarchical aggregation structures, not with grouped structures. They involve first making forecasts for the Total level $y_t$, and next disaggregating these down the hierarchy [17]. Let $p_1, \ldots, p_m$ be a set of disaggregation proportions that deliver the forecasts of the Total series, which are to be distributed in order to obtain forecasts for all series at the bottom-level of the structure. To illustrate, concerning our hierarchy by applying proportions to Figure 1, we get $p_1, \ldots, p_{14}$:

$$
\begin{aligned}
\widetilde{y}_{AA,t} &= p_1 \hat{y}_t, \ \widetilde{y}_{AB,t} = p_2 \hat{y}_t, \ \widetilde{y}_{AC,t} = p_3 \hat{y}_t. \\
\widetilde{y}_{BA,t} &= p_4 \hat{y}_t, \ \widetilde{y}_{BB,t} = p_5 \hat{y}_t, \ \widetilde{y}_{BC,t} = p_6 \hat{y}_t, \ \widetilde{y}_{BD,t} = p_7 \hat{y}_t. \\
\widetilde{y}_{CA,t} &= p_8 \hat{y}_t, \ \widetilde{y}_{CB,t} = p_9 \hat{y}_t, \ \widetilde{y}_{CC,t} = p_{10} \hat{y}_t, \ \widetilde{y}_{CD,t} = p_{11} \hat{y}_t. \\
\widetilde{y}_{DA,t} &= p_{12} \hat{y}_t, \ \widetilde{y}_{DB,t} = p_{13} \hat{y}_t, \ \widetilde{y}_{DC,t} = p_{14} \hat{y}_t.
\end{aligned}
\tag{5}
$$

This can be rewritten using matrix notation. If we stack the set of proportions in an $m$-dimensional vector $p = (p_1, \ldots, p_m)'$, we have the bottom-level $h$-step-ahead predictions. Overall, for a given set of proportions, top-down approaches can be written as:

$$
\begin{aligned}
\widetilde{b}_t &= p_j \hat{y}_t. \\
\widetilde{y}_t &= S p_j \hat{y}_t.
\end{aligned}
\tag{6}
$$

The main TD models stipulate disaggregation proportions according to the historical proportions of the data. Among the main models of this approach, we highlight the following three: (i) top-down Gross–Sohl method A (TDGSA), (ii) top-down Gross–Sohl method F (TDGSF), and (iii) Top-down forecast proportions (TDFP) (Table 2). Additional details and demonstrations of Table 2 can be obtained from [18,31].

**Table 2.** TD disaggregation proportions according to the historical proportions of the data.

| TD Gross-Sohl Method A<br>TDGSA | TD Gross-Sohl Method F<br>TDGSF | TD Forecast Proportions<br>TDFP |
|---|---|---|
| $p_j = \frac{1}{T} \sum\limits_{t=1}^{T} \frac{y_{j,t}}{y_t}$ | $p_j = \sum\limits_{t=1}^{T} \frac{y_{j,t}}{T} \bigg/ \sum\limits_{t=1}^{T} \frac{y_t}{T}$ | $p_j = \prod\limits_{l=0}^{K-1} \frac{\hat{y}_{j,t}^{(l)}}{\hat{S}_{j,t}^{(l+1)}}$ |
| for $j = 1, \ldots, m$. Each proportion $p_j$ reflects the average of the historical proportions of the bottom-level series $y_{j,t}$, $t$ over the period $t = 1, \ldots, T$ relative to the total aggregate $y_t$. | for $j = 1, \ldots, m$. Each proportion $p_j$ takes the average historical value of the bottom-level series $y_{j,t}$ related to the average value of the total aggregate $y_t$. | where $j = 1, \ldots, m$, $\hat{y}_{j,h}^{(l)}$ is the $h$-step-ahead forecast and $\hat{S}_{j,t}^{(l)}$ is the sum of the $h$-step-ahead forecasts below the node that is $l$ levels above node $j$. |

### 2.3. The Optimal Reconciliation Approaches

The optimal reconciliation approach proposed by [19] consists of an ordinary least squares problem based on the calculation of independent projections for all hierarchical levels, then applying a regression model to optimize the combination of these forecasts. According to [32], we can write the base prediction as:

$$
\hat{y}_{t+h|t} = S \beta_{t+h|t} + \varepsilon_h,
\tag{7}
$$

where $\beta_{t+h|t}$ represents the unknown conditional mean of the most disaggregated series, and $\varepsilon_h$ is the error with mean of zero and covariance matrix $\Sigma_h$. If $\Sigma_h$ were known, the estimator of $\beta_{t+h|t}$ would lead to the following weighted least squares, producing reconciled forecasts, as follows:

$$
\widetilde{y}_{t+h|t} = S \hat{\beta}_{t+h|t} = S \left( S' \sum\nolimits_h{}^{-1} S \right)^{-1} S' \sum\nolimits_h{}^{-1} \hat{y}_{t+h|t} = S P \hat{y}_{t+h|t},
\tag{8}
$$

where $P = \left( S' \sum_h^{-1} S \right)^{-1} S' \sum_h^{-1} S$. If the base forecasts $\hat{y}_{t+h|t}$ are unbiased, then the reconciled forecasts $\widetilde{y}_{t+h|t}$ will be unbiased, provided that $SPS = S$ [19]. This condition is valid for this reconciliation procedure for the bottom-up, although not for the top-down, methods. Consequently, the top-down approaches will never give unbiased reconciled forecasts, even if the base forecasts are unbiased.

Additionally, [27] proved that, in general, $\sum h$ is not known and not identifiable. The covariance matrix of the $h$-step-ahead reconciled forecast errors is given by the following expression:

$$Var_{(y_{t+h}-\tilde{y}_{t+h|t})} = SPW_hP'S', \tag{9}$$

for any $P$ such that $SPS = S$, then $W_h = Var_{(y_{t+h}-\hat{y}_{t+h|t})} = E(\hat{e}_{t+h|t}\hat{e}'_{t+h|t})$ is the covariance matrix of the corresponding $h$-step ahead base forecast errors. The purpose is to get the matrix $P$ that minimizes the error variances of the reconciled forecasts which are on the diagonal of the covariance matrix $Var_{(y_{t+h}-\tilde{y}_{t+h|t})}$. Finally, [27] demonstrated that the optimal reconciliation matrix $P$ that minimizes the trace of $SPW_hP'S' =$, such that $SPS = S$, and the optimal reconciled forecasts, respectively, are given by:

$$P = (S'W_h^{-1}S)^{-1}S'W_h^{-1}$$
$$\tilde{y}_{t+h|t} = S(S'W_h^{-1}S)^{-1}S'W_h^{-1}\hat{y}_{t+h|t}, \tag{10}$$

which is introduced as the MinT (minimum trace) estimator. The next step consists of estimating $W_h$, a matrix of order $n$. Wickramasuriya, Athanasopoulos and Hyndman [27] proposed the following procedures (Table 3) to obtain the matrix:

**Table 3.** Hierarchical forecasting for electricity generation based on the ARIMA procedure.

| Procedure | Description |
| --- | --- |
| **OLS** | $W_h = k_hI$, $\forall h$ where $k_h > 0$. This is the most simplifying premise, and collapses the MinT estimator to the OLS estimator, proposed by Hyndman et al. [19]. This is optimal when the base forecast errors are uncorrelated and equivariant. |
| **WLSv** | $W_h = k_h\text{diag}(\widehat{W}_1)$, $\forall h$ where $k_h > 0$ and: $\widehat{W} = \frac{1}{T}\sum_{t=1}^{T}\hat{e}_t(1)\hat{e}_t(1)'$, is the unbiased sample covariance estimator of the in-sample one-step-ahead base forecast errors. In this case, we can describe MinT as a WLS estimator applying variance scaling [27]. |
| **WLSs** | $W_h = k_h\Lambda$, $\forall h$ where $k_h > 0$ and $\Lambda = \text{diag}(S1)$ with 1 being a unit column vector of dimension $n$. We assume that each of the bottom-level base forecast errors has a variance $k_h$ and is uncorrelated between nodes. Consequently, every element of the diagonal $\Lambda$ matrix receives the number of forecast error variances contributing to that aggregation level [27]. This estimator depends only on the grouping structure of the hierarchy. |
| **MinT (Sample)** | $W_h = k_w\widehat{W}_1$, $\forall h$ where $k_h > 0$, the unrestricted sample covariance estimator for $h = 1$ [27]. In the results section, we denote this as MinT (Sample). |
| **MinT (Shrink)** | $W_h = k_w\widehat{W^*_{1,D}}$; $\forall h$; $k_h > 0$; $W^*_{1,D} = \lambda_D\widehat{W}_{1,D} + (1 - \lambda_D)\widehat{W}_1$, is a shrinkage estimator with diagonal target, $\widehat{W}_{1,D}$, which is a diagonal matrix comprising the diagonal entries of $\widehat{W}_1$, and $\lambda_D$ is the shrinkage intensity parameter. Thus, off-diagonal elements of $\widehat{W}_1$ are shrunk toward zero and diagonal elements (variances) remain unchanged [27]. Wickramasuriya, Athanasopoulos and Hyndman [27] suggested a scale and location invariant shrinkage estimator by parameterizing the shrinkage in terms of variances and correlations: $$\hat{\lambda}_D = \frac{\sum_{i\neq j}\hat{var}_{(\hat{r}_{ij})}}{\sum_{i\neq j}\hat{r}_{ij}^2},$$ where $\hat{r}_{ij}$ is the $ij$th element of $\hat{R}_1$, the 1-step-ahead sample correlation matrix to shrink it toward an identity matrix. |

Source: adapted by authors from: [27].

### 2.4. ARIMA and ETS Formulation

ARIMA is one of the most-widely-used time series approaches for forecasting power generation [33]. Although studies have shown that ETS outperforms ARIMA [34], it is recommended to keep ARIMA as a reference model during the forecasting process. Moreover, several statistical software packages, like R®, provide automatic model identification and parameter estimation skills for both ARIMA and ETS [17]. Professor Hyndman [19] developed the HTS package initially based on these predictive models. The present paper aims to test different approaches to optimal forecast reconciliation and, to do so, only the ARIMA and ETS models will be used. It is recommended that future studies extend these forecasting procedures using different predictive models, such as machine learning ones.

ARIMA was proposed by [33]. It is a linear forecasting method for dealing with stationary time series [34]. In the initial step, a time series is built stationary by differencing $d$ times along with some nonlinear transformations, such as logging [34]. The consequential data are recognized as a linear function of past $p$ data values and $q$ errors (11), i.e., modeled as an autoregressive moving average (ARMA) model,

$$y_t = \varnothing_1 y_{t-1} + \varnothing_2 y_{t-2} + \ldots + \varnothing_p y_{t-p} + \Theta_1 \varepsilon_{t-1} + \Theta_2 \varepsilon_{t-2} + \ldots + \Theta_q \varepsilon_{q-1}, \tag{11}$$

where $y_t$ denotes real value at time $t$, $\varepsilon_t$ describes the error sequence: it is supposed to be white noise and Gaussian distributed $(0, \sigma^2)$. $\varnothing_i$ for $(i = 1, 2, \ldots, p)$ are autoregressive $(AR)$ coefficients and $\Theta_j$ for $(j = 1, 2, \ldots, q)$ are moving average $(MA)$ coefficients. $p$ and $q$ are integers referred to as model orders. The time series model is denoted as $ARIMA(p, d, q)$ [35,36].

According to [34], the group of exponential smoothing methods utilizes the principle of weighted averages of past information for making forecasts. Since its formulation in 1950, a variety of exponential smoothing methods have been developed. All exponential smoothing methods were initially classified by [37], which has been continued by [38–40]. ETS stands for error, trend, and seasonality elements. As pointed by [34], the usual representation for these patterns involves a state vector $x_t = (l_t, b_t, s_t, s_{t-1}, \ldots, s_{t-m+1})'$, and the state space equations [39] have the resulting structure:

$$\begin{aligned} y_t &= w(x_{t-1}) + r(x_{t-1})\varepsilon_t \\ x_t &= f(x_{t-1}) + g(x_{t-1})\varepsilon_t, \end{aligned} \tag{12}$$

where $(\varepsilon_t)$ denotes a Gaussian white noise $(0, \sigma^2)$ and $\mu_t = w(x_t - 1)$. The model with additive error has $r_t(x_t - 1) = 1$, so $y_t = \mu_t + \varepsilon_t$. The model with multiplicative errors has $r_t(x_t - 1) = \mu_t = \mu_t$, so $y_t = \mu_t(1 + \varepsilon_t)$. Consequently, $\varepsilon_t = (y_t - \mu_t)/\mu_t$ is a relative error for the multiplicative model and any value of $r_t(x_t - 1)$ will lead to the identical point forecast for $y_t$ [34,39].

## 2.5. Evaluating Forecast Accuracy

According to [20], there are several accuracy metrics, such as mean absolute percentage error (*MAPE*), mean absolute error (*MAE*), mean absolute scaled error (*MASE*), or root-mean-square error (*RMSE*), to evaluate the performance of point prediction methods, defined as follows:

$$MAPE = \frac{1}{T} \sum_{t=1}^{T} \left| \frac{y_t - \hat{y}_t}{y_t} \right|. \tag{13}$$

$$MAE = \frac{1}{T} \sum_{t=1}^{T} |y_t - \hat{y}_t|. \tag{14}$$

$$MASE = \frac{MAE}{MAE_{in-sample,naive}} \tag{15}$$

$$RMSE = \sqrt{\frac{1}{T} \sum_{t=1}^{T} (y_t - \hat{y}_t)^2}, \tag{16}$$

where $y_t$ is the amount of power generation at time t, $\hat{y}_t$ is the fitted value for power generation, and $MAE_{in-sample,naive}$ is the MAE generated by a naive forecast.

Specifically, in studies of hierarchical time series, the *MAPE* indicator appears the most frequently in the literature [41–43]. *MAPE* was also the selected metric for the present paper (Figures 2 and 3). Complementarily, *MAE*, *MASE*, and *RMSE* were estimated, and the results can be found in the Appendix A (Figures A3 and A4). The values of the *MAPE*, *MAE*, *MASE* and *RMSE* statistics were obtained using a weighted average, with proportions from Table 1.

| | | Predictive model: Autoregressive integrated moving average (ARIMA) | | | | | | | | | |
|---|---|---|---|---|---|---|---|---|---|---|---|
| **MAPE** | | Forecast horizon (h) | | | | | | | | | |
| | | 1 | 2 | 3 | 4 | 5 | 6 | 7 | 8 | 9 | Mean |
| | | Hierarchical level 0 : Total - Brazil | | | | | | | | | |
| Method | BU | 2.00 | 3.53 | 5.62 | 8.04 | 10.17 | 11.45 | 11.17 | 10.76 | 10.48 | 8.14 |
| | TDGSA | 2.07 | 3.76 | 6.10 | 8.79 | 11.25 | 12.87 | 12.95 | 12.72 | 12.63 | 9.24 |
| | TDGSF | 2.07 | 3.76 | 6.10 | 8.79 | 11.25 | 12.87 | 12.95 | 12.72 | 12.63 | 9.24 |
| | TDFP | 2.07 | 3.76 | 6.10 | 8.79 | 11.25 | 12.87 | 12.95 | 12.72 | 12.63 | 9.24 |
| | OLS | 1.98 | 3.65 | 5.94 | 8.59 | 10.99 | 12.54 | 12.55 | 12.23 | 12.06 | 8.95 |
| | WLSv | 1.91 | 3.51 | 5.70 | 8.24 | 10.51 | 11.93 | 11.79 | 11.32 | 10.99 | 8.43 |
| | WLSs | 1.88 | 3.46 | 5.63 | 8.15 | 10.39 | 11.77 | 11.59 | 11.09 | 10.76 | 8.30 |
| | MintT (Sample) | 1.68 | 3.29 | 5.50 | 8.02 | 10.24 | 11.57 | 11.31 | 10.85 | 10.60 | 8.12 |
| | MinT (Shrink) | 1.74 | 3.36 | 5.59 | 8.12 | 10.35 | 11.69 | 11.44 | 10.94 | 10.69 | 8.21 |
| | | Hierarchical level 1 : Electrical subsystems | | | | | | | | | |
| Method | BU | 1.97 | 3.64 | 6.12 | 8.75 | 10.78 | 11.93 | 11.90 | 11.70 | 11.88 | 8.74 |
| | TDGSA | 31.97 | 31.74 | 30.37 | 28.93 | 28.12 | 27.49 | 26.71 | 26.04 | 25.36 | 28.53 |
| | TDGSF | 32.38 | 32.14 | 30.71 | 29.21 | 28.21 | 27.46 | 26.71 | 26.06 | 25.41 | 28.70 |
| | TDFP | 1.86 | 3.88 | 6.68 | 9.89 | 9.89 | 12.52 | 14.19 | 14.45 | 14.34 | 9.75 |
| | OLS | 1.90 | 3.55 | 6.30 | 9.20 | 11.70 | 13.36 | 13.64 | 13.56 | 13.66 | 9.65 |
| | WLSv | 1.77 | 3.35 | 5.84 | 8.62 | 10.84 | 12.38 | 12.56 | 12.40 | 12.41 | 8.91 |
| | WLSs | 1.81 | 3.41 | 5.92 | 8.74 | 11.00 | 12.57 | 12.79 | 12.68 | 12.75 | 9.07 |
| | MintT (Sample) | 1.64 | 3.20 | 5.66 | 8.50 | 10.76 | 12.23 | 12.40 | 12.21 | 12.20 | 8.76 |
| | MinT (Shrink) | 1.66 | 3.28 | 5.75 | 8.57 | 10.84 | 12.33 | 12.50 | 12.31 | 12.28 | 8.83 |
| | | Hierarchical level 2 : Energy sources | | | | | | | | | |
| Method | BU | 2.66 | 5.05 | 6.53 | 7.71 | 8.88 | 9.46 | 9.40 | 9.22 | 9.11 | 7.56 |
| | TDGSA | 46.33 | 44.34 | 41.72 | 40.35 | 39.87 | 39.29 | 38.44 | 37.52 | 36.58 | 40.49 |
| | TDGSF | 47.66 | 45.70 | 42.87 | 41.24 | 40.42 | 39.64 | 38.80 | 37.90 | 36.96 | 41.24 |
| | TDFP | 2.83 | 5.51 | 7.53 | 9.45 | 9.45 | 11.46 | 12.79 | 13.20 | 13.33 | 9.50 |
| | OLS | 2.51 | 5.07 | 6.78 | 8.29 | 9.78 | 10.62 | 10.73 | 10.63 | 10.56 | 8.33 |
| | WLSv | 2.60 | 5.11 | 6.74 | 8.09 | 9.42 | 10.18 | 10.21 | 10.07 | 9.97 | 8.04 |
| | WLSs | 2.56 | 4.98 | 6.64 | 8.00 | 9.31 | 10.00 | 9.95 | 9.70 | 9.63 | 7.86 |
| | MintT (Sample) | 2.48 | 4.96 | 6.58 | 7.91 | 9.27 | 10.10 | 10.22 | 10.10 | 10.00 | 7.96 |
| | MinT (Shrink) | 2.52 | 5.04 | 6.68 | 8.02 | 9.38 | 10.20 | 10.30 | 10.19 | 10.10 | 8.05 |

**Figure 2.** Hierarchical forecasting for electricity generation based on the ARIMA procedure (MAPE). (Note: The performance was indicated into a color scale, where green means better values for calculated accuracy, and red means worse accuracy. The intermediate values are colored yellow.).

| MAPE | | Predictive model: Error, trend, seasonality (ETS) | | | | | | | | | |
|---|---|---|---|---|---|---|---|---|---|---|---|
| | | Forecast horizon (h) | | | | | | | | | |
| | | 1 | 2 | 3 | 4 | 5 | 6 | 7 | 8 | 9 | Mean |
| | | Hierarchical level 0 : Total - Brazil | | | | | | | | | |
| Method | BU | 2.60 | 4.89 | 7.73 | 10.90 | 13.79 | 15.78 | 16.20 | 16.28 | 16.48 | 11.63 |
| | TDGSA | 2.11 | 4.21 | 6.90 | 9.95 | 12.73 | 14.64 | 14.98 | 14.99 | 15.12 | 10.62 |
| | TDGSF | 2.11 | 4.21 | 6.90 | 9.95 | 12.73 | 14.64 | 14.98 | 14.99 | 15.12 | 10.62 |
| | TDFP | 2.11 | 4.21 | 6.90 | 9.95 | 12.73 | 14.64 | 14.98 | 14.99 | 15.12 | 10.62 |
| | OLS | 2.13 | 4.24 | 6.93 | 9.98 | 12.77 | 14.68 | 15.03 | 15.03 | 15.17 | 10.66 |
| | WLSv | 2.26 | 4.42 | 7.16 | 10.25 | 13.06 | 15.00 | 15.37 | 15.39 | 15.55 | 10.94 |
| | WLSs | 2.29 | 4.46 | 7.20 | 10.29 | 13.11 | 15.05 | 15.42 | 15.45 | 15.61 | 10.99 |
| | MintT (Sample) | **2.06** | **4.14** | **6.80** | **9.84** | **12.61** | **14.51** | **14.84** | **14.84** | **14.96** | **10.51** |
| | MinT (Shrink) | 2.07 | 4.15 | 6.82 | 9.86 | 12.63 | 14.53 | 14.87 | 14.87 | 14.99 | 10.53 |
| | | Hierarchical level 1 : Electrical subsystems | | | | | | | | | |
| Method | BU | 2.59 | 4.94 | 8.09 | 11.43 | 14.21 | 16.13 | 16.71 | 16.93 | 17.36 | 12.04 |
| | TDGSA | 31.98 | 31.78 | 30.44 | 29.05 | 28.29 | 27.76 | 27.08 | 26.51 | 25.91 | 28.76 |
| | TDGSF | 32.38 | 32.17 | 30.78 | 29.32 | 28.37 | 27.72 | 27.08 | 26.53 | 25.96 | 28.92 |
| | TDFP | 2.17 | 4.37 | 7.37 | 10.58 | **10.58** | **13.24** | 15.05 | 15.54 | 15.68 | 10.51 |
| | OLS | 2.15 | 4.34 | 7.35 | 10.55 | 13.22 | 15.04 | 15.53 | 15.68 | 16.04 | 11.10 |
| | WLSv | 2.30 | 4.55 | 7.60 | 10.85 | 13.56 | 15.42 | 15.94 | 16.11 | 16.49 | 11.42 |
| | WLSs | 2.27 | 4.50 | 7.54 | 10.79 | 13.49 | 15.34 | 15.85 | 16.03 | 16.40 | 11.36 |
| | MintT (Sample) | **1.89** | **3.98** | **6.92** | **10.09** | 12.74 | 14.55 | **15.03** | **15.20** | **15.56** | **10.66** |
| | MinT (Shrink) | 1.94 | 4.04 | 7.00 | 10.17 | 12.83 | 14.64 | 15.13 | 15.29 | 15.64 | 10.74 |
| | | Hierarchical level 2 : Energy sources | | | | | | | | | |
| Method | BU | 3.13 | 6.56 | 9.01 | 11.16 | 13.24 | 14.40 | 14.74 | 15.21 | 15.63 | 11.45 |
| | TDGSA | 46.34 | 44.42 | 41.84 | 40.48 | 40.00 | 39.45 | 38.69 | 37.87 | 37.00 | 40.68 |
| | TDGSF | 47.66 | 45.77 | 42.98 | 41.35 | 40.54 | 39.77 | 38.99 | 38.17 | 37.32 | 41.40 |
| | TDFP | 2.90 | 6.31 | 8.74 | 10.97 | **10.97** | **13.05** | 14.29 | 14.58 | 14.94 | 10.75 |
| | OLS | 3.14 | 6.60 | 9.07 | 11.30 | 13.38 | 14.62 | 14.92 | 15.31 | 15.74 | 11.56 |
| | WLSv | 2.76 | 6.08 | 8.42 | 10.56 | 12.53 | 13.72 | 13.97 | 14.18 | 14.56 | 10.75 |
| | WLSs | 3.13 | 6.57 | 9.02 | 11.21 | 13.29 | 14.49 | 14.79 | 15.21 | 15.63 | 11.48 |
| | MintT (Sample) | 2.67 | 5.88 | 8.10 | **10.10** | 11.99 | 13.05 | **13.20** | **13.47** | **13.81** | **10.25** |
| | MinT (Shrink) | **2.63** | **5.89** | 8.17 | 10.21 | 12.14 | 13.25 | 13.43 | 13.72 | 14.09 | 10.39 |

**Figure 3.** Hierarchical forecasting for electricity generation based on the ETS procedure. (Note: The performance was indicated into a color scale, where green means better values for calculated accuracy, and red means worse accuracy. The intermediate values are colored yellow.).

## 3. Results and Discussion

Figure 2, below, shows the predictive result obtained, using the ARIMA model, considering a predictive window of nine hours ($h = 1, \ldots, 9$). Note that the model was estimated, taking the main hierarchical adjustment approaches into account, for the following levels: (i) total power generation in Brazil (Level 0), (ii) total energy generation by electrical subsystem (Level 1), and (iii) total energy generation by the energy generating source (Level 2). For Level 1, four forecasts (one for each electrical subsystem) were estimated. For Level 2, 14 forecasts (one for each energy source) were estimated.

Therefore, we estimated 1539 predictive models satisfying the following proportions: (i) 81 models for Level 0, (ii) 324 models for Level 1, and (iii) 1134 models for Level 2. The MAPE calculation for Levels 1 and 2 was based on a weighted average of the predictive errors. The weighting factors used are shown in Table 1.

The performance of each predictive model, divided by the forecast horizon, is illustrated by a color scale. The green colors indicate the most accurate forecasts, while the red colors symbolize less accurate forecasts. The best forecasts, for each of the predictive horizons, are highlighted in bold. The last column of Table 1 presents the average performance for each forecast horizon ($h$) for each hierarchical approach.

As pointed by [27], the MinT procedure has a useful feature: it systematizes results into a unique analytical solution that incorporates information about the correlation structure of the entire dataset. Additionally, the minimum trace reconciliation, with or without regularization, presented the best results of all linear reconciliation methods, such as OLS and WLS, with variations. Moreover, the MinT (Sample) approach returns the most accurate, coherent forecasts for all levels considering just the first forecast horizons. However, as the predictive window grows, the BU method becomes more accurate. Furthermore, the performance of the BU model increases as the time series disaggregate.

As expected, the results obtained using the top-down technique did not present good predictive results, since it is intended to generate forecasts for level 0, with worse accuracy for the other levels. Both BU and TD present disadvantages: they do not take the correlation among the series at each level into account.

The other accuracy metrics presented in the Appendix A (*MAE*, *MAE*, and *RMSE*) reinforce the results found. In general, the performance of the optimal reconciliation models, by trace minimization, provides more uniform estimates and better predictive potential for the first hours of the predictive horizon (Figures A3 and A4).

In addition to the ARIMA predictive model, Figure 3 presents the same forecasting procedures. However, they are based on the ETS automatic adjustment model. The objective is to show the influence of different forecasting methods for each hierarchical reconciliation model. In general, the error percentage produced by the ETS model was slightly higher than that produced by the ARIMA model. Figure 3 also shows the influence of trace minimization procedures (MinT) on the improvement of predictive performance. In particular, the MinT models have good predictive performance, even with the increase of the forecast horizon hours.

The average performance of the trace minimization (MinT) models shows stability, considering all hierarchical levels. As shown in Figure 2, the ETS-based predictive model shares some similarities with the ARIMA model. The BU technique is better for the most disaggregated levels, whereas the TD technique stands out only at the more aggregated levels. Note that the trace minimization procedures show significant gains over the classic linear models, namely OLS and WLS.

Figures 2 and 3 present some limitations. In general, it is not possible to test the predictive influence of each of the subsystems within the established forecast horizon. To show this problem, Figure 4 presents a predictive comparison (MAPE) for each of the Brazilian electrical subsystems, considering the nine-hour predictive horizon. On the left is the technique with the best aggregation/disaggregation performance (BU) for the ARIMA model. On the right is the technique with the best average performance (MinT) for the ETS automatic selection model.

Figure 4 thus shows a negative influence of the "south" electrical subsystem in the global measures of accuracy, especially from a predictive horizon of three hours onward. This system should be analyzed more thoroughly to identify energy sources located in the "south" subsystem that contributed most to the predictive instability of this system. Simultaneously, the use of individualized predictive models for this "south" system can be a good strategy, since unique climatic conditions exist in southern Brazil.

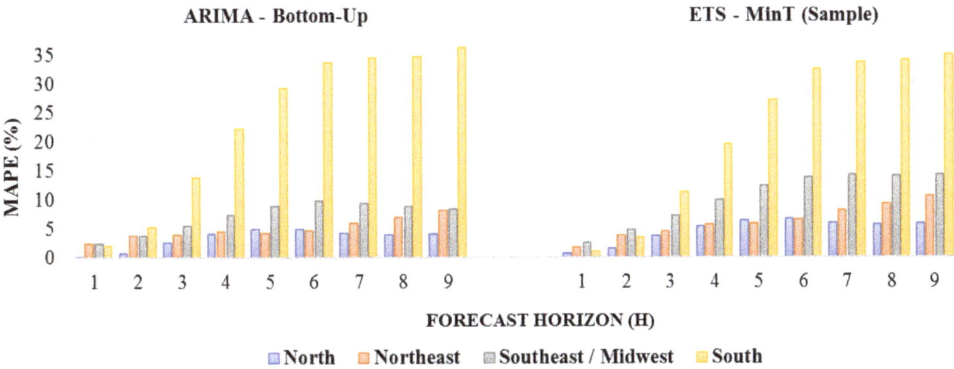


**Figure 4.** Hierarchical forecasting for power generation: electrical subsystem versus forecast horizon.

Figures A1 and A2 (Appendix A) present the accuracy measure of the ARIMA and ETS models in detail, considering energy sources versus electrical subsystems. These results reinforce those in Figure 4, indicating instability in the southern subsystem, especially wind energy data.

Finally, some limitations of the present paper are recognized here. First, predictive models are based on past information evaluable, so the presented results cannot be extrapolated for different contexts and other time periods. Additionally, it is necessary to incorporate other predictive models to make the results more robust. In future research, it is recommended that models which integrate high-frequency data, e.g., the Wavelet approach, be adopted.

## 4. Conclusions

Analysis of the energy market is complicated. It involves the relationship between forecasting models and uncertainty, distinctly regarding the stochastic behavior of variables. The present paper is aimed at policymakers, offering a forecasting tool that deals with grouped time series. It also proposes a new forecasting approach, based on hierarchical modeling of the energy generation in Brazil.

The present paper introduces the use of trace minimization procedures (MinT) to aggregate and disaggregate forecasts based on the ARIMA and ETS models. MinT models performed better than the classic linear approaches, such as OLS and WLS. The MinT models also have high reliability for short predictive horizons. It is noteworthy that both hierarchical procedures and forecasting methods influence the predictive values of power generation in Brazil. Despite its advantages, the optimal reconciliation approach also has some limitations. This method could be unduly influenced by the sample period, and thus its ranking might change for other periods.

Therefore, the use of other predictive models, such as those based on analogs, machine learning, and other hybrid techniques, for example, is recommended. For future research, fine-tuning forecasts of the "south" electrical subsystem, as well as testing the accuracy of the hierarchal methods by using new forecasting approaches, is also recommended.

Finally, the present study contributes to the energy planning processes of different agents, given that understanding energy generation patterns is singularly important for minimizing risks and supporting reliable production planning. Good forecasts for future energy generation can support operational arrangements since energy supply and demand impact spot market sales prices.

**Author Contributions:** Both authors made substantial contributions to the analysis presented in the paper. T.S.G. took lead responsibility for proposing the methodology and for drafting the manuscript and M.A.C. for revising it critically. M.A.C. supervised the project. All authors have read and agreed to the published version of the manuscript.

**Funding:** This research was funded by [National Council for Scientific and Technological Development—CNPq] grant number [141740/2019-1].

**Acknowledgments:** The authors would like to thank the National Council for Scientific and Technological Development (CNPq) and Companhia Energética Integrada (CEI) for supporting this research.

**Conflicts of Interest:** The authors declare that there is no conflict of interest.

## Abbreviations

The following abbreviations are used in this manuscript:

| | |
|---|---|
| ARIMA | Autoregressive integrated moving average model |
| BU | Bottom-up |
| ETS | Error, trend, and seasonality model |
| GWh | Gigawatt hours |
| MAPE | Mean absolute percentage error |
| MinT | Minimum trace reconciliation |
| OLS | Ordinary least squares |
| ONS | Operator of the National System |
| TD | Top-down |
| TDFP | Top-down forecast proportions |
| TDGSA | Top-down Gross-Sohl method A |
| TDGSF | Top-down Gross-Sohl method F |
| WLS | Weighted least squares |

## Nomenclature

The following nomenclature is used in this manuscript:

| | |
|---|---|
| $k$ | Level of disaggregation |
| $h$ | Forecast horizon |
| $\hat{b}_t$ | $m$-dimensional vector of $h$-step-ahead forecasts |
| $\beta_{t+h\|t}$ | Unknown conditional mean of the most disaggregated series |
| $\varepsilon_h$ | Error for each forecast horizon |
| $\Sigma_h$ | Covariance matrix |
| $p$ | Set of proportions in an $m$-dimensional vector |
| $p_j$ | The average of the historical proportions |
| $S$ | Summing matrix |
| $\hat{S}_{j,t}^{(l)}$ | The sum of the $h$-step-ahead forecasts for TD |
| $W_h$ | Covariance matrix of the corresponding $h$-step ahead base forecast errors |
| $y_t$ | Total level of power generation |
| $\widetilde{y}_t$ | an $n$-dimensional vector of $h$-step-ahead forecasts |
| $\hat{y}_{j,h}^{(l)}$ | The $h$-step-ahead forecast for TD |
| $\widetilde{y}_{t+h\|t}$ | Reconciled forecasts |
| $\hat{\lambda}_D$ | Shrinkage estimator |

## Appendix A

**ARIMA**

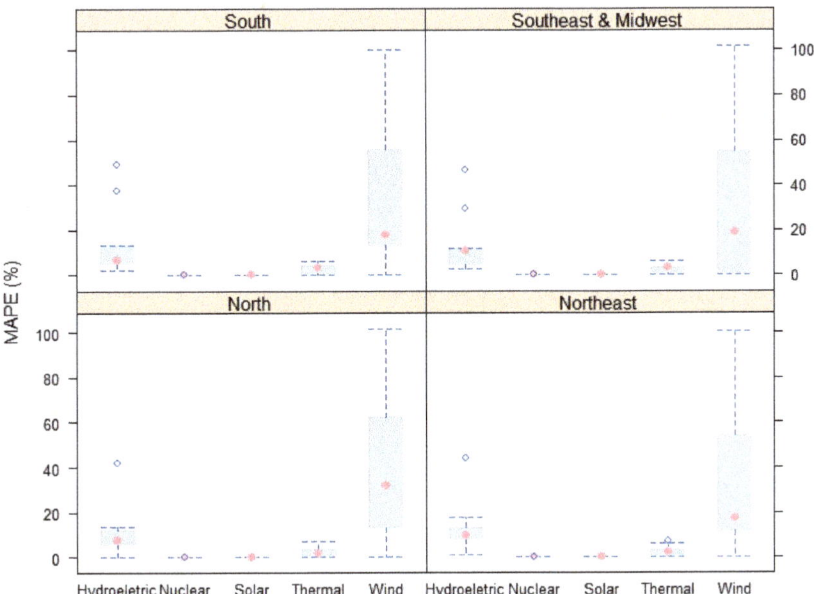

**Figure A1.** Hierarchical forecasting for power generation: electrical subsystem *versus* generating source.

**ETS**

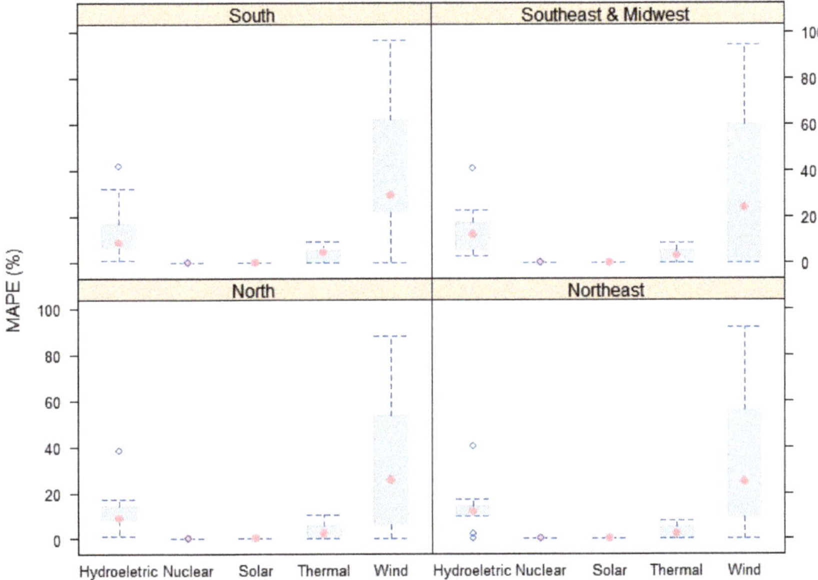

**Figure A2.** Hierarchical forecasting for power generation: electrical subsystem *versus* generating source.

| | | Predictive model: Autoregressive integrated moving average (ARIMA) | | | | | | | | | | | |
|---|---|---|---|---|---|---|---|---|---|---|---|---|---|
| | | RMSE | | | | MAE | | | | MASE | | | |
| | | Forecast horizon (h) - mean of the interval | | | | | | | | | | | |
| | | [1-3] | [4-6] | [7-9] | Mean | [1-3] | [4-6] | [7-9] | Mean | [1-3] | [4-6] | [7-9] | Mean |
| | | Hierarchical level 0 : Total - Brazil | | | | | | | | | | | |
| Method | BU | 2.36 | 7.77 | 8.50 | 6.21 | 2.20 | 6.46 | 7.14 | 5.27 | 1.11 | 3.27 | 3.62 | 2.67 |
| | TDGSA | 2.47 | 8.07 | 9.33 | 6.62 | 2.35 | 7.11 | 8.35 | 5.93 | 1.19 | 3.60 | 4.23 | 3.01 |
| | TDGSF | 2.47 | 8.07 | 9.33 | 6.62 | 2.35 | 7.11 | 8.35 | 5.93 | 1.19 | 3.60 | 4.23 | 3.01 |
| | TDFP | 2.47 | 8.07 | 9.33 | 6.62 | 2.35 | 7.11 | 8.35 | 5.93 | 1.19 | 3.60 | 4.23 | 3.01 |
| | OLS | 2.41 | 7.97 | 9.12 | 6.50 | 2.28 | 6.95 | 8.05 | 5.76 | 1.15 | 3.52 | 4.08 | 2.92 |
| | WLSv | 2.34 | 7.82 | 8.75 | 6.31 | 2.19 | 6.66 | 7.49 | 5.45 | 1.11 | 3.37 | 3.80 | 2.76 |
| | WLSs | 2.32 | 7.79 | 8.67 | 6.26 | 2.16 | 6.59 | 7.36 | 5.37 | 1.09 | 3.34 | 3.73 | 2.72 |
| | MintT (Sample) | 2.22 | 7.71 | 8.56 | 6.16 | 2.06 | 6.49 | 7.22 | 5.26 | 1.05 | 3.29 | 3.65 | 2.66 |
| | MinT (Shrink) | 2.25 | 7.75 | 8.63 | 6.21 | 2.10 | 6.56 | 7.28 | 5.31 | 1.07 | 3.32 | 3.69 | 2.69 |
| | | Hierarchical level 1 : Electrical subsystems | | | | | | | | | | | |
| Method | BU | 1.04 | 2.99 | 3.18 | 2.40 | 0.97 | 2.52 | 2.66 | 2.05 | 1.04 | 2.74 | 2.97 | 2.25 |
| | TDGSA | 5.48 | 6.88 | 7.06 | 6.47 | 5.46 | 6.72 | 6.88 | 6.35 | 6.30 | 7.44 | 7.45 | 7.06 |
| | TDGSF | 5.51 | 6.91 | 7.08 | 6.50 | 5.50 | 6.75 | 6.90 | 6.38 | 6.28 | 7.40 | 7.41 | 7.03 |
| | TDFP | 1.16 | 2.95 | 3.99 | 2.70 | 1.13 | 2.75 | 3.71 | 2.53 | 1.20 | 2.93 | 3.95 | 2.70 |
| | OLS | 1.08 | 3.20 | 3.65 | 2.64 | 1.04 | 2.89 | 3.27 | 2.40 | 1.12 | 3.14 | 3.64 | 2.64 |
| | WLSv | 1.05 | 3.10 | 3.44 | 2.53 | 1.00 | 2.73 | 3.01 | 2.24 | 1.06 | 2.91 | 3.26 | 2.41 |
| | WLSs | 1.05 | 3.10 | 3.44 | 2.53 | 1.00 | 2.73 | 3.00 | 2.24 | 1.07 | 2.95 | 3.33 | 2.45 |
| | MintT (Sample) | 1.01 | 3.07 | 3.41 | 2.50 | 0.96 | 2.70 | 2.95 | 2.20 | 1.00 | 2.87 | 3.18 | 2.35 |
| | MinT (Shrink) | 1.02 | 3.08 | 3.43 | 2.51 | 0.97 | 2.72 | 2.97 | 2.22 | 1.02 | 2.90 | 3.22 | 2.38 |
| | | Hierarchical level 2 : Energy sources | | | | | | | | | | | |
| Method | BU | 0.89 | 2.32 | 2.38 | 1.87 | 0.83 | 1.98 | 2.04 | 1.62 | 1.22 | 2.70 | 2.95 | 2.29 |
| | TDGSA | 3.39 | 4.70 | 4.93 | 4.34 | 3.37 | 4.57 | 4.79 | 4.24 | 14.02 | 14.92 | 15.04 | 14.66 |
| | TDGSF | 3.36 | 4.68 | 4.90 | 4.32 | 3.35 | 4.55 | 4.76 | 4.22 | 14.79 | 15.66 | 15.77 | 15.41 |
| | TDFP | 0.98 | 2.30 | 2.97 | 2.08 | 0.95 | 2.15 | 2.77 | 1.95 | 1.62 | 3.57 | 4.91 | 3.37 |
| | OLS | 0.90 | 2.37 | 2.49 | 1.92 | 0.85 | 2.07 | 2.15 | 1.69 | 1.85 | 5.25 | 7.12 | 4.74 |
| | WLSv | 0.91 | 2.42 | 2.63 | 1.98 | 0.86 | 2.18 | 2.33 | 1.79 | 1.25 | 2.88 | 3.22 | 2.45 |
| | WLSs | 0.90 | 2.35 | 2.44 | 1.89 | 0.84 | 2.04 | 2.09 | 1.66 | 1.60 | 4.22 | 5.37 | 3.73 |
| | MintT (Sample) | 0.88 | 2.40 | 2.62 | 1.97 | 0.84 | 2.17 | 2.31 | 1.77 | 1.23 | 2.87 | 3.24 | 2.45 |
| | MinT (Shrink) | 0.89 | 2.41 | 2.64 | 1.98 | 0.85 | 2.18 | 2.33 | 1.79 | 1.24 | 2.89 | 3.26 | 2.46 |

**Figure A3.** Hierarchical forecasting for electricity generation based on the ARIMA procedure (RMSE, MAE, MASE). (Note: The performance was indicated into a color scale, where green means better values for calculated accuracy, and red means worse accuracy. The intermediate values are colored yellow.).

| Predictive model: Error, trend, seasonality (ETS) | | | | | | | | | | | |
| RMSE | | | | MAE | | | | MASE | | | |
| Forecast horizon (h) - mean of the interval | | | | | | | | | | | |
| [1-3] | [4-6] | [7-9] | Mean | [1-3] | [4-6] | [7-9] | Mean | [1-3] | [4-6] | [7-9] | Mean |
|---|---|---|---|---|---|---|---|---|---|---|---|
| **Hierarchical level 0 : Total - Brazil** | | | | | | | | | | | |
| BU — 3.02 | 9.14 | 11.26 | 7.81 | 2.98 | 8.62 | 10.57 | 7.39 | 1.51 | 4.37 | 5.35 | 3.74 |
| TDGSA — 2.65 | 8.62 | 10.51 | 7.26 | 2.59 | 7.98 | 9.76 | 6.78 | 1.31 | 4.04 | 4.94 | 3.43 |
| TDGSF — 2.65 | 8.62 | 10.51 | 7.26 | 2.59 | 7.98 | 9.76 | 6.78 | 1.31 | 4.04 | 4.94 | 3.43 |
| TDFP — 2.65 | 8.62 | 10.51 | 7.26 | 2.59 | 7.98 | 9.76 | 6.78 | 1.31 | 4.04 | 4.94 | 3.43 |
| OLS — 2.67 | 8.64 | 10.54 | 7.28 | 2.61 | 8.01 | 9.78 | 6.80 | 1.32 | 4.06 | 4.96 | 3.44 |
| WLSv — 2.77 | 8.78 | 10.75 | 7.43 | 2.72 | 8.19 | 10.01 | 6.97 | 1.38 | 4.15 | 5.07 | 3.53 |
| WLSs — 2.78 | 8.81 | 10.78 | 7.46 | 2.73 | 8.22 | 10.05 | 7.00 | 1.38 | 4.16 | 5.09 | 3.54 |
| MintT (Sample) — 2.61 | 8.56 | 10.43 | 7.20 | 2.55 | 7.91 | 9.66 | 6.71 | 1.29 | 4.01 | 4.89 | 3.40 |
| MinT (Shrink) — 2.62 | 8.58 | 10.45 | 7.21 | 2.56 | 7.92 | 9.68 | 6.72 | 1.30 | 4.01 | 4.90 | 3.40 |
| **Hierarchical level 1 : Electrical subsystems** | | | | | | | | | | | |
| BU — 1.30 | 3.66 | 4.48 | 3.15 | 1.29 | 3.47 | 4.18 | 2.98 | 1.44 | 3.84 | 4.66 | 3.31 |
| TDGSA — 5.54 | 7.09 | 7.45 | 6.70 | 5.54 | 6.98 | 7.30 | 6.61 | 6.37 | 7.65 | 7.81 | 7.28 |
| TDGSF — 5.58 | 7.12 | 7.48 | 6.73 | 5.57 | 7.00 | 7.32 | 6.63 | 6.34 | 7.62 | 7.77 | 7.24 |
| TDFP — 1.13 | 2.95 | 4.07 | 2.72 | 1.11 | 2.77 | 3.79 | 2.56 | 1.22 | 3.03 | 4.14 | 2.80 |
| OLS — 1.15 | 3.42 | 4.15 | 2.91 | 1.13 | 3.21 | 3.84 | 2.73 | 1.23 | 3.50 | 4.22 | 2.99 |
| WLSv — 1.20 | 3.50 | 4.25 | 2.98 | 1.18 | 3.29 | 3.95 | 2.81 | 1.31 | 3.61 | 4.37 | 3.10 |
| WLSs — 1.20 | 3.50 | 4.26 | 2.98 | 1.18 | 3.29 | 3.95 | 2.81 | 1.30 | 3.60 | 4.36 | 3.08 |
| MintT (Sample) — 1.20 | 3.50 | 4.25 | 2.98 | 1.17 | 3.28 | 3.93 | 2.79 | 1.25 | 3.52 | 4.26 | 3.01 |
| MinT (Shrink) — 1.18 | 3.47 | 4.22 | 2.96 | 1.16 | 3.26 | 3.91 | 2.77 | 1.25 | 3.51 | 4.25 | 3.00 |
| **Hierarchical level 2 : Energy sources** | | | | | | | | | | | |
| BU — 1.15 | 2.93 | 3.53 | 2.54 | 1.14 | 2.83 | 3.34 | 2.44 | 1.64 | 3.87 | 4.68 | 3.40 |
| TDGSA — 3.43 | 4.85 | 5.21 | 4.50 | 3.42 | 4.75 | 5.08 | 4.42 | 14.21 | 15.56 | 16.08 | 15.28 |
| TDGSF — 3.41 | 4.83 | 5.18 | 4.47 | 3.40 | 4.73 | 5.05 | 4.39 | 14.97 | 16.29 | 16.78 | 16.01 |
| TDFP — 1.05 | 2.47 | 3.28 | 2.27 | 1.03 | 2.36 | 3.11 | 2.17 | 1.87 | 3.88 | 5.06 | 3.60 |
| OLS — 1.13 | 2.89 | 3.48 | 2.50 | 1.11 | 2.79 | 3.29 | 2.40 | 2.64 | 5.55 | 6.80 | 5.00 |
| WLSv — 1.07 | 2.81 | 3.36 | 2.41 | 1.05 | 2.69 | 3.17 | 2.30 | 1.55 | 3.74 | 4.50 | 3.26 |
| WLSs — 1.13 | 2.91 | 3.50 | 2.51 | 1.12 | 2.80 | 3.31 | 2.41 | 2.31 | 4.99 | 6.09 | 4.46 |
| MintT (Sample) — 1.01 | 2.71 | 3.23 | 2.32 | 0.99 | 2.59 | 3.03 | 2.20 | 1.44 | 3.38 | 3.99 | 2.93 |
| MinT (Shrink) — 1.03 | 2.75 | 3.28 | 2.35 | 1.01 | 2.63 | 3.08 | 2.24 | 1.43 | 3.49 | 4.19 | 3.03 |

**Figure A4.** Hierarchical forecasting for electricity generation based on the ETS procedure (RMSE, MAE, MASE). (Note: The performance was indicated into a color scale, where green means better values for calculated accuracy, and red means worse accuracy. The intermediate values are colored yellow.).

## References


1.  Medojevic, M.; Medic, N.; Marjanovic, U.; Lalic, B.; Majstorovic, V. Exploring the impact of industry 4.0 concepts on energy and environmental management systems: Evidence from Serbian manufacturing companies. In Proceedings of the IFIP International Conference on Advances in Production Management Systems, Austin, TX, USA, 1–5 September 2019; pp. 355–362.
2.  Alcácer, V.; Cruz-Machado, V. Scanning the industry 4.0: A literature review on technologies for manufacturing systems. *Eng. Sci. Technol. Int. J.* **2019**, *22*, 899–919. [CrossRef]
3.  Bourdeau, M.; Zhai, X.Q.; Nefzaoui, E.; Guo, X.; Chatellier, P. Modeling and forecasting building energy consumption: A review of data-driven techniques. *Sustain. Cities Soc.* **2019**, *48*, 101533. [CrossRef]
4.  Hammad, M.A.; Jereb, B.; Rosi, B.; Dragan, D. Methods and models for electric load forecasting: A comprehensive review. *Logist. Sustain. Transp.* **2020**, *11*, 51–76. [CrossRef]

5.	Runge, J.; Zmeureanu, R. Forecasting energy use in buildings using artificial neural networks: A review. *Energies* **2019**, *12*, 3254. [CrossRef]

6.	Choi, Y.B. *Paradigms and Conventions: Uncertainty, Decision Making, and Entrepreneurship*; University of Michigan Press: Ann Arbor, MI, USA, 1993.

7.	Jiang, W.; Yan, Z.; Feng, D.H.; Hu, Z. Wind speed forecasting using autoregressive moving average/generalized autoregressive conditional heteroscedasticity model. *Eur. Trans. Electr. Power* **2012**, *22*, 662–673. [CrossRef]

8.	Hao, C.H.E.N. A new method of load forecasting based on generalized autoregressive conditional heteroscedasticity model. *Autom. Electr. Power Syst.* **2007**, *15*, 012.

9.	Stefenon, S.F.; Ribeiro, M.H.D.M.; Nied, A.; Mariani, V.C.; dos Santos Coelho, L.; da Rocha, D.F.M.; Grebogif, R.B.; de Barros Ruano, A.E. Wavelet group method of data handling for fault prediction in electrical power insulators. *Int. J. Electr. Power Energy Syst.* **2020**, *123*, 106269. [CrossRef]

10.	Frizzo Stefenon, S.; Silva, M.C.; Bertol, D.W.; Meyer, L.H.; Nied, A. Fault diagnosis of insulators from ultrasound detection using neural networks. *J. Intell. Fuzzy Syst.* **2019**, *37*, 6655–6664. [CrossRef]

11.	Gupta, S.; Srinivasan, D.; Reindl, T. Forecasting solar and wind data using dynamic neural network architectures for a micro-grid ensemble. In Proceedings of the 2013 IEEE Computational Intelligence Applications in Smart Grid (CIASG), Singapore, 16–19 April 2013; IEEE: New York, NY, USA, 2013; pp. 87–92.

12.	Frizzo Stefenon, S.; Zanetti Freire, R.; dos Santos Coelho, L.; Meyer, L.H.; Bartnik Grebogi, R.; Gouvêa Buratto, W.; Nied, A. Electrical insulator fault forecasting based on a wavelet neuro-fuzzy system. *Energies* **2020**, *13*, 484. [CrossRef]

13.	Moghaddam, A.A.; Seifi, A.R. Study of forecasting renewable energies in smart grids using linear predictive filters and neural networks. *IET Renew. Power Gener.* **2011**, *5*, 470–480. [CrossRef]

14.	Barbounis, T.G.; Theocharis, J.B. A locally recurrent fuzzy neural network with application to the wind speed prediction using spatial correlation. *Neurocomputing* **2007**, *70*, 1525–1542. [CrossRef]

15.	Xia, J.; Zhao, P.; Dai, Y. Neuro-fuzzy networks for short-term wind power forecasting. In Proceedings of the 2010 International Conference on Power System Technology, Hangzhou, China, 24–28 October 2010; IEEE: New York, NY, USA, 2010; pp. 1–5.

16.	Dawan, P.; Sriprapha, K.; Kittisontirak, S.; Boonraksa, T.; Junhuathon, N.; Titiroongruang, W.; Niemcharoen, S. Comparison of power output forecasting on the photovoltaic system using adaptive neuro-fuzzy inference systems and particle swarm optimization-artificial neural network model. *Energies* **2020**, *13*, 351. [CrossRef]

17.	Hyndman, R.J.; Kandahar, Y. Automatic time series forecasting: The forecast package for R. *J. Stat. Softw.* **2008**, *26*, 1–22.

18.	Athanasopoulos, G.; Ahmed, R.A.; Hyndman, R.J. Hierarchical forecasts for Australian domestic tourism. *Int. J. Forecast.* **2009**, *25*, 146–166. [CrossRef]

19.	Hyndman, R.J.; Ahmed, R.A.; Athanasopoulos, G.; Shang, H.L. Optimal combination forecasts for hierarchical time series. *Comput. Stat. Data Anal.* **2011**, *55*, 2579–2589. [CrossRef]

20.	Almeida, V.; Ribeiro, R.; Gama, J. Hierarchical time series forecast in electrical grids. In *Information Science and Applications (ICISA)*; Springer: Singapore, 2016; pp. 995–1005.

21.	Panamtash, H.; Zhou, Q. Coherent probabilistic solar power forecasting. In Proceedings of the 2018 IEEE International Conference on Probabilistic Methods Applied to Power Systems (PMAPS), Boise, ID, USA, 24–28 June 2018; IEEE: New York, NY, USA, 2018; pp. 1–6.

22.	Abouarghoub, W.; Nomikos, N.K.; Petropoulos, F. On reconciling macro and micro energy transport forecasts for strategic decision making in the tanker industry. *Transp. Res. Part E Logist. Transp. Rev.* **2018**, *113*, 225–238. [CrossRef]

23.	Auder, B.; Cugliari, J.; Goude, Y.; Poggi, J.M. Scalable clustering of individual electrical curves for profiling and bottom-up forecasting. *Energies* **2018**, *11*, 1893. [CrossRef]

24.	Silva, F.L.; Souza, R.C.; Oliveira, F.L.C.; Lourenco, P.M.; Calili, R.F. A bottom-up methodology for long term electricity consumption forecasting of an industrial sector-Application to pulp and paper sector in Brazil. *Energy* **2018**, *144*, 1107–1118. [CrossRef]

25.	Ghedamsi, R.; Settou, N.; Gouareh, A.; Khamouli, A.; Saifi, N.; Recioui, B.; Dokkar, B. Modeling and forecasting energy consumption for residential buildings in Algeria using bottom-up approach. *Energy Build.* **2016**, *121*, 309–317. [CrossRef]

26. Kosiorowski, D.; Mielczarek, D.; Rydlewski, J. Forecasting of a hierarchical functional time series on example of macromodel for day and night air pollution in silesia region: A critical overview. *arXiv* **2017**, arXiv:1712.03797.

27. Wickramasuriya, S.L.; Athanasopoulos, G.; Hyndman, R.J. Optimal forecast reconciliation for hierarchical and grouped time series through trace minimization. *J. Am. Stat. Assoc.* **2019**, *114*, 804–819. [CrossRef]

28. National System Operator. Operation History (Report of Power Generation). 2020. Available online: http://www.ons.org.br/paginas/resultados-da-operacao/historico-da-operacao (accessed on 15 May 2020).

29. R Core Team. *R: A Language and Environment for Statistical Computing*; R Foundation for Statistical Computing: Vienna, Austria, 2020; Available online: https://www.R-project.org/ (accessed on 15 May 2020).

30. Orcutt, G.H.; Watts, H.W.; Edwards, J.B. Data aggregation and information loss. *Am. Econ. Rev.* **1968**, *58*, 773–787.

31. Gross, C.W.; Sohl, J.E. Disaggregation methods to expedite product line forecasting. *J. Forecast.* **1990**, *9*, 233–254. [CrossRef]

32. Oliveira, J.M.; Ramos, P. Assessing the performance of hierarchical forecasting methods on the retail sector. *Entropy* **2019**, *21*, 436. [CrossRef]

33. Yang, D.; Kleissl, J.; Gueymard, C.A.; Pedro, H.T.; Coimbra, C.F. History and trends in solar irradiance and PV power forecasting: A preliminary assessment and review using text mining. *Sol. Energy* **2018**, *168*, 60–101. [CrossRef]

34. Panigrahi, S.; Behera, H.S. A hybrid ETS–ANN model for time series forecasting. *Eng. Appl. Artif. Intell.* **2017**, *66*, 49–59. [CrossRef]

35. Dong, Z.; Yang, D.; Reindl, T.; Walsh, W.M. Short-term solar irradiance forecasting using exponential smoothing state space model. *Energy* **2013**, *55*, 1104–1113. [CrossRef]

36. Box, G.E. Jenkins. In *Time Series Analysis: Forecasting and Control*; Holden-Day Inc.: New York, NY, USA, 1976.

37. Pegels, C.C. Exponential forecasting: Some new variations. *Manag. Sci.* **1969**, 311–315.

38. Gardner, E.S., Jr. Exponential smoothing: The state of the art. *J. Forecast.* **1985**, *4*, 1–28. [CrossRef]

39. Hyndman, R.J.; Koehler, A.B.; Snyder, R.D.; Grose, S. A state space framework for automatic forecasting using exponential smoothing methods. *Int. J. Forecast.* **2002**, *18*, 439–454. [CrossRef]

40. Taylor, J.W. Exponential smoothing with a damped multiplicative trend. *Int. J. Forecast.* **2003**, *19*, 715–725. [CrossRef]

41. Liu, Z.; Yan, Y.; Yang, J.; Hauskrecht, M. Missing value estimation for hierarchical time series: A study of hierarchical Web traffic. In Proceedings of the 2015 IEEE International Conference on Data Mining, Atlantic City, NJ, USA, 14–17 November 2015; IEEE: New York, NY, USA, 2015; pp. 895–900.

42. Weiss, C. Essays in Hierarchical Time Series Forecasting and Forecast Combination. Ph.D. Thesis, University of Cambridge, Cambridge, UK, 2018.

43. Hong, T.; Xie, J.; Black, J. Global energy forecasting competition 2017: Hierarchical probabilistic load forecasting. *Int. J. Forecast.* **2019**, *35*, 1389–1399. [CrossRef]

Article

# Application of IRT Models to Selection of Bidding Paths in Financial Transmission Rights Auction: U.S. New England

**Peter Y. Jang [1], Kwanghee Jung [2,*] and Mario G. Beruvides [1]**

[1] Department of Industrial, Manufacturing, and Systems Engineering, Texas Tech University, Lubbock, TX 79409, USA; peter.y.jang@ttu.edu (P.Y.J.); mario.beruvides@ttu.edu (M.G.B.)
[2] Department of Educational Psychology and Leadership, Texas Tech University, Lubbock, TX 79409, USA
* Correspondence: kwanghee.jung@ttu.edu

Received: 2 June 2020; Accepted: 26 June 2020; Published: 30 June 2020

**Abstract:** This paper explores a way to apply Item Response Theory (IRT), one of the popular statistical methodologies in measurement and psychometrics, to evaluate Financial Transmission Rights (FTR) paths in the U.S. electricity market. FTR is an energy derivative product to hedge congestion cost risks inherent in constrained transmission lines. In New England, with about 1200 pricing locations, the theoretical combinations of FTR paths amount to 1.4 million in prevailing flows alone. With capital constraints, it is imperative that FTR market participants build the capability to evaluate FTR paths to bid on. IRT provides a framework of how well tests work, and how individual items work on tests, estimating respondents' latent abilities, and individual item parameters. IRT is utilized to analyze historical electricity data of 2019 for a daily congestion cost of eight customer load zones and one hub in the U.S., New England, for the evaluation of FTR paths. In the analysis, an item represents an FTR path, while item difficulty, item discrimination, and a latent trait variable for the path correspond to the path profitability, risk level, and daily congestion ability, respectively. This paper explores the experimental procedures by which IRT, a psychometric tool, may also be applicable in complex energy markets, providing a consistent and standardized analytical framework to address the issues of selection and prioritization among multiple opportunities. FTR path evaluation is conducted in three steps to determine bid priority paths in FTR auctions: parameter significance tests, ranking on path profitability and risk level, and weighting scores of individual rankings on the two criteria.

**Keywords:** energy economics; financial transmission rights (FTR); FTR auction; FTR path evaluation; electric market; item response theory (IRT); psychometrics

---

## 1. Introduction

Financial Transmission Rights (FTR) is an energy derivative that allows market participants to receive an annual or monthly share of congestion cost revenues collected in settled electricity prices, or locational marginal price (LMP), by Independent System Operators (ISO) [1,2]. The ISO is a third-party organization to ensure electric systems reliability in generation resources and transmission lines. Congestion cost at a pricing location is a price difference between the least expensive electricity available in the ISO region and more expensive options due to transmission system constraints.

FTR holders are paid a congestion cost difference (credit) settled on a transmission path when it is positive (prevailing flow FTR) and must pay the difference (charge) when negative (counterflow FTR). The above FTR is called the FTR Obligation, compared to the FTR Option where FTR Option holders do not have to pay the difference even when the settled value is negative [1]. As of April 2020, FTR Option products are not available in ISO New England (ISO-NE) in the U.S., but exist in other electric markets, such as the Pennsylvania–New Jersey–Maryland Interconnection LLC (PJM Interconnection).

As FTR values are derived from the pairs of pricing locations in an electric market, the possible combinations of FTR paths could be very large, providing multiple FTR bidding opportunities for market participants. For example, ISO New England has an existing generating capacity of 31,200 MW from 1976 generators in its six member states, with a generation mix of natural gas (40%), nuclear (25%), net imports (19%), renewable (9%), and hydro (7%) [3]. According to the Day-Ahead Energy Market Hourly LMP Report for 14 April 2020 published by the ISO-NE, the total number of pricing locations in the region is 1192, comprising 1125 network nodes, 32 hub nodes, 20 demand response locations, 8 load zones, 6 external nodes, and 1 hub. The total pricing locations could theoretically be translated into 1.4 million FTR paths in the ISO auction for prevailing flows, and when added by counterflows, the total number could be doubled.

There are two types of FTR market participants: hedgers and speculators. Hedging participants, with electricity supply obligations, want to hedge against congestion costs by purchasing FTRs on the paths from their supply sources to their customer load zones, while speculating participants, without any physical supply obligations, may purchase FTRs to arbitrage differences between expected and actual settled values of FTR paths [1]. With such numerous choices of potential FTR paths available, FTR market participants need to reasonably evaluate which FTR paths to bid in the auctions, subject to their limited capital budgets. Consistent and standardized methodology is crucial in evaluating interested paths, in terms of profitability and risks associated with the paths.

Item Response Theory (IRT) is one of the most influential methods in the field of educational and psychological measurement, to understand the behaviors of individual test items or variables [4]. IRT models provide information about item parameters and latent traits of test respondents, helping gain insights and assessments about their performance as well as the items. It is also useful for test development, item analysis, equating, item banking, and computer aided test (CAT) [5]. As a group of statistical models with probabilistic and stochastic procedures, IRT connects the pattern of responses to a group of items to predict a latent trait/ability, and then, converts discrete item responses into the levels or locations of probability estimates which respondents possess underlying the latent trait [6,7].

The most basic model is the One-Parameter Logistic model (1PL), or the Rasch model named after Georg Rasch, a Danish mathematician. In the model, the probability of correct response (denoted as $X_i = 1$) to each item (labeled $i$) is a function of the item's difficulty level (labeled $b_i$) and the respondent's trait level (labeled $\theta$), with a mathematical expression as in Equation (1) [7]:

$$P_i(\theta) = P_i\left(X_i = 1 | \theta\right) = \frac{e^{(\theta - b_i)}}{1 + e^{(\theta - b_i)}} = \frac{1}{1 + e^{-(\theta - b_i)}} \tag{1}$$

In the Equation (1), $X_i = 1$ indicates that a respondent endorsed an item i or provided a correct response. A horizontal line at $P_i(\theta) = 0.5$ on the y-axis in Figure 1 denotes a mid-probability of the correct response to item *i* being correct. That is, it indicates that the respondent has 50% chance of providing a correct response to the item. The difficulty coefficient ($b_i$) of an item are the value of a latent trait level ($\theta$) on the x-axis which is an intersection point between the mid-probability, shown on the horizontal line, and an individual characteristic curve (ICC) of the item. Figure 1 illustrates three items in Rasch 1PL model, with the values of difficulty coefficients with $b_i = -1, 0$, and 1. Both the latent trait level ($\theta$) and difficulty coefficient ($b_i$) are on the same z-score metric, with the latent trait level ($\theta$) typically in the range of $[-2, 2]$ [6].

IRT models are logistic regression models to predict dichotomous, or binary, outcomes, with a monotonically increasing S-shaped curve, called the Item Characteristic Curves (ICCs) [8]. ICCs display the relationship between a latent trait level and the probability of correct response. Figure 1 illustrates three ICCs of probability of correct response ($P_i(\theta)$), with assumptions of three items' difficulty parameters of $-1, 0$, and 1, respectively, given a range of the latent trait levels $[-4, 4]$. This paper will use the term latent trait and ability interchangeably in describing the IRT and its application to FTR path evaluation in the U.S. electricity market.

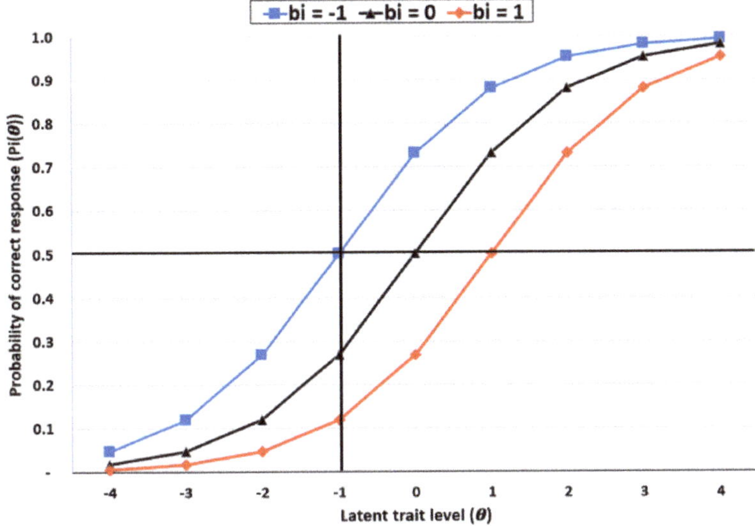


**Figure 1.** Item characteristic curves, Item Response Theory (IRT) 1PL model.

The ICCs in Figure 1 may be interpreted such that an item is more difficult to the right of the chart, and easier to the left of the chart, and that low $P_i(\theta)$ of the item implies highly unlikeliness of providing correct responses at a given latent trait level [7]. For example, in Figure 1 at the level of $-1$ for the latent trait ($\theta$), the probability of correct response varies with each of the three items, 50%, 27%, and 12% as indicated on a vertical line on the Figure 1.

While IRT is popular in the psychometrics discipline, there have been several studies on the applications of the IRT theory in the fields of health behavior research [6,7], as well as in financial literacy measurement [9]. The Two-Parameter Logistic model (2PL) is an extension of the Rasch 1PL model, with additional parameter of an item's discrimination parameter. Item discrimination represents how well an item differentiates among respondents on the latent trait continuum, e.g., differentiating respondents of different ability levels. The IRT 2PL model and ICCs will be discussed in greater detail later in this study.

In summary, the IRT 2PL model provides analytical advantages in terms of parameter parsimony, easier parameter interpretation, distinguishability among multiple items, and visual presentation. The parameters, difficulty ($b_i$) and differentiation ($a_i$), may be derived from historical data observation, and the magnitude of the parameters may be used to easily interpret how difficult or different each item is relative to other items. The parameters also provide a foundation to build the probability function of a right response ($P_i(\theta)$ or $P_i(X_i = 1|\theta)$) in visual presentation of ICCs on a level of latent trait variable ($\theta$). When there are multiple opportunities available in the marketplace, essential decision-making factors involve estimation and comparison of their return and risk profiles. With such capability, IRT may be applied to FTR markets, where the astonishing number of paths are available, 1.4 million paths in U.S. New England ISO alone, and a consistent and standardized evaluation model is required for FTR participants to understand return and risk profiles of path they are interested in.

This paper is the first experiment to apply IRT, particularly the IRT 2PL model, to the U.S. energy market, in evaluating and selecting the FTR paths to bid in market auctions. This paper is organized as follows: Section 2. Literature Review; Section 3. Data and Methodology; Section 4. Results and Discussion; and Section 5. Conclusion and Implications.

## 2. Literature Review

### 2.1. Financial Transmission Rights

The first FTR auction took place in 1999, in the PJM Interconnection in the U.S. In the auctions, ISOs have a goal of maximizing FTR revenues, subject to the constraints of transmission capacity and contingencies [10]. Electric suppliers calculate FTR values of the paths to bid, based on their own forecasts of future LMP prices in the interested locations. The FTR calculation may have analytical frameworks of game theoretic models, with multiple participants, or network contingencies in the ISO systems [11,12].

FTRs are defined in U.S. Dollar ($) per mega-watts (MWs), from a source (receipt, inject) pricing point to a sink (delivery, withdrawal) pricing point on a transmission line path. In New England, FTR products are offered in monthly and annual auctions, for two categories, onpeak hours (weekdays hours ending 0800–2300) and offpeak hours (weekdays hours ending 2400–0700, and 24 h on weekends and NERC holidays). Available pricing locations in ISO-NE include generator nodes, external nodes, hub (specified set of predefined pricing nodes), load zones (aggregate of pricing nodes in a specific area), and DRR (demand response resources) aggregate zones [13]. Figure 2 presents a flow chart that summarizes typical FTR auction procedures involving several entities in terms of exchanging data and information [13].

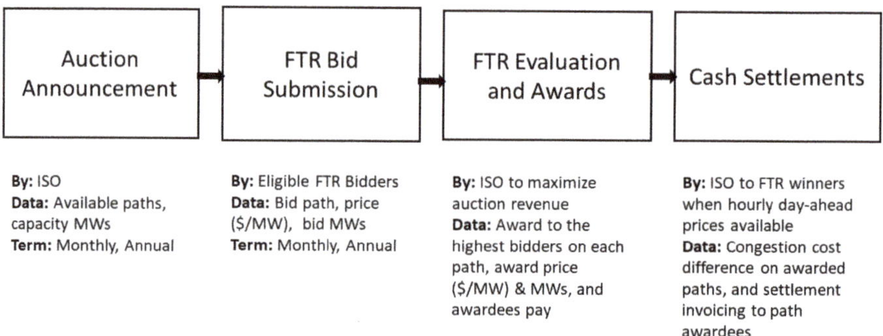


**Figure 2.** Financial Transmission Rights (FTR) auction flow chart.

FTR auction results in ISO-NE provide the magnitude of the auctions and major FTR participants [14]. For the April 2020 auction, a total of 5625 FTR paths were cleared, with onpeak at 2719 and offpeak at 2906, on a total of 25,586 MWs. There was a total of 28 FTR winners, with the top five companies accounting for 71% of total MWs cleared: NextEra Energy Marketing, Vitol, MAG Energy Solutions, Exelon Generation, Transgrid Midwest. In the annual FTR auction for 2020, the number of cleared FTR paths was 3348 (onpeak 1593 and offpeak 1755), on a total of 17,514 MWs, onpeak and offpeak combined. The top five participants accounted for 75% of total MWs cleared: Vitol, Mercuria Energy America, Castleton Commodities Merchant, Citigroup Energy, NextEra Energy Marketing.

ISO operates the wholesale electricity market that consists of two markets, Day-Ahead Market (DAM) and Real Time Market (RTM). DAM is a forward sport market, where DAM LMP are settled in day-ahead auctions. Generators submit offers, and customer loads submit bids to the ISO with hourly MWs for each hour of the next day. The ISO calculates a nodal price, or a locational marginal price (LMP) of a location, based on all the submitted offers and bids, subject to the Lagrange multipliers, or constraints of active power balance and transmission [10].

Settled LMP is made up of three components, energy, congestion cost, and loss [15]. Congestion cost is created by binding constraints of transmission lines and generation resources in the auction, resulting in incremental cost for some points and different LMPs. FTR is the difference in the congestion cost component of the LMPs between two locations. RTM is a balancing market to DAM that addresses actual power systems and generated MWs. Energy sellers in DAM get paid real time prices for the MWs generated in real time over the MWs sold in day-ahead market with cleared DAM prices [16].

There is uncertainty in future LMPs that may be significantly different from the LMPs that FTR market participants expected when estimating the values of FTRs for bidding. In this context, FTR holders may face the risks of liability at the time of LMP settles, due to counterflows on the awarded paths, unexpected outages, severe economic congestions, and transmission losses [17]. Load serving utilities (LSE) or suppliers to retail customers are exposed to major risks of LMP, comprising energy (fuel), transmission constraint cost at a given time, and line losses. The transmission constraint cost may be called a transmission opportunity cost, or difference between the clearing LMPs on a given path [18]. Other reliability services costs include capacity for adequate resources, and ancillary services to maintain the electric systems.

In this context, the biggest challenge for FTR participants is how to simulate market participants' behavior [10], as well as to calculate expected FTR payoffs and financial risks associated with the FTR path [19]. A realistic view is that it is practically difficult to formulate all the electricity prices and market behaviors, given thousands of pricing nodes available in the ISO [11]. Due the complexity of predicting clearing LMPs and estimating the value of FTR derivative product, the research studies that addressed FTR bidding strategies usually involved simulation approaches or problem formulation with a limitation of two to four pricing nodes [10,19].

There have been some studies related to FTR from bidder and generator standpoints, but none of them addressed the question of how to evaluate and select FTR paths to bid among multiple choices. Hogan [18] also noted that the U.S. energy market design has been successful with bid-based LMP and FTRs, but still has remaining challenges with both theory and implementation.

Li and Shahidehpour [19] illustrated a three-bus system with four FTR bidders, subject to the ISO's goal of maximizing auction revenues, as well as the impacts of transmission constraints, forecasted LMP differences, and bidder's risk tolerance on FTR bidding strategies. Das et al. [12] experimented with a matrix-game model to analyze FTR bidding strategies. This study involved multiple FTR participants on a sample network, with assumptions that bidders have forecasts of LMPs, and assessed impacts of various bid prices.

From a power plant generator's perspective, Liu and Wu [11] investigated an FTR position by exploring the interaction between generator's bidding and transmission rights holding. The study suggested that transmission rights helped reconfigure a generator's behavior in bidding their electricity into the ISO. Liu and Gross [20] proposed a way, based on simulation approaches, to integrate bi-lateral transaction with a centralized pool market, or ISO, for the efficient allocation of transmission services affecting FTR evaluations.

### 2.2. Item Response Theory in Psychometrics

IRT model has been developed as a new way of data analysis for categorical data to measure a latent trait variable (also called ability, denoted as $\theta$), as well as to model the item responses ($X_i$) of respondents. The data may be dichotomous (binary) or polytomous [4,21]. Major assumptions in the IRT models include monotonicity of the latent trait variable and the probability of an item correct response, unidimensionality of measuring one single latent ability with a set of items, and local independence among the item responses.

There are basic IRT parameter logistic models: Rasch 1PL (1-parameter logistic model) and 2PL (2-parameter), depending on the number of parameters used in modeling for items, and a parameter of a single latent variable underlying the item responses of a respondent [21]. Two basic parameters in IRT are item difficulty ($b_i$, location index), and item discrimination ($a_i$, differentiation). A latent trait (ability, $\theta$) parameter is a construct or a factor measured by the item responses.

Further to the Rasch 1PL model described in Equation (1) and Figure 1, 2PL model is introduced here as an extension of 1PL. The IRT 2PL model has one more parameter, discrimination ($a_i$), than the Rasch model, and may be interpreted in a way that the higher discrimination ($a_i$) of an item is, the more discriminating the item is with a steeper slope on an ICC. Conversely, a flatter ICC of an item indicates the item is less likely to discriminate among respondents than other items. The discrimination coefficient ($a_i$) typically takes the value of [−0.5, 2] [6].

The IRT 2PL model is expressed as in Equation (2):

$$P_i(\theta) \;=\; P_i\,(X_i \;=\; 1|\,\theta) \;=\; \frac{e^{a_i(\theta-b_i)}}{1+e^{a_i(\theta-b_i)}} \;=\; \frac{1}{1+e^{-a_i(\theta-b_i)}} \tag{2}$$

where exp (e) is the constant 2.718, $b_i$ = difficulty parameter for an item i, $a_i$ = discrimination parameter for an item i, and $\theta$ = ability level.

Figure 3 presents an illustration of 2PL-based item characteristic curves (ICCs) for three items, built on Equation (2). It displays the impacts of varying discrimination coefficients of $a$ = 0.5, 1, and 2 for three items, given a latent trait level, and the same difficulty coefficient ($b_i$) of 0 for all items. Item discrimination parameter represents a slope on an inflection point of each ICC. As the discrimination coefficients ($a_i$) describe a sharp distinction between respondents to each item, the corresponding latent trail levels ($\theta$) also vary in a given range of probability of correct response (P($\theta$)) from 25% to 75%, as referenced in two horizontal lines. When discrimination coefficient ($a_i$) = 0.5, the range of the trait ($\theta$) is [−2.1, 2.1], at $a$ = 1, the range is [−1.3, 1.3], and at $a$ = 2 the range is narrower to [−0.8, 0.8]. The results indicate that a greater discrimination coefficient ($a_i$) produces a tighter range of a latent trail level ($\theta$) with a steeper slope, and more discrimination power among the respondents between lower and upper groups.

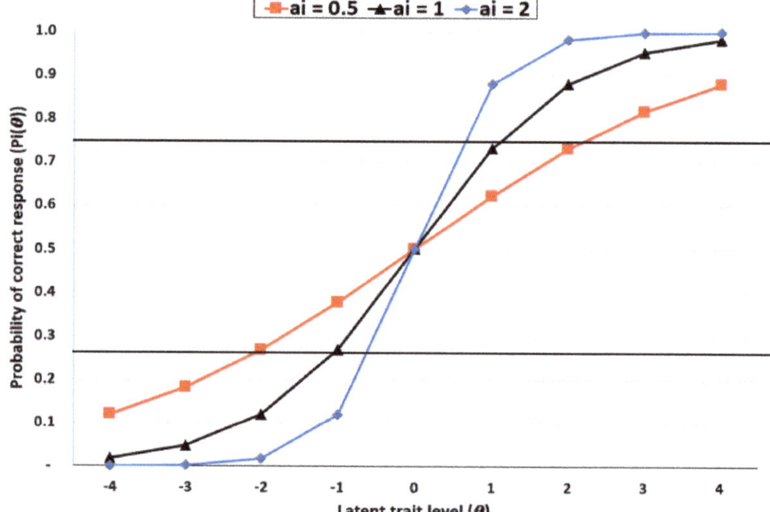


**Figure 3.** Item characteristic curves, IRT 2PL model: difficulty coefficient b = 0 for all items.

While there are more advanced IRT models to add a feature of weighting scores in survey responses in education and psychology [22], there have been applications of IRT in other disciplines of health care and financial sectors. Hays et al. [6] experimented IRT models in health outcome measurements. Their study involved analysis of the 9-item measure from study participants in the HIV Cost and Services Utilization Study (HCSUS) [23]. The 9-items included physical functioning, including basic activities, instrumental activities, and mobility, in the past four weeks, and interpreted the difficulty parameter ($b_i$) and discrimination parameter ($a_i$) of each activity in terms of physical functioning level and distinction among the activities.

Warne et al. [7] introduced the IRT 2PL model to health behavior research on life-time substance-use data from the Adolescent Risk Health Behavior Questionnaire [24]. The study analyzed 1,360 responses on health behaviors related to 23 substance items, i.e., alcohol, tobacco, and other drug use, and interpreted difficulty parameter ($b_i$) of each substance as the likelihood that a respondent had tried it. The difficulty parameters ($b_i$) of the substances helped identify two groups of substances that respondents are likely to endorse. Himefarb [25] also introduced IRT models to chiropractic and health educators as a standard way of standardized assessments in their practice

In financial sector, Knoll and Houts [9] developed a measure of financial knowledge components in financial literacy by applying the IRT 2PL to narrow down the items from three national surveys: the ALP, a RAND-managed Internet-based panel, The Health and Retirement Study (HRS) conducted by the University of Michigan since 1992, and The National Survey (NS-NFCS) portion of the 2009 National Financial Capability Study. The study suggested that the index based on their selected twenty items would be useful to compare financial knowledge among programs and populations.

### 3. Data and Methodology

#### 3.1. Data

Historical electricity price data is collected from the pricing reports of ISO New England Inc. (https://www.iso-ne.com/isoexpress/web/reports/pricing/-/tree/zone-info), where historical Standard Market Design (SMD) hourly data files are available for LMP by month and year. This paper used a file of 2019 SMD Hourly Data in the reports for analysis.

Each spreadsheet in the file has price information for eight zones and one hub, including hourly LMPs, the components of energy, congestion cost, and marginal loss, as well as hourly demand and weather. This research will use the DAM congestion costs because an FTR value is settled at the DAM. Besides, hourly prices will be broken down to two categories, onpeak and offpeak. Analysis focus will be on daily onpeak prices only, with onpeak hours being from hours ending 0800 to 2300 during weekdays, while offpeak hours include hours ending 2400 to 0700 during weekdays and all 24 h during weekends and NERC holidays.

#### 3.2. Methodology

ISO New England has one internal hub and eight customer load zones in its six member states. The trading hub, called Massachusetts Hub (MassHub), is actively traded at electricity futures markets. Market hedging participants are interested in the FTRs between the MassHub, a liquid pricing point, and a customer zone that they are obligated to serve, while speculators look to capture any arbitrage opportunities associated with active pricing locations. In this context, this creates eight FTR paths, with Massachusetts Hub as a source pricing point and each of the eight customer zones as a sink pricing point. The number of Massachusetts Hub-related FTRs awarded in 2020 annual auction was significant at 276 (as a source), or 8% of total paths, and at 165 (as a sink), while the number for April 2020 auction was 332 (as a source), or 2% of total, and 313 (as a sink) (https://www.iso-ne.com/isoexpress/web/reports/auctions/-/tree/auction-results-ftr).

The eight customer load zones for analysis designated as sink points, are: northeastern Massachusetts (NEMA), Vermont (VT), New Hampshire (NH), Maine (ME), Rhode Island (RI), Southeast Massachusetts (SEMA), Connecticut (CT), and West/Central Massachusetts (WCMA), with Mass Hub being a source point. Figure 4 presents a map of the eight load zones, and hourly day-ahead LMP prices for the zones, Mass Hub, and import interface locations. (Sources: https://www.iso-ne.com/about/key-stats/maps-and-diagrams/ and https://www.iso-ne.com/isoexpress/web/charts).

The R-software package was used to build the IRT model, and the analysis package is "ltm" package, or Latent Trait Models under IRT Analyses, dated 17 April 2018. The detail and functions are available on https://cran.r-project.org/web/packages/ltm/ltm.pdf.

In this study, an item (i), or a variable in IRT, is an FTR path for analysis to evaluate the profitability and risk levels of the FTR path relative to other candidate paths. Hourly FTR path values for each path during each of onpeak hours are first calculated as the difference between congestion costs at Massachusetts Hub and each of customer delivery zones, and then summarized as average FTR values for each of the onpeak days in 2019. When the average daily FTR values on each path is positive, it is coded as 1, else 0, to build a dichotomous variable as an item response ($X_i$).

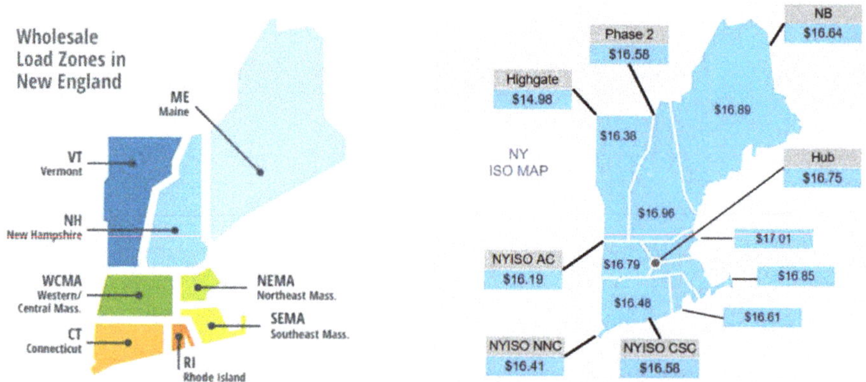


**Figure 4.** Eight load zonal map (**left**); and hourly day-ahead LMP Prices for zones, hub, and import interface locations (**right**). Reprinted with permission from ISO New England.

The binary FTR value ($X_i$) of 1 in this analysis represents positive profitability of an FTR path on an onpeak day. Analysis will begin with data summary statistics of the eight FTR paths in binary format (0, 1), or from Mass Hub to each customer load zone. Each onpeak date is treated as an individual respondent identifier (ID) in this analysis. In 2019, there were 255 onpeak days, resulting in 255 IDs, and daily binary items (FTR profitability) of all the eight FTR paths may be used to derive a latent trait variable (ability, $\theta$). A latent trait (ability) will be labeled as congestion ability.

The IRT 2PL model, as specified in Equation (2), is used to estimate difficulty coefficient ($b_i$), discrimination coefficient ($a_i$), as well as to build the ICCs. The ICC, that is, $P_i(\theta)$ or $P_i(X_i = 1|\theta)$ derived from the Equation, will provide visual comparisons of path profitability and risk levels, based on the parameters ($b_i$, $a_i$) and congestion ability level ($\theta$) in the range of [−4, 4].

Item difficulty coefficient ($b_i$) will represent profitability in this paper for each FTR path, and be interpreted that the greater the coefficient, the lower profitability, or vice versa. The item discrimination coefficient ($a_i$) will represent the risk level of an FTR path, translating into the level of differentiating among candidate FTR paths on different congestion levels. The greater the discrimination coefficient for an FTR path, the wider the FTR value distribution is, indicating riskiness itself of an FTR path. The two parameters will be compared among the paths, to evaluate and select bidding paths in FTR auctions. Congestion ability ($\theta$) refers to an underlying latent trait measured by daily responses to each FTR path, and it may be interpreted as a daily congestion ability. The parameter would indicate how often congested transmission situation took place in daily electricity market, potentially resulting in creating more congestion cost difference or values on FTR paths.

## 4. Results and Discussion

### 4.1. Summary Statistics for Eight FTR Paths

Table 1 summarizes proportions of positive profits, or FTR profitability days, for eight FTR paths, from the Mass Hub to each customer load zone, during 255 onpeak days in 2019. Among the eight

FTR paths, two FTR paths of MassHub_RI (0.2392 and MassHub_SEMA (0.2235) took the lead in FTR profitability days, with FTR paths of MassHub_VT (0.0392) and MassHub_WCMA (0.0431) at the bottom. Notably, standard deviations were also the greatest on the top two FTR paths with about 0.42, while the rest of FTR paths stayed in the range of 0.20 to 0.28. In other words, sink points in Rhode Island and Southeast Massachusetts, sourcing electricity from Massachusetts Hub, recorded profits at 23.9% and 22.4% of onpeak days in 2019, while the profitability distribution is wider than other zones, as indicated by their higher standard deviations.

**Table 1.** Summary statistics of profitable onpeak days for FTR path in 2019.

| FTR Path | Mean | SD |
|---|---|---|
| MassHub_ME | 0.0706 | 0.2566 |
| MassHub_NH | 0.0510 | 0.2204 |
| MassHub_VT | 0.0392 | 0.1945 |
| MassHub_CT | 0.0627 | 0.2430 |
| MassHub_RI | 0.2392 | 0.4274 |
| MassHub_SEMA | 0.2235 | 0.4174 |
| MassHub_WCMA | 0.0431 | 0.2036 |
| MassHub_NEMA | 0.0824 | 0.2754 |

*4.2. IRT 2PL Model Results*

Table 2 presents a summary of estimated coefficients of FTR profitability (difficulty, $b_i$) and risk level (discrimination, $a_i$), estimated by the IRT 2PL model. As a smaller difficulty coefficient on an FTR path represents the greater probability of profitability, the FTR paths, MassHub_RI (0.56) and MassHub_SEMA (0.61), showed the highest profitability among the eight candidate paths, with the least profitable paths of MassHub_VT (3.67) and MassHub_WCMA (3.32). These profitability results are consistent with the results of proportions of positive profits, as shown in Table 1. Item discrimination coefficient, a measure of risk level in this paper, is also the greatest on the two FTR paths, MassHub_RI (26.80) and MassHub_SEMA (26.25), implying higher risks than other paths. The two paths with the smallest discrimination coefficients were MassHub_VT (0.92) and MassHub_WCMA (1.00), implying the least risks among all the candidate paths.

**Table 2.** 2PL model—path difficulty (profitability) and discrimination (risk) coefficients.

| Parameter | FTR Path | Estimate | SE | z-Value | p-Value |
|---|---|---|---|---|---|
| Difficulty | MassHub_ME | 2.00 | 0.37 | 5.44 | 0.000 |
| (profitability, $b_i$) | MassHub_NH | 1.85 | 0.27 | 6.76 | 0.000 |
| | MassHub_VT | 3.67 | 1.30 | 2.81 | 0.002 |
| | MassHub_CT | 2.02 | 0.37 | 5.46 | 0.000 |
| | MassHub_RI * | 0.56 | 2.16 | 0.26 | 0.397 |
| | MassHub_SEMA * | 0.61 | 2.04 | 0.30 | 0.382 |
| | MassHub_WCMA | 3.32 | 1.03 | 3.23 | 0.001 |
| | MassHub_NEMA | 1.30 | 0.16 | 8.04 | 0.000 |
| Discrimination | MassHub_ME | 1.55 | 0.37 | 4.16 | 0.000 |
| (risk level, $a_i$) | MassHub_NH | 2.28 | 0.56 | 4.10 | 0.000 |
| | MassHub_VT | 0.92 | 0.37 | 2.50 | 0.006 |
| | MassHub_CT | 1.65 | 0.41 | 3.99 | 0.000 |
| | MassHub_RI * | 26.80 | 498.14 | 0.05 | 0.479 |
| | MassHub_SEMA * | 26.25 | 791.95 | 0.03 | 0.487 |
| | MassHub_WCMA | 1.00 | 0.35 | 2.82 | 0.002 |
| | MassHub_NEMA | 3.51 | 0.84 | 4.17 | 0.000 |

* Statistically not significant at the 0.05 level.

The two zones of Rhode Island and Southeast Massachusetts, as sink points from Massachusetts Hub, recorded the highest profitability, and at the same time, the highest risk profiles, representing high-return, high-risk opportunities. On the other hand, the least risky sink zones, Vermont and West/Central Massachusetts, from Massachusetts Hub, did not necessarily display the highest profitability.

When parameters of difficulty (profitability, $b_i$) and discrimination (risk level, $a_i$) are estimated from IRT 2PL model, the estimates need to be tested if they are different from 0. This research designed the priority ranking system to evaluate FTR paths, after accounting for statistical significance of the two parameter estimates, FTR profitability ($b_i$), and FTR path risk level ($a_i$).

First, the p-value criteria of statistical significance at the 0.05 level, for difficulty and discrimination coefficients are compared. Two FTR paths, MassHub_RI and MassHub_SEMA, are not statistically significant at the 0.05 level, meaning that the difficult and discrimination estimates could be unreliably zero, and be excluded for further evaluation. Table 3 presents a summary of evaluation processes, after exclusion of two insignificant paths, to determine bidding priority among FTR paths.

**Table 3.** Evaluation processes to determine bidding priority among FTR paths.

| FTR Path | Difficulty ($b_i$) | Discrimination ($a_i$) | Rank by ($b_i$) | Rank by ($a_i$) | Weighted Sum: ($b_i$) and ($a_i$) | Bidding Priority |
|---|---|---|---|---|---|---|
| MassHub_ME | 2.00 | 1.55 | 3 | 3 | 3.0 | 1 |
| MassHub_NH | 1.85 | 2.28 | 2 | 5 | 3.2 | 3 |
| MassHub_VT | 3.67 | 0.92 | 6 | 1 | 4.0 | 5 |
| MassHub_CT | 2.02 | 1.65 | 4 | 4 | 4.0 | 6 |
| MassHub_WCMA | 3.32 | 1.00 | 5 | 2 | 3.8 | 4 |
| MassHub_NEMA | 1.30 | 3.51 | 1 | 6 | 3.0 | 2 |
| MassHub_RI * | 0.56 | 26.80 | excluded | excluded | excluded | excluded |
| MassHub_SEMA * | 0.61 | 26.25 | excluded | excluded | excluded | excluded |

* Excluded for further evaluation, since the coefficients are not statistically significant at the 0.05 level.

The second step is ranking the remaining six FTR paths with the difficulty and discrimination coefficients. The FTR paths are ranked by two individual categories, in the ascending order of the two individual coefficients, since the lower difficulty of an FTR path stands for higher profitability and lower discrimination for less risk level, as shown columns of "rank by ($b_i$) and rank by ($a_i$)" in Table 3. As an FTR auction bidder, more profitable and less risky paths are favorable target paths to bid on.

The third and last step is to obtain weighted scores of the two ranks of each path, in this example, 60% for profitability (difficulty, $b_i$) and 40% for risk level (discrimination, $a_i$), laying the foundation to determine FTR bidding priority for the candidate paths. As a result, the FTR bidding priority is set up in the ranking order of the paths: MassHub_ME, MassHub_NEMA, MassHub_NH, MassHub_WCMA, MassHub_VT, and MassHub_CT.

*4.3. Item (Path) Characteristic Curves (ICCs)*

The FTR path profitability (item difficulty, $b_i$) and risk level (discrimination, $a_i$) in Tables 2 and 3 may be translated and illustrated on graphical forms (ICCs), representing the probability of profitability ($P_i(\theta)$) on the y-axis on a given congestion ability ($\theta$) on the x-axis. Figure 5 presents ICCs for each of the eight FTR paths for discussion purpose, where FTR paths, MassHub_RI ($a_i = 26.80$) and MassHub_SEMA ($a_i = 26.25$), show steep slopes, indicating high risk levels, due to their higher discrimination coefficients. FTR paths, MassHub_VT ($a_i = 0.92$) and MassHub_WCMA ($a_i = 1.00$), show the smallest slopes, indicating the least discrimination, or the least risks among all the FTR paths.

As the item (path) difficulty ($b_i$) coefficients represent a scale on the x-axis of a latent trait variable (congestion ability $\theta$) at a mid-probability (0.50) on the y-axis, their implication is that the further right an FTR path is, the less profitable (more difficult) it is. For example, the two FTR paths with the greatest difficulty coefficients are MassHub_VT (3.67) and MassHub_WCMA (3.32), as shown in Tables 2 and 3, and Figure 5.

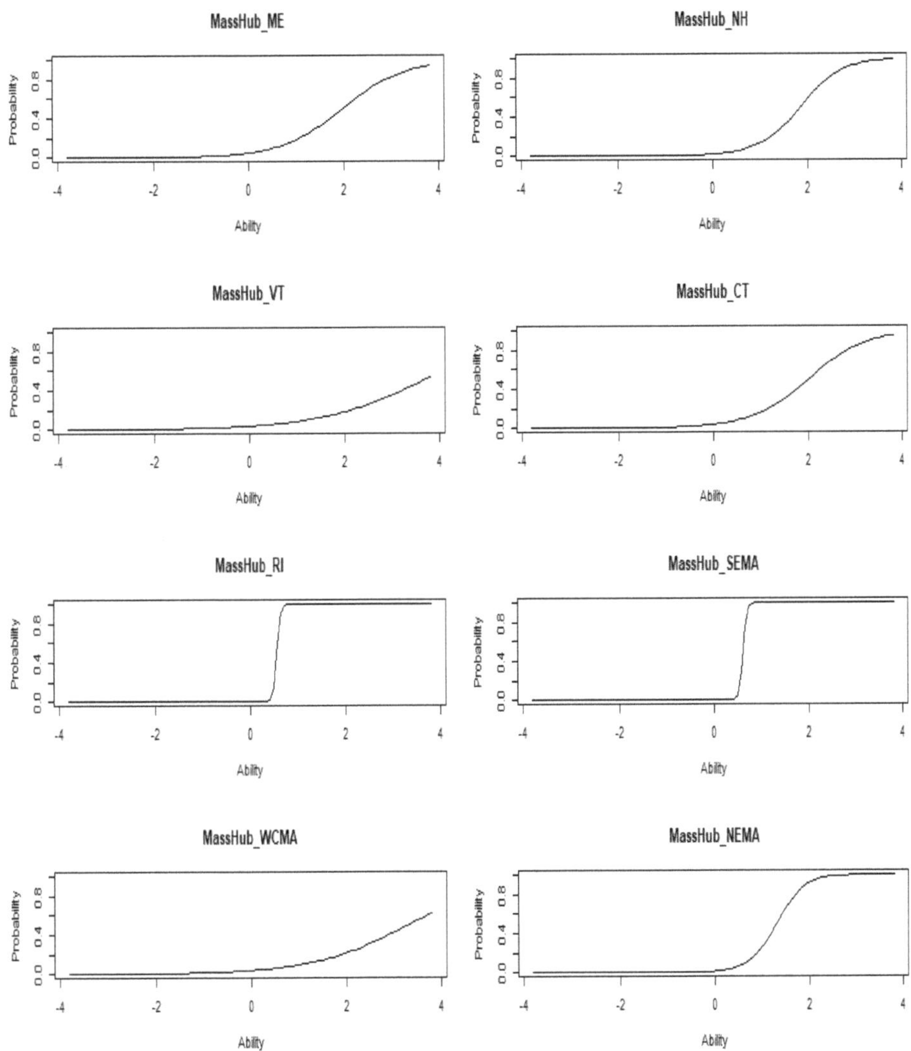

**Figure 5.** Individual item characteristic curves: eight paths.

## 5. Conclusion and Implications

Financial Transmission Rights (FTR) is an energy derivative and financial instrument in electricity markets. Transmission system constraints are one of the risks for market participants, called congestion cost risk as part of LMP, particularly for those with supply obligations to serve customer loads from a different pricing location. The U.S. New England ISO, with six members states, has about 1200 pricing nodes, including load zones, hubs, and generating plant nodes. The number of the pricing nodes may be translated into 1.4 million potential FTR paths on prevailing flows only, which created tremendous opportunities for FTR bidders, as well as challenges of making decisions which paths to bid in FTR auctions. It is essential that FTR participants have a standardized and consistent model to evaluate those paths, given the complexity and magnitude of FTR path choices available to them.

Item response theory (IRT), a popular statistical model in psychometrics, measures information of an item (e.g., item difficulty and item discrimination) with a latent variable estimated by all the item responses. This paper examined a way to apply the IRT 2PL model to evaluate and select the FTR paths to bid in market auctions, with a historical price data for eight candidate paths in 2019 in U.S. New England. The parameters in IRT were defined in a way that an FTR path is an item, the FTR value (binary) for an item response, the difficulty parameter for path profitability, and the discrimination for its risk level.

This study selected eight FTR paths, with Massachusetts Hub as a source and eight load zones in six states as a sink, with several steps of evaluation. Balancing FTR profitability and risk level was prudently considered in the whole process of applying the IRT model. For each of the eight paths, the IRT 2PL model produced difficulty parameter ($b_i$) for FTR path profitability and discrimination ($a_i$) for risk level. As a first step, significance of the parameters for the candidate paths was calculated based on a p-value hurdle of 0.05. Two paths of MassHub_SEMA and MassHub_RI were removed for further consideration because their p-values of both profitability ($b_i$) and discrimination ($a_i$) were greater than 0.05. As a result, six remaining FTR paths were selected for second evaluation process, that is, ranking the paths based on two criteria of profitability ($b_i$) and discrimination ($a_i$). The two rankings of each path were finally evaluated with weighting factor of 60% and 40% each on profitability and discrimination, resulting in priority order among the paths. Results show that FTR paths of Mass_Maine and Mass_NEMA took the top two spots, followed by Mass_NH, Mass_WCMA, Mass_VT, and Mass_CT.

This experiment shows that the IRT model may provide a standardized analytical framework, with three parameters, in the evaluation of FTR paths, and may be implemented to address the path choice challenges for FTR participants. The model could also be useful and applicable in other energy markets, with proper definitions of terms for analysis and interpretation of estimated parameters. In this study, the lowest item difficulty coefficient of an FTR path was interpreted as a greatest profitability path, and the greatest item discrimination coefficients, or the steepest slope on the ICCs, as the riskiest path.

We note that there are a couple of limitations in this research. One limitation is that it has not developed rigorous interpretation and utilization of the latent trait variable ($\theta$), or congestion ability, measured by the FTR path item (binary) responses ($X_i$), in the path evaluation process. Another limitation is that the paper focused on evaluation and selection aspect of FTR paths to bid under the analytical framework of item response theory (IRT). Future research may include topics of how to utilize the latent trait level (congestion ability) in the FTR path evaluation, and how to determine FTR bid prices for the auction under uncertainty of electric prices in future. Another extension of IRT applications may involve using more FTR paths for analysis, longer time horizons across multi-years, and testing results with other future time periods.

**Author Contributions:** Conceptualization, P.Y.J.; Data curation, P.Y.J.; Formal analysis, P.Y.J.; Investigation, P.Y.J.; Methodology, P.Y.J. and K.J.; Project administration, K.J. and M.G.B.; Resources, P.Y.J.; Software, P.Y.J.; Supervision, K.J. and M.G.B.; Validation, P.Y.J., K.J. and M.G.B.; Visualization, P.Y.J.; Writing – original draft, P.Y.J.; Writing – review & editing, P.Y.J., K.J. and M.G.B. All authors have read and agreed to the published version of the manuscript.

**Funding:** This research received no external funding.



## References

1.  ISO New England Inc. Financial Transmission Rights (FTR). Available online: https://www.iso-ne.com/markets-operations/settlements/understand-bill/item-descriptions/ftr (accessed on 9 April 2020).
2.  Opgrand, J.J. The Role of Auction Revenue Rights in Markets for Financial Transmission Rights. Ph.D. Thesis, Purdue University, West Lafayette, Indiana, 2019.
3.  ISO New England Inc., "New England Power Grid State Profiles 2019–2020," 2020. Available online: https://www.iso-ne.com/static-assets/documents/2020/02/ (accessed on 12 April 2020).

4.  De Ayala, R.J. *The Theory and Practice of Item Response Theory*; Guilford Publications: New York, NY, USA, 2013.

5.  Albano, T. *Introduction to Educational and Psychological Measurement Using R*; Tony Albano: Lincoln, NE, USA, 2018; pp. 145–162.

6.  Hays, R.D.; Morales, L.S.; Reise, S.P. Item Response Theory and Health Outcomes Measurement in the 21st Century. *Med. Care* **2000**, *39* (Suppl. 9), II28–II42. [CrossRef]

7.  Warne, R.T.; McKyer, E.L.J.; Smith, M.L. An Introduction to Item Response Theory for Health Behavior Researchers. *Am. J. Health Behav.* **2012**, *36*, 31–43. [CrossRef]

8.  Columbia Mailman School of Public Health. Available online: https://www.mailman.columbia.edu/research/population-health-methods/item-response-theory (accessed on 23 April 2020).

9.  Knoll, M.A.Z.; Houts, C.R. The Financial Knowledge Scale: An Application of Item Response Theory to the Assessment of Financial Literacy. *J. Consum. Aff.* **2012**, *46*, 381–410. [CrossRef]

10. Ziogos, N.P.; Tellidou, A.C.; Gountis, V.P.; Bakirtzis, A.G. A Reinforcement Learning Algorithm for Market Participants in FTR Auctions. In Proceedings of the 2007 IEEE Lausanne Power Tech, Lausanne, Switzerland, 1–5 July 2007.

11. Liu, Y.; Wu, F.F. Transmission Rights and Generator's Strategic Bidding in Electricity Markets. In Proceedings of the 2006 IEEE Power Engineering Society General Meeting, Montreal, Que., Canada, 18–22 June 2006; pp. 1–6.

12. Das, T.K.; Rocha, P.; Babayigit, C. A matrix game model for analyzing FTR bidding strategies in deregulated electric power markets. *Electr. Power Energy Syst.* **2010**, *32*, 760–768. [CrossRef]

13. ISO New England Inc. ISO New England Manual for Financial Transmission Rights Manual M-06. Available online: https://www.iso-ne.com/static-assets/documents/2018/10/ (accessed on 9 April 2020).

14. ISO New England Inc. FTR Auction Results. Available online: https://www.iso-ne.com/isoexpress/web/reports/auctions/-/tree/auction-results-ftr (accessed on 12 April 2020).

15. Baig, S.M.; Imran, K.; Rafique, M.N.; Khan, H.H. Combined Simulation of Auction Revenue Rights Allocation and Financial Transmission Rights Auction for Electricity Markets. In Proceedings of the 4th International Conference on Power Generation Systems and Renewable Energy Technologies (PGSRET), Islamabad, Pakistan, 10–12 September 2018.

16. ISO New England Inc. *Introduction to Wholesale Electricity Markets (WEM 101)*; ISO New England Inc.: Northampton, MA, USA, 2019.

17. ISO New England Inc. Wholesale vs. Retail Electricity Costs. Available online: https://www.iso-ne.com/about/what-we-do/in-depth/wholesale-vs-retail-electricity-costs (accessed on 9 April 2020).

18. Hogan, W.W. Electricity Market Design Interactions of Multiple Market. Available online: https://media.rff.org/documents/170914_PowerMarkets_WilliamHogan.pdf (accessed on 9 April 2020).

19. Li, T.; Shahidehpour, M. Risk-Constrained FTR Bidding Strategy in Transmission Markets. *IEEE Trans. Power Syst.* **2005**, *20*, 1014–1021. [CrossRef]

20. Liu, M.; Gross, G. Congestion rents and FTR evaluations in mixed-pool-bilateral systems. *Int. J. Electr. Power Energy Syst.* **2008**, *30*, 447–454. [CrossRef]

21. Baker, F.B.; Kim, S.-H. *The Basics of Item Response Theory Using R*; Springer: New York, NY, USA, 2017.

22. Camilli, G. IRT Scoring and Test Blueprint Fidelity. *Appl. Psychol. Meas.* **2018**, *42*, 393–400. [CrossRef] [PubMed]

23. Hays, R.D.; Spritzer, K.; McCaffrey, D.F.; Cleary, P.; Collins, R.L.; Sherbourne, C.D.; Wu, A.W.; Bozzette, S.A.; Shapiro, M.F.; Cunningham, W.; et al. *The HIV Cost and Services Utilization Study (HCSUS) Measures of Health-Related Quality of Life*; RAND: Santa Monica, CA, USA, 1998.

24. Smith, M.L.; McKyer, E.L.J.; Larsen, R.A. Factor structure and psychometrics of the adolescent health risk behavior survey instrument. *Am. J. Health Behav.* **2010**, *34*, 328–339. [CrossRef] [PubMed]

25. Himelfarb, I. A primer on standardized testing: History, measurement, classical test theory, item response theory, and equating. *J. Chiropr. Educ.* **2009**, *33*, 151–163. [CrossRef] [PubMed]

 *energies*

*Article*

# Renewable Energy Utilization Analysis of Highly and Newly Industrialized Countries Using an Undesirable Output Model

**Chia-Nan Wang [1],*, Hector Tibo [1,2],* and Duy Hung Duong [3]**

[1] Department of Industrial Engineering and Management, National Kaohsiung University of Science and Technology, Kaohsiung 80778, Taiwan

[2] Department of Electrical Engineering, Technological University of the Philippines Taguig, Taguig 1630, Philippines

[3] Department of Mechanical Engineering, National Kaohsiung University of Science and Technology, Kaohsiung 80778, Taiwan; F107142154@nkust.edu.tw

* Correspondence: cnwang@nkust.edu.tw (C.-N.W.); i107143104@nkust.edu.tw (H.T.)

Received: 13 May 2020; Accepted: 20 May 2020; Published: 21 May 2020

**Abstract:** In the fight against climate change, the utilization of renewable energy resources is being encouraged in every country all over the world to lessen the emissions of greenhouse gases. However, not all countries are able to efficiently utilize these resources, and instead of providing solutions, the inefficient use of renewable energy may lead to even more damage to the environment. Data from eight countries belonging to the highly industrialized countries (HIC) group and nine from newly industrialized countries (NIC) group were used to evaluate the energy utilization of these groups. Factors such as total renewable energy capacity, the labor force, and total energy consumption were considered to be the input factors, while, $CO_2$ emission and gross domestic product are the output factors. These factors were used to calculate efficiency scores of every country from 2013 to 2018 using the undesirable output model of Data envelopment analysis (DEA). The grey prediction model was also used to measure the forecasted values of the input and output factors for the year 2019 to 2022, and measure again the future efficiency scores of the HICs and NICs. The combination of grey prediction and DEA undesirable output model made this paper unusual and the most appropriate method in dealing with data that contains both desired and undesired outputs. The results show that the United Kingdom, Germany, France, and the United States continuously top the efficiency ranking among the HIC group, with a perfect 1.0 efficiency score from 2013 to 2022. Russia demonstrates the lowest score of 0.1801 and is expected to perform the same low-efficiency score in the future. Within the NIC group, Indonesia can be highlighted for performing with perfect efficiency starting from the year 2015 and even through 2022. Other NICs are performing at a very low-efficiency, with scores ranging from 0.2278 to 0.2734 on average, with Turkey displaying the lowest rank. This study recommends some useful strategies to improve the utilization of renewable energy resources such as improvements in the political and legal structure surrounding their use and regulation, tax incentives or exemptions to private power producers to encourage shifting away from conventional energy production, partnerships with non-governmental and international organizations that can provide assistance in managing renewable energies, strengthening of the energy sector's research and development activities and long-term strategic plans for the development in renewable energy with considerations to the social, environmental, and economic impact on each country.

**Keywords:** renewable energy; utilization efficiency; data envelopment analysis; undesirable output model; bad output; grey prediction model

---

## 1. Introduction

Global warming, as it is called, is simply the heating up of the atmosphere attributed mainly to the greenhouse effect caused by the rising concentrations of carbon dioxide, chlorofluorocarbons (CFCs), and other air pollutants. This extreme phenomenon is the primary reason for the climate change that is considered to be an environmental crisis. During the United Nations Framework Convention on Climate Change, several state parties agreed to pursue efforts to reduce this century's average increase of the Earth's temperature from 2 degrees Celsius to 1.5 degrees Celsius. The consensus entered force last November 2016 and is referred to as the Paris Agreement [1]. One foreseen solution is to slowly transform from conventional energy production based on fossil fuel combustion to renewable energy resources. Fuel combustion is a traditional way to generate electric power through the burning of coal and fossil fuels. Many countries are dependent on this practice. Even some developed ones are major contributors to the presence of $CO_2$ in the atmosphere caused by this process [2].

In the fight against climate change, other nations are now making use of different renewable energy sources. In this way, they can lessen the use of $CO_2$-emitting methods in the production of electricity which is very essential to human living standards. Renewable energies are sources of energy which are continuously being replenished naturally by the Earth itself. These energies can be obtained straight from the Sun (solar power and thermal), wind power, hydroelectric power, tidal or wave energy, and geothermal and biomass, but the transition to these energy sources can be difficult and costly. Though many potential benefits can ensue, there are also some technical limitations that must be considered. Also, instead of providing an additional solution to the climate change problem, the improper utilization of renewable energy may lead to more damage to the environment. These environmental damages may be caused by the disadvantages of using renewable energies such as air pollution caused by biomass burning, corrosion problems when using geothermal energy, risk of flooding in the communities surrounding a hydropower plant, negative impact on marine wildlife in using marine energy, and impact on the environmental landscape of using solar and wind energies [3].

There are international agencies and organizations that aim to provide guidance and assistance to those countries that are making their way to the use of renewable energies. One is the International Renewable Energy Agency (IRENA) which is an association of world governments that provides support to countries as they shift to a future with sustainable energy. They also function as a primary platform for global collaboration and as an archive of policies, resources, technologies, and economic know-how on renewable energy [4]. This study will use data from IRENA, Enerdata, and the World Bank.

To evaluate the relative efficiencies of the renewable energy utilization of seventeen (17) nations belonging from highly industrialized countries group (HIC—Russia, Canada, the United States, Japan, the United Kingdom, Italy, Germany, and France) which are also being referred to as the G8 or the Group of Eight Industrialized Nations [5] and newly industrialized countries group (NIC—Turkey, Thailand, Malaysia, Indonesia, India, China, Brazil, Mexico, and South Africa) [6] is the main goal of this study. These HICs and NICs are expected to have highly developed and developing economies. The HICs and NICs are chosen by the authors to be the subject countries for this study because of their high potential in investing in renewable energies since they have more developed economies. The aim is to identify which of the countries are performing efficiently as they progress in the use of their renewable energy resources. Energy consumption has an essential effect to the country's gross domestic product (GDP) as the ratio between the two factors affects the economic output of several countries since energy is a major input in continuous consumption of goods from energy-demanding sectors such as in production and manufacturing [7].

Three input and two output factors during the six-year periods will be considered for forecasting future values using the grey forecasting method or GM (1,1). The data from the past and the future will then be analyzed using the data envelopment analysis (DEA) undesirable output model.

The combination of these two models makes this study different from other papers, especially the use of the undesirable output model in the energy sector. This model will give consideration to the presence of good and bad outputs, treating the bad or undesired outputs as less important contrary to good outputs. This paper will make use of these two methods to evaluate the past to future efficiencies of seventeen countries.

The whole paper is divided into five sections. Reviews of previous literature related to the study are found in the Section 2. The proposed approach in forecasting future values and evaluation of efficiencies is in the Section 3. The Section 4 presents the interpretations and analyses of data gathered using GM (1,1) and the results of the DEA undesirable model. Concluding statements are described in the Section 5.

## 2. Literature Review

The energy sector has been a very important aspect of human life and has a strong impact on the economic, social, institutional, and environmental conditions of every country. Cirstea et al. [8] calculated the renewable energy sustainable index (RESI) using the normalization and multivariate analysis which affects the said conditions. The goal of the index is to provide a framework that can be used by the renewable energy sector's potential investors to aid their decision making. Another study conducted by Iddrisu and Bhattacharyya [9] made use of the arithmetic mean of the four normalized indicators for the measurement of the sustainable energy development index (EDI) which was devised to evaluate, rate, and rank countries according to the calculated energy indices. Lee and Zhong [10] by using min-max normalization combined with multivariate analysis were able to draft the renewable energy responsible investment index (RERII) wherein the primary intention is to help energy investors to decide effectively and proactively and also, to establish an investment framework for energy stakeholders in developing or revising current approaches for investing in the renewable energy field. The ecological factor was considered by Schlör et al. [11] to form the sustainable development index (SDI). In their study, the methods of selecting variables, normalization, and weighting to analyze whether the German energy sector is on a sustainable development track even under the pressure of sustainability goals. A general sustainability indicator for the consumption of renewable energy resources was established by Liu [12] using weighing, quantification, and evaluation of theoretical criteria. The framework incorporated a multicriteria decision-making model (MCDM) called the analytic hierarchy process (AHP) to provide a precise measurement of sustainability. The same index model was developed beforehand by Doukas et al. [13] applying a multivariate technique called the principal component analysis (PCA) for the analysis of nine different indicators to quantitatively measure the energy sustainability of rural communities. Due to the integration of MCDM techniques in various efficiency analysis, the method has become popular to evaluate the energy sector. Štreimikienė et al. [14] combined AHP with the additive ratio assessment technique (ARAS) to analyze the environmental impact criteria and rate the electricity generation technologies in Lithuania. Another method popularly known as the Technique for Order Preference by Similarity to Ideal Solution (TOPSIS) was merged with AHP to provide a comparison and rank the five low carbon energy resources in China in a study conducted by Ren and Sovacool [15] in which they found that wind and hydroelectric power have the most potential for improvement. Troldborg et al. [16] make use of the Preference Ranking Organization Method of Enrichment Evaluation (PROMETHEE) method to formulate an assessment model of sustainability in a national-level and ranks Scotland's technological capabilities for the renewable energy sector. The complex proportional assessment (COPRAS) technique is used by Yazdani-Chamzini et al. [17] for an effective selection of the most pertinent renewable energy project in comparison to the current available options. Another project selection method for renewable energy programs in Spain was applied by San Cristóbal [18] by using the compromise ranking method which allows the Spanish government decision-makers to provide weights of importance to different criteria according to their own preferences. Kabak and Dağdeviren [19] used a hybrid MCDM framework—Benefits, Opportunities, Costs, and Risks (BOCR)—combined with an analytic network process (ANP) method, to aid Turkish policymakers in decision-making in terms of choosing which of five alternative energy sources should be given priority for development.

Aside from the studies which make use of advance statistical methods, several past studies have used DEA as a mathematical method to evaluate the energy production and consumption of several countries. The method is very essential to the assessment since it can provide a comparison of systems with multiple input and output factors.

Halkos and Tzeremes [20] used total capital stock and total labor as input variables and GDP as desirable output and $CO_2$ emissions as undesirable output for the analysis of 110 countries using data from the year 2007. In this study, DEA constant return-to-scale (CRS) scores provided an analytical result showing that only five countries most likely performed efficiently. However, the outcome of the bias-corrected scores are larger than the scores of the standard deviation, so the authors were able to rank the countries accordingly to get a list of ten countries with the highest and the ten lowest scores. The authors then concluded that the environmental efficiencies of each country have shown positive effects over the first six years after signing the Kyoto Protocol Agreement, but this performance did not last long, as the efficiencies of the countries appeared to decline in the following years. This can be reflected from the countries' avoidance of compliance with the impositions of the agreement and their inability to accordingly adjust their $CO_2$ emission reduction to a value that is relative to the growth rate of their economies.

Wang et al. [21] make use of combined DEA models, the super slack-based model, and the Malmquist productivity index for evaluation and selection of sustainable logistics providers in the US. Using financial indicators such as the equity, liabilities, operating expense, and assets as input factors and revenue, net income, and earnings-per-share as outputs, their paper established a list of rankings among different decision making units (DMUs). With the combined methods used, the paper was able to determine the nine perfectly efficient logistic providers among the sixteen DMUs.

Oggioni et al. [22] measure the ecological efficiency of the cement industry from 21 countries. Using CCR (Charnes-Cooper-Rhodes) and BCC (Banker-Charnes-Cooper) DEA models, the study was able to determine which of these countries performed efficiently in terms of disposability of undesirable outputs. Two outputs were used, cement production and $CO_2$ emission, first being the desirable one and the second otherwise. These data—including the total labor, installed capacity, energy consumption, and raw materials—were collected from the years 2005 to 2008 as inputs for the application of DEA. Analyzing the scores for every country, the outcome points out that during the period under study there are countries like Canada, Brazil, Turkey, and the United States that performed the very worst in terms of eco-efficiency. This is due to the absence of strong environmental protection regulations. Emerging countries like China and India, at that time, were showing high-efficiency levels in cement production. This is attributed to their investment in more efficient technologies and the production of low-quality cement which emits lower $CO_2$ levels.

Woo et al. [23] applied the DEA-based Malmquist productivity index (MPI) to evaluate the environmental efficiency of renewable energy of 31 country members of the Organisation for Economic Co-operation and Development (OECD). This study makes use of total labor, total capital, and total renewable energy supply as input factors, while renewable electricity generation and GDP are used as output factors. Carbon emissions are also part of the study as they were considered to be the undesirable output for the analysis. The MPI model is deployed to measure the technical efficiency change, frontier change, and productivity of the involved countries with and without consideration of carbon emissions. The results of this study concluded that in the evaluation of efficiencies for the energy sector, the undesirable output must always be considered as they have a significant relationship to energy performance. The paper also encourages policymakers to support the development of technologies related to the use of renewable energy in their own country as it also has a significant impact on the energy market.

The same DEA model, MPI, was used by Zhou et al. [24] to measure the carbon emission performance from the world's top 18 $CO_2$-producing countries. Capital stock, total energy consumption, and total labor force were used as the input variables while GDP and $CO_2$ emissions were the outputs. The MPI scores display the performance of these countries in terms of carbon emission productivity, efficiency change, and technological change over the period of 1997 to 2004. The results showed that there was a 24% increase in carbon emissions during the study period and cites technological progress as the primary reason. Germany turned out to be the number one carbon emitter while Indonesia and China displayed negative productivity.

Wang et al. [25] made a forecast of energy efficiency or the period from 2018 to 2023 using data from 25 countries. The study method used the DEA Slack-Based Model (SBM) to determine the efficiencies using historical data from 2008 to 2017 and then applied grey forecasting to aid the SBM to produce future efficiency scores. The countries were chosen according to the sufficiency of data that is available from different sources. Two commonly used economic indicators (labor force and capital stock) and one energy-related factor (energy consumption) were used as inputs. The desirable output is GDP and the undesirable one is $CO_2$ emissions. This combination of variables used for the study is enough to provide appropriate results to achieve the goal of the paper. After the result analysis, the authors found out that during the past period, only eight out of 25 countries performed efficiently and this performance will be maintained for the future period. This indicates the proper balance in their growing economies while protecting the environment due to a deliberate reduction of $CO_2$ emissions. The findings also suggest that European countries have higher efficiency scores compared to those in Asia and America. The paper further recommended a substantial government policy intervention for every country that should focus on improving the energy production and environmental sectors.

More studies have suggested different considerations for input and output factors to be used for the analysis of countries' energy efficiency scores, as listed in Table 1 below.

**Table 1.** List of commonly used input and output factors for several past literature.

| Authors, Year [References] | Factors | | Method/s | No. of Counties |
|---|---|---|---|---|
| | **Inputs** | **Outputs** | | |
| Zofio and Prieto, 2001 [26] | Energy Consumption Capital stock Labor | GDP $CO_2$ emission | DEA | 18 |
| Xie et al., 2014 [27] | Labor Installed capacity Fuel and nuclear | Power generation $CO_2$ emission | DEA-SBM | 26 |
| Cicea et al., 2014 [28] | GDP capita Energy intensity Investment to renewables | $CO_2$ emission | DEA | 22 |
| Wang et al., 2018 [29] | Energy Consumption Population | GDP $CO_2$ $CH_4$ methane $N_2O$ nitrous oxide | DEA-Undesirable model | 42 |
| Chien and Hu, 2009 [30] | Capital stock Energy consumption Labor | GDP | DEA | 45 |

These studies are proof that DEA is an effective way to measure the efficiencies of the energy sector from different countries using diverse combinations of inputs and outputs. This method has played a very important role in efficiency analyses since it was introduced by Charnes et al. [31] in 1978. The Charnes, Cooper, Rhodes (CCR) model became the first traditional method to calculate the relative efficiency scores of several numbers of DMUs which represents the technical efficiency. Banker et al. [32] in 1984 presented another model called Banker, Charnes, Cooper (BCC) to evaluate efficiencies of the DMUs that are not or not yet operating at an optimum scale in which CCR is incapable of. Another DEA-based model was proposed in 1982 by Caves et al. [33] and is called the Malmquist Productivity Index (MPI). It was later been split into two segments by Fare et al. [34] to represent catch-up and frontier-shift efficiencies. Furthermore, non-radial DEA types such as the additive (ADD) [35] and slacks-based measure (SBM) [36] models were also introduced.

These developments to DEA involve the introduction of a model that will be able to consider the presence of some bad outputs or unwanted factors. Cooper et al. [37] modified the SBM model to be able to provide a more precise measurement of efficiency and is called the undesirable outputs model.

DEA models are used to measure the efficiency coming from data currently available. There can be no measurement of future efficiencies but only can measure the past up to the present scores. Wang et al. [38] use GM (1,1) along with DEA to measure the future efficiency performances of some Vietnamese ICT firms. The grey forecasting method is a time-series prediction model that does not require enormous amounts of data to be able to produce highly accurate results [39]. This capability of GM (1,1) have made it become a popular forecasting tool for many studies which this paper will also utilize and combine with the DEA undesirable output model.

## 3. Materials and Methods

### 3.1. Research Process

To be able to reach the goal of this paper, this study is divided into four stages as shown in Figure 1. This will serve as the guide for the authors in finalizing the study.

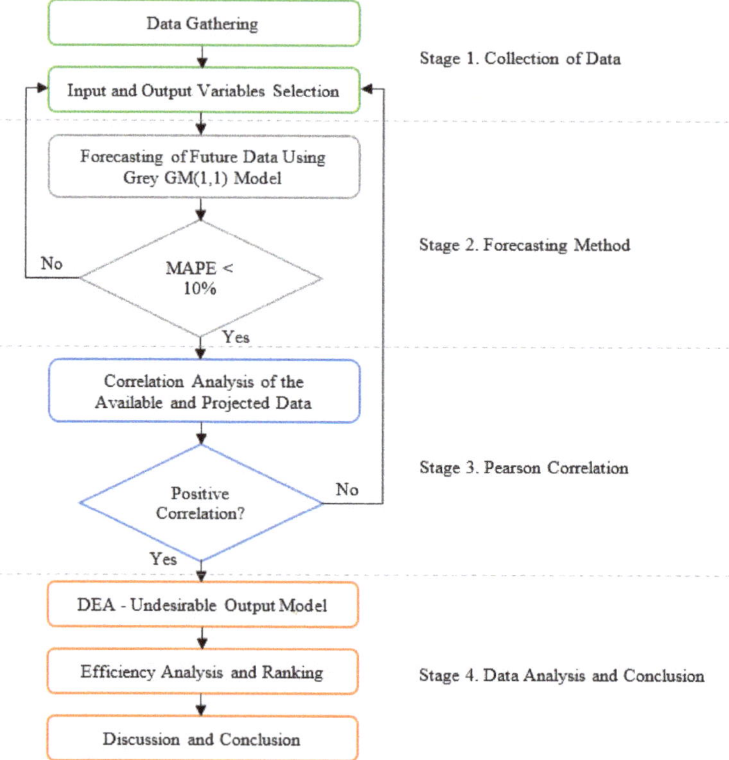


**Figure 1.** The research process.

### Stage 1. Collection of Data

Data were collected through IRENA, Enerdata, and World Bank. Based on several pieces of cited literature, the authors selected the appropriate input and output factors suitable for this study. These factors are also commonly used by previous studies related to this paper.

### Stage 2. Grey Forecasting Method

The grey forecasting method is used to predict the values of the input and output factors for the future period. The method uses historical data. The mean absolute percentage error (MAPE) determines whether the predicted value is acceptable or not. Lower values of MAPE means higher accuracy.

### Stage 3. Pearson Correlation

To check if the selected input and output factors and the predicted values are suitable for DEA processing, the calculation of the Pearson coefficient of correlation is very necessary. This method was widely used in previous studies. It is used to confirm the isotonic relationship between factors and a positive correlation is a requirement for DEA.

### Stage 4. Data Analysis and Conclusion

Since this paper focuses on the efficiency of renewable energy programs, the presence of carbon emissiosn suggests the use of the DEA undesirable output model. The DEA result will show which countries performed efficiently and those did not. The ranking will be based on the output efficiencies. This is to determine which countries among HICs and NICs have better renewable energy capabilities. The conclusions will provide a summary and addresses the objective of the study. The authors will specify some recommendations and information valuable for decision-making by policymakers.

### 3.2. GM (1,1) Grey Prediction Model

The GM (1,1) grey prediction model is a widely used forecasting method associated with time series and differential equations. One advantage of this method is the requirement of few historical data, at least four successive data with intervals that are equally distributed in a timely manner, to generate an acceptable prediction and calculated efficiently as discussed by Julong [39] and supported by Tseng et al. [40]. The procedure for grey prediction is shown in Figure 2 below.

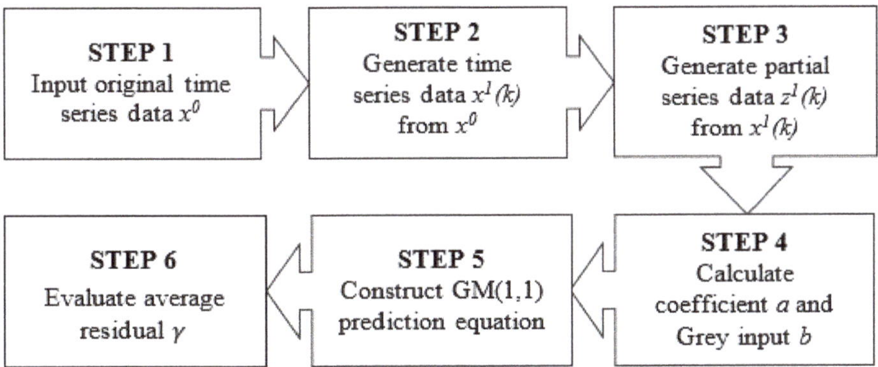


**Figure 2.** The grey prediction model procedure.

Given the variable primitive series $X^{(0)}$ in Equation (1), a more detailed procedure of prediction using the GM (1,1) grey model is discussed:

$$X^{(0)} = \left[ X^{(0)}(1),\ X^{(0)}(2), \ldots,\ X^{(0)}(n) \right], \qquad n \geq 4 \tag{1}$$

where $X^{(0)}$ is a positive sequence and $n$ is the total number of historical observations [39].

One very necessary property of a grey model is the accumulating generation operator (AGO) that is used for elimination of uncertainties from the primitive data. The equation for AGO is presented in Equation (2):

$$X^{(1)} = \left[\, X^{(1)}(1),\ X^{(1)}(2), \ldots,\ X^{(1)}(n) \,\right], \qquad n \geq 4 \tag{2}$$

where $X^{(1)}(1) = X^{(0)}(1)$, $X^{(1)}(k) = \sum_{i=1}^{k} X^{(0)}(i)$, and $k = 1, 2, \ldots, n$ [39].

The partial data series is described in Equation (3):

$$Z^{(0)} = \left[\, Z^{(1)}(1),\ Z^{(1)}(2), \ldots,\ Z^{(1)}(n) \,\right] \tag{3}$$

where $Z^{(1)}(k)$ is the value of the mean from the adjacent data described in Equation (4):

$$Z^{(1)}(k) = \frac{1}{2} \times \left[X^{(1)}(k) + X^{(1)}(k-1)\right],\ k = 2,\ 3, \ldots, n, \tag{4}$$

Through $X^{(1)}$, the first order differential equation $X^{(1)}(k)$ of grey prediction model can be derived from Equation (5) [39]:

$$\frac{dX^{(1)}(k)}{dk} + aX^{1}(k) = b \tag{5}$$

wherein $a$ is the developing coefficient and $b$ is the grey input.

In general, Equation (5) does not generate the values for parameters $a$ and $b$. The above equation is solved through the least square methods (Equation (6)):

$$\hat{X}^{(1)}(k+1) = \left(X^{(0)}(1) - \frac{b}{a}\right)e^{-ak} + \frac{b}{a} \tag{6}$$

where $\hat{X}^{(1)}(k+1)$ depicts the prediction value of $X$ at a $k + 1$ point in time. Using the method of ordinary least square (OLS), the values of $[a,b]^{T}$ can be acquired as described by Equations (7)–(9) [39]:

$$[a, b]^{T} = \left(B^{T}B\right)^{-1}B^{T}Y \tag{7}$$

$$Y = \begin{bmatrix} x^{(0)}(2) \\ x^{(0)}(3) \\ \cdots \cdots \\ \cdots \cdots \\ x^{(0)}(n) \end{bmatrix} \tag{8}$$

$$B = \begin{bmatrix} -z^{(1)}(2) & 1 \\ -z^{(1)}(3) & 1 \\ \cdots \cdots & \cdots \\ \cdots \cdots & \cdots \\ -z^{(1)}(n) & 1 \end{bmatrix} \tag{9}$$

wherein $[a, b]^{T}$ is referred to as the parameter series, $Y$ is the data series and $B$ is the data matrix.

The values for $\hat{X}^{(1)}(k)$ will be calculated by letting $\hat{X}^{(0)}$ become the predicted series:

$$\hat{X}^{(0)}\ X^{(0)}(1),\ \hat{X}^{(0)}(2),\ \ldots, \hat{X}^{(0)}(n) \tag{10}$$

where $\hat{X}^{(0)}(1) = X^{(0)}(1)$

Equation (11) is obtained through the application of inverse accumulated generation operation [39]:

$$X^{(0)}(k+1) = \left(X^{(0)}(1) - \frac{b}{a}\right)e^{-ak}(1 - e^{a}) \tag{11}$$

The accuracy of the predicted values can be measured using the actual and predicted data. The measurement is called the mean absolute percentage error or *MAPE* and is described in the formula below:

$$MAPE = \frac{1}{n}\sum\left(\frac{x^{(0)}(k) - \hat{x}^{(0)}(k)}{x^{(0)}(k)}\right) \times 100\% \tag{12}$$

The acceptability of the predicted data depends on the values of the *MAPE*. Small values for *MAPE* means a higher rate of accuracy. Ju-long [41] also categorized the reliability classes into four as listed in Table 2.

**Table 2.** Equivalent forecast category for every *MAPE* percentage score.

| *MAPE* | Forecast Categories |
|---|---|
| <10% | High Accuracy |
| 10–20% | Good |
| 20–50% | Reasonable |
| >50% | Inaccurate |

*3.3. Data Envelopment Analysis—Undesirable Output Model*

The undesirable output model is one of the many widely used DEA models. One thing that makes this model special and different from the others, is that this model considers the presence of bad output factors in the data set. Cooper et al. [37] modified the slack-based model (SBM) to be able to take account of the undesirable outputs during efficiency analysis. This research contains data involving the presence of a bad output making it more suitable for the study. The undesirable model will be described in the following paragraphs.

The input and output matrix of the DMUs will be denoted as $(x_0, y_0)$. The output parameters of the matrix $y$ will be decomposed into two: the desirable outputs are $Y^g$ (good matrices) and the undesirable outputs are $Y^b$ (bad matrices). The whole decomposition will become $\left(x_0, y_0^g, y_0^b\right)$.

The set for production possibility is described by:

$$P = \left\{ \left(x, y^g, y^b\right) \mid x \geq X\lambda, \ y^g \leq Y^g\lambda, \ y^b \geq Y^b\lambda, \ L \leq e\lambda \leq U, \ \lambda \geq 0\right\} \tag{13}$$

wherein the intensity factor is $\lambda$, $L$ is the lower boundary and $U$ is the upper boundary for $\lambda$.

In a presence of bad output, a DMU is efficient if there is no vector $\left(x, y^g, y^b\right) \in P$ in such $x_0 \geq x$, $y_0^g \leq y^g$, $y_0^b \geq y^b$ having at least one inequality.

The modification of SBM to attain the objective of the undesirable model is described as:

$$\rho^* = \min \frac{1 - \frac{1}{m}\sum_{i=1}^{m}\frac{s_{io}^-}{x_{io}}}{1 + \frac{1}{s}\left(\sum_{r=1}^{s_1}\frac{s_r^g}{y_{ro}^g} + \sum_{r=1}^{s_2}\frac{s_r^b}{y_{ro}^b}\right)} \tag{14}$$

subject to $x_0 = X\lambda + s^-$; $y_0^g = Y\lambda - s^g$; $y_0^b = Y\lambda + s^b$; $L \leq e\lambda \leq U$; $s^-, s^g, s^b, \lambda \geq 0$. The excesses in inputs are denoted by the vector $s^-$ and bad outputs is $s^b$. In contrast, $s^g$ denotes the lack of good outputs. $s_1$ and $s_2$ express the number of elements in $s^b$, and $s^g$, and $s = s_1 + s_2$.

According to Cooper et al. [37], the DMU $\left(x_0,\, y_0^g,\, y_0^b\right)$ is efficient even under a condition of any undesirable outputs if $\rho^* = 1$. An inefficient DMU, $\rho^* < 1$, can be enhanced by removing the excesses in inputs and bad outputs and intensifying the deficiencies in good outputs with the following projection:

$$
\begin{aligned}
x_0 &\Leftarrow x_0 - s^{-*}\\
y_0^g &\Leftarrow y_0^g - s^{g*}\\
y_0^b &\Leftarrow y_0^b - s^{b*}
\end{aligned}
\tag{15}
$$

Through the Charnes–Cooper transformation method as described by Tone in 2001 [36], the fractional formula can be converted into a linear program with the following variables for the constant return to scale.

Whereas:

$$
\begin{aligned}
&v,\, u^g,\, u^b\\
&L = 0,\; U = \infty\\
&\max u^g y_o^g - v - u^b y_o^b
\end{aligned}
\tag{16}
$$

subject to:

$$
u^g Y^g - vX - u^b y^b \le 0 \tag{17}
$$

$$
v \ge \frac{1}{m}\left[\frac{1}{x_o}\right] \tag{18}
$$

$$
u^g \ge \frac{1 + u^g y_o^g - vx_o - u^b y_o^b}{s}\; [1/y_o^g] \tag{19}
$$

$$
u^b \ge \frac{1 + u^g y_o^g - vx_o - u^b y_o^b}{s}\; [1/y_o^b] \tag{20}
$$

The $v$ and $u^b$ variables, are respectively referred to as the values of inputs and bad outputs while $u^g$ refers to the value of good outputs.

Cooper et al. [37], set out the weights for bad and good outputs to be encode before running the undesirable output model. The weight variables must satisfy the $w_1, w_2 \ge 0$ conditions for the bad and good outputs so that the calculated relative weights will become $W_1 = sw_1/(w_1 + w_2)$ and $W_2 = sw_2/(w_1 + w_2)$. The final function will be transformed into:

$$
\rho^* = \min \frac{1 - \frac{1}{m}\sum_{i=1}^{m}\frac{s_{io}^-}{x_{io}}}{1 + \frac{1}{s}\left(W_1 \sum_{r=1}^{s_1}\frac{s_r^g}{y_{ro}^g} + W_2 \sum_{r=1}^{s_2}\frac{s_r^b}{y_{ro}^b}\right)} \tag{21}
$$

The default value for $w_1$ and $w_2$ is 1. To give importance to the degree of emphasis for the evaluation of bad outputs, the value of $w_2$ can be larger than $w_1$ or vice versa. In this model, the heading (O) refers to good outputs while (OBAD) is for bad outputs.

## 4. Results

### 4.1. Data Analysis of the Input and Output Factors

The mere difference between the NIC and HIC is that NICs are those countries in which their economic development is said to be in between those classified to be as developing and highly developed. Substantial growth in their own gross domestic product (GDP) is a key indicator in transitioning from one classification to another. However, the authors choose to process the data combining NICs and HICs due to the small number of HICs in the world. For DEA to come up with a highly reliable result, there should be sufficient number of DMUs for comparative analysis. Table 3 below lists down the names of countries belonging to NICs and HICs with their respective DMU representation.

**Table 3.** Name of the countries in their respective DMU number and group.

| DMU No. | Highly Industrialized Countries (HIC) | DMU No. | Newly Industrialized Countries (NIC) |
|---|---|---|---|
| CTRY1 | France | CTRY9 | South Africa |
| CTRY2 | Germany | CTRY10 | Mexico |
| CTRY3 | Italy | CTRY11 | Brazil |
| CTRY4 | United Kingdom | CTRY12 | China |
| CTRY5 | Japan | CTRY13 | India |
| CTRY6 | United States | CTRY14 | Indonesia |
| CTRY7 | Canada | CTRY15 | Malaysia |
| CTRY8 | Russia | CTRY16 | Thailand |
|  |  | CTRY17 | Turkey |

The primary objective of data envelopment analysis is to calculate the efficiency using multiple inputs and outputs. From the diverse combination of factors used by previous studies as presented in Table 1, the authors decided to use total renewable energy consumption (TREC), the labor force (LF) and total energy consumption (TEC) as input factors, while carbon dioxide emission ($CO_2$) and gross domestic product (GDP) are the outputs. Table 4 below summarizes the descriptive statistical values and the coefficient of correlation from the 2015 period with reference to the input and output factors.

**Table 4.** Descriptive statistics and Pearson correlation coefficients of the year 2015.

| | Input Factors | | | Output Factors | |
|---|---|---|---|---|---|
| | Total Renewable Energy Capacity in GW (TREC) | Labor Force In millions (LF) | Total Energy Consumption in Mtoe (TEC) | Carbon Dioxide Emission in $MtCO_2$ ($CO_2$) | Gross Domestic Product in $ Million (GDP) |
| | | | Descriptive Statistics | | |
| Max | 479.11 | 785,372.42 | 2993.90 | 9061.26 | 18,219.3 |
| Min | 3.43 | 14,589.35 | 86.33 | 228.53 | 296.64 |
| Average | 80.98 | 124,337.42 | 560.54 | 1401.99 | 3229.5 |
| SD | 110.48 | 197,968.12 | 780.22 | 2226.36 | 4468.96 |
| | | | Correlation Scores | | |
| TREC | 1 | 0.8274 | 0.9191 | 0.9467 | 0.7080 |
| LF | 0.8274 | 1 | 0.8104 | 0.8515 | 0.4637 |
| TEC | 0.9191 | 0.8104 | 1 | 0.9888 | 0.8581 |
| $CO_2$ | 0.9467 | 0.8515 | 0.9888 | 1 | 0.7941 |
| GDP | 0.7080 | 0.4637 | 0.8581 | 0.7941 | 1 |

Note: The scores of the correlation coefficient are all positive values from 2013–2018. 2015 data is used to represent the other year periods.

### 4.2. GM (1,1) Grey Prediction Model Results

Acquiring a positive correlation using the data from 2013 to 2018, indicates that the input and output factors used complied with the homogeneity and isotonicity requirement of DEA. For this reason, the data is suitable for Grey prediction to obtain the future factors for 2019 to 2022 periods. Table 5 provides the summary of gained mean absolute percentage error (*MAPE*) from 2013 to 2022.

The results above show that most of the *MAPE* from HIC and NIC are below 10%, aside from Russia which gained 12.82% for the GDP factor. This can be a result of a tremendous decline in their GDP output from the year 2015 to 2018. However, this score is still considerably "good" as far as a grey prediction is concerned. Since most of the average MAPE scores are less than 10%, this study can proceed to the next phase using the predicted values for 2019 to 2022.

Table 6 below displays no negative coefficient using the projected data for the year 2021. This also indicates that the other projected data for the year 2019, 2020, and 2022 do not contain any negative coefficients. Thus, the authors can use these data for further analysis using DEA.

**Table 5.** Summary of the average mean absolute percentage error (*MAPE*) of HICs and NICs.

| DMU No. | Country | TREC | LF | TEC | $CO_2$ | GDP | Average |
|---|---|---|---|---|---|---|---|
| CTRY1 | France | 0.21% | 0.09% | 0.39% | 1.02% | 4.81% | 1.30% |
| CTRY2 | Germany | 0.34% | 0.20% | 0.85% | 0.99% | 4.60% | 1.39% |
| CTRY3 | Italy | 0.37% | 0.25% | 0.59% | 0.71% | 4.65% | 1.31% |
| CTRY4 | United Kingdom | 2.85% | 0.06% | 0.32% | 0.72% | 3.03% | 1.40% |
| CTRY5 | Japan | 2.28% | 0.19% | 0.35% | 0.44% | 2.51% | 1.15% |
| CTRY6 | United States | 0.77% | 0.14% | 1.00% | 1.37% | 0.46% | 0.75% |
| CTRY7 | Canada | 1.24% | 0.14% | 0.80% | 0.63% | 4.57% | 1.48% |
| CTRY8 | Russia | 0.13% | 0.21% | 1.42% | 1.29% | 12.82% | 3.17% |
| CTRY9 | South Africa | 5.17% | 0.43% | 1.14% | 1.38% | 5.29% | 2.68% |
| CTRY10 | Mexico | 1.51% | 0.09% | 0.62% | 1.39% | 4.51% | 1.62% |
| CTRY11 | Brazil | 0.31% | 0.14% | 0.89% | 1.26% | 8.10% | 2.14% |
| CTRY12 | China | 0.52% | 0.07% | 0.94% | 0.83% | 2.12% | 0.90% |
| CTRY13 | India | 1.06% | 0.08% | 0.47% | 1.07% | 2.14% | 0.96% |
| CTRY14 | Indonesia | 0.40% | 0.28% | 0.96% | 1.49% | 2.01% | 1.03% |
| CTRY15 | Malaysia | 3.33% | 0.08% | 1.21% | 0.53% | 5.40% | 2.11% |
| CTRY16 | Thailand | 1.96% | 0.17% | 0.34% | 0.55% | 2.84% | 1.17% |

**Source:** Calculated by the authors.

**Table 6.** Descriptive statistics and Pearson correlation coefficients using the projected values for the year 2021.

| | Input Factors | | | Output Factors | |
|---|---|---|---|---|---|
| | **Total Renewable Energy Capacity in GW (TREC)** | **Labor Force In millions (LF)** | **Total Energy Consumption in Mtoe (TEC)** | **Carbon Dioxide Emission in $MtCO_2$ ($CO_2$)** | **Gross Domestic Product in $ Million (GDP)** |
| | | | Descriptive Statistics | | |
| Max | 1025.678 | 788,934.32 | 3265.25 | 9,616.34 | 22,840.18 |
| Min | 9.79 | 16,459.70 | 94.97 | 240.41 | 347.94 |
| Average | 142.74 | 130,441.84 | 604.62 | 1475.28 | 3965.78 |
| SD | 233.78 | 203,616.31 | 839.58 | 2342.19 | 5920.01 |
| | | | Correlation Scores | | |
| TREC | 1 | 0.8531 | 0.9110 | 0.9512 | 0.7194 |
| LF | 0.8531 | 1 | 0.8212 | 0.8597 | 0.5379 |
| TEC | 0.9110 | 0.8212 | 1 | 0.9888 | 0.8774 |
| $CO_2$ | 0.9512 | 0.8597 | 0.9888 | 1 | 0.8226 |
| GDP | 0.7194 | 0.5379 | 0.8774 | 0.8226 | 1 |

Note: The scores of the correlation coefficient are all positive values from 2019–2022. The data from 2021 is used to represent the other year periods.

### 4.3. Results of the DEA Undesirable Model for the Period 2013–2018

#### 4.3.1. Efficiency Scores of HICs and NICs

Since the requirement for DEA has been met from the previous analysis, the DEA undesirable output model will be used to calculate the efficiencies of NICs and HICs as well as their rankings according to their corresponding country category. Table 7 below shows how every country performed in terms of technical efficiency as well as which country is on top for every year in the period.

As seen in the table, three countries—France, the United Kingdom, and the United States—are the most efficient among the HICs which recorded a score of 1 in all year periods. Germany was able to follow through in 2018. While countries from NICs have lower efficiency scores compare to HICs, one country—Indonesia—has shown improvement by obtaining a score of 1 starting from the year 2015 to 2018. South Africa was recorded to be the most efficient NIC from 2013 to 2014.

Since the United Kingdom, the United States, and France have shown consistency in obtaining an efficiency score of 1, there is no need to include them in the line graph as shown in Figure 3 below.

It can be observed that Russia has the lowest scores among the groups. Russia's lowest point score of 0.14 efficiency was during 2015, wherein most HIC countries (except the US, UK, and France) also experienced the same decline.

**Table 7.** Efficiency scores of countries and group rankings from period from 2013 to 2018.

| Countries | Year Periods and Rankings | | | | | | | | | | | |
|---|---|---|---|---|---|---|---|---|---|---|---|---|
| | 2013 | Rank | 2014 | Rank | 2015 | Rank | 2016 | Rank | 2017 | Rank | 2018 | Rank |
| Highly Industrialized Countries (HICs) | | | | | | | | | | | | |
| CTRY1 | 1.0000 | 1 | 1.0000 | 1 | 1.0000 | 1 | 1.0000 | 1 | 1.0000 | 1 | 1.0000 | 1 |
| CTRY2 | 0.7030 | 6 | 0.6277 | 5 | 0.5518 | 6 | 0.6498 | 6 | 0.7326 | 5 | 1.0000 | 1 |
| CTRY3 | 0.7783 | 4 | 0.6829 | 4 | 0.5860 | 4 | 0.6794 | 5 | 0.7358 | 4 | 0.7325 | 5 |
| CTRY4 | 1.0000 | 1 | 1.0000 | 1 | 1.0000 | 1 | 1.0000 | 1 | 1.0000 | 1 | 1.0000 | 1 |
| CTRY5 | 0.7341 | 5 | 0.5928 | 6 | 0.5647 | 5 | 0.7187 | 4 | 0.7153 | 6 | 0.6770 | 6 |
| CTRY6 | 1.0000 | 1 | 1.0000 | 1 | 1.0000 | 1 | 1.0000 | 1 | 1.0000 | 1 | 1.0000 | 1 |
| CTRY7 | 0.5117 | 7 | 0.4115 | 7 | 0.3648 | 7 | 0.3945 | 7 | 0.4341 | 7 | 0.4157 | 7 |
| CTRY8 | 0.2215 | 8 | 0.1887 | 8 | 0.1417 | 8 | 0.1531 | 8 | 0.1852 | 8 | 0.1904 | 8 |
| Newly Industrialized Countries (NICs) | | | | | | | | | | | | |
| CTRY9 | 1.0000 | 1 | 1.0000 | 1 | 0.3000 | 3 | 0.2600 | 4 | 0.2760 | 4 | 0.2748 | 4 |
| CTRY10 | 0.3574 | 2 | 0.3439 | 2 | 0.3455 | 2 | 0.3518 | 2 | 0.4020 | 2 | 0.3873 | 2 |
| CTRY11 | 0.3362 | 4 | 0.2718 | 5 | 0.2050 | 6 | 0.2338 | 5 | 0.2791 | 3 | 0.2266 | 6 |
| CTRY12 | 0.1338 | 8 | 0.1315 | 8 | 0.1500 | 8 | 0.1668 | 8 | 0.1811 | 8 | 0.1840 | 8 |
| CTRY13 | 0.1001 | 9 | 0.0993 | 9 | 0.1150 | 9 | 0.1351 | 9 | 0.1537 | 9 | 0.1411 | 9 |
| CTRY14 | 0.2993 | 5 | 0.2877 | 3 | 1.0000 | 1 | 1.0000 | 1 | 1.0000 | 1 | 1.0000 | 1 |
| CTRY15 | 0.2285 | 6 | 0.2185 | 6 | 0.2049 | 7 | 0.2309 | 6 | 0.2568 | 7 | 0.2810 | 3 |
| CTRY16 | 0.2055 | 7 | 0.1876 | 7 | 0.2104 | 5 | 0.2291 | 7 | 0.2626 | 6 | 0.2716 | 5 |
| CTRY17 | 0.3363 | 3 | 0.2793 | 4 | 0.2634 | 4 | 0.2806 | 3 | 0.2653 | 5 | 0.2152 | 7 |

**Source:** Calculated by the authors.

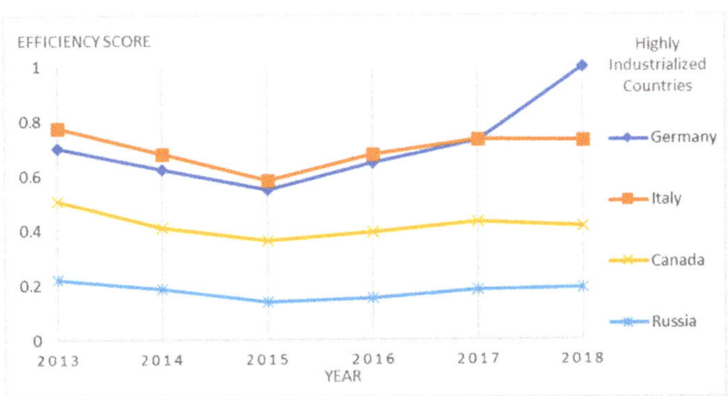


**Figure 3.** Graphical Presentation of other HICs Efficiency Scores (<1.0) from 2013–2018.

It can also be noticed that all of them have declining efficiencies from 2013 to 2015. After this period, it can be observed that most of these countries have increased their efficiencies from 2015 to 2017. However, Japan's score dropped again until 2018. Germany shows a huge increase in their technical efficiency, achieving a score of 1 at the end of 2018.

South Africa started in a high score during the 2013 to 2014 period as seen in Figure 4 below. Unfortunately, the country's efficiency dropped from 2015 and this trend continued until 2018. In contrast, Indonesia improved at the same time South Africa's score fell. Indonesia was able to maintain a score of 1.0 efficiency until 2018, making the country the most efficient among the NICs. India performed with the lowest efficiency. India scores only 0.09 in 2014 and reached its highest point of 0.15 in 2017. With almost the same performance with India, China placed second lowest in terms of efficiencies during the whole 2013 to 2018 period. Some other NICs performed very low, with efficiency scores below 0.5.

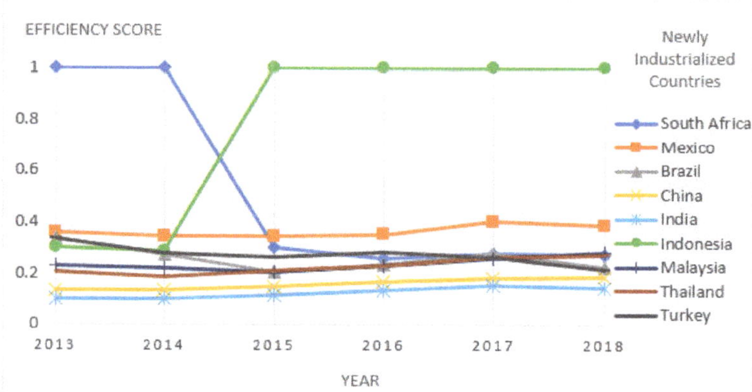


**Figure 4.** Graphical Presentation NICs Efficiency Scores (<1.0) from 2013–2018.

### 4.3.2. Average Efficiency Scores and Overall Ranking

Table 8 below arranges the HICs and NICs according to their average efficiency scores and ranking. The average efficiency scores were calculated from the values of efficiency scores per individual year 2013 to 2018 as shown in Table 7 from the previous section.

**Table 8.** Average Efficiency Scores and Overall Rankings of HICs and NICs.

| DMU No. | Countries | Average Efficiency Score | Overall Ranking |
|---|---|---|---|
| **Highly Industrialized Countries** | | | |
| CTRY1 | France | 1.0000 | 1 |
| CTRY4 | United Kingdom | 1.0000 | 1 |
| CTRY6 | United States | 1.0000 | 1 |
| CTRY2 | Germany | 0.7108 | 4 |
| sCTRY3 | Italy | 0.6992 | 5 |
| CTRY5 | Japan | 0.6671 | 6 |
| CTRY7 | Canada | 0.4221 | 7 |
| CTRY8 | Russia | 0.1801 | 8 |
| **Newly Industrialized Countries** | | | |
| CTRY14 | Indonesia | 0.7645 | 1 |
| CTRY9 | South Africa | 0.5185 | 2 |
| CTRY10 | Mexico | 0.3647 | 3 |
| CTRY17 | Turkey | 0.2734 | 4 |
| CTRY11 | Brazil | 0.2588 | 5 |
| CTRY15 | Malaysia | 0.2368 | 6 |
| CTRY16 | Thailand | 0.2278 | 7 |
| CTRY12 | China | 0.1579 | 8 |
| CTRY13 | India | 0.1241 | 9 |

**Source:** Calculated by the authors.

Since France, the UK and the US are consistently obtaining a 1.0 score for the whole year periods, their average scores are the highest among the others and therefore, the three countries ranked first while Germany, with a score of 0.711 is not too far from acquiring the highest score in the future. Russia remains the least efficient in the HIC group. With the Indonesia obtaining a score of 1.0 from 2015 to

2018, the country was able to take the lead in terms of efficiency among the NIC group with an average score of 0.765. Not too far from Indonesia's score is South Africa in second with 0.519 efficiency. With the rest of the countries from the NIC group obtaining an efficiency score below 0.5, India appeared to be the most inefficient, ranking 9th with a very low score of 0.124.

*4.4. Projected Efficiency Scores for the Period 2019–2022*

The projected efficiency scores are calculated using the values obtained from the grey prediction method. These values are used as inputs and outputs for the computation of technical efficiencies using the undesirable output model of DEA. Table 9 shows the efficiency scores for the period of 2019 to 2022.

**Table 9.** Projected efficiency scores of countries and group rankings from period from 2019 to 2022.

| Countries | Year Periods and Rankings | | | | | | | |
|---|---|---|---|---|---|---|---|---|
| | 2019 | Rank | 2020 | Rank | 2021 | Rank | 2022 | Rank |
| **Highly Industrialized Countries (HICs)** | | | | | | | | |
| CTRY1 | 1.0000 | 1 | 1.0000 | 1 | 1.0000 | 1 | 1.0000 | 1 |
| CTRY2 | 1.0000 | 1 | 1.0000 | 1 | 1.0000 | 1 | 1.0000 | 1 |
| CTRY3 | 0.7687 | 5 | 0.8030 | 5 | 0.8208 | 5 | 0.8381 | 6 |
| CTRY4 | 1.0000 | 1 | 1.0000 | 1 | 1.0000 | 1 | 1.0000 | 1 |
| CTRY5 | 0.7577 | 6 | 0.7977 | 6 | 0.8207 | 6 | 0.8493 | 5 |
| CTRY6 | 1.0000 | 1 | 1.0000 | 1 | 1.0000 | 1 | 1.0000 | 1 |
| CTRY7 | 0.3909 | 7 | 0.3686 | 7 | 0.3482 | 7 | 0.3295 | 7 |
| CTRY8 | 0.1580 | 8 | 0.1498 | 8 | 0.1423 | 8 | 0.1356 | 8 |
| **Newly Industrialized Countries (NICs)** | | | | | | | | |
| CTRY9 | 0.2342 | 5 | 0.2152 | 6 | 0.1983 | 7 | 0.1834 | 7 |
| CTRY10 | 0.3958 | 2 | 0.3717 | 2 | 0.3495 | 2 | 0.3288 | 2 |
| CTRY11 | 0.2332 | 6 | 0.2308 | 5 | 0.2286 | 5 | 0.2261 | 5 |
| CTRY12 | 0.2101 | 8 | 0.2126 | 7 | 0.2149 | 6 | 0.2173 | 6 |
| CTRY13 | 0.1627 | 9 | 0.1639 | 9 | 0.1651 | 9 | 0.1664 | 8 |
| CTRY14 | 1.0000 | 1 | 1.0000 | 1 | 1.0000 | 1 | 1.0000 | 1 |
| CTRY15 | 0.2576 | 4 | 0.2532 | 4 | 0.2489 | 4 | 0.2447 | 4 |
| CTRY16 | 0.2720 | 3 | 0.2728 | 3 | 0.2737 | 3 | 0.2747 | 3 |
| CTRY17 | 0.2248 | 7 | 0.1999 | 8 | 0.1748 | 8 | 0.1534 | 9 |

**Source:** Calculated by the authors.

The result of the forecasted efficiencies shows that in the HIC group, France, Germany, United Kingdom, and the United States will continue topping the ranks, while Canada and Russia remain in the 7th and 8th rank, respectively. Japan and Italy can be seen switching their ranks in the last period of 2022 with the former going up from 6th to 5th.

Figure 5 does not include the countries that obtain an efficiency score of 1 through the whole period of 2019 to 2022 as they are understood to be highly efficient already. It can be observed that Italy and Japan display a quite positive trend, with a slight increase in efficiency, while Canada and Russia have a slightly negative trend.

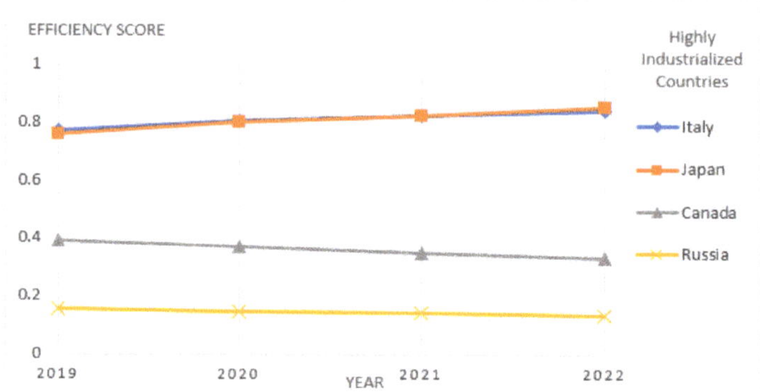


**Figure 5.** Graphical Presentation of other HICs' Efficiency Scores (<1.0) from 2019–2022.

Efficiency scores of the NICs are mostly around of below a 0.5 level, except for Indonesia which garnered a score of 1.0 from 2019 to 2022. None of the countries show any remarkable positive trends. Instead, some countries like China, India, and Thailand are expected to display stable performance or very little improvement in efficiency. Mexico, Thailand, and Turkey will have declining efficiencies during the projected period, as seen in Figure 6.

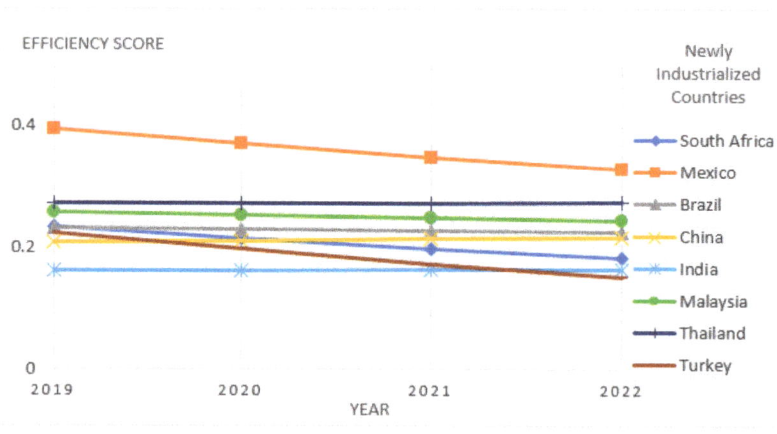


**Figure 6.** Graphical Presentation of NICs' Efficiency Scores from 2019–2022.

As presented in Figure 7, the calculated average efficiencies of HICs during the past and projected periods are comparatively higher than the NICs. However due to the existence of the very low efficiency scores of Russia and Canada, the effect to the average efficiency scores of the HIC group reached the 0.7099 level for the past period and is expected to increase by 7.76% to reach the 0.765 projected efficiency level. A different scenario is expected from the NIC group which will exhibit a decline of 1.23% from a 0.325 score down to a projected average level of 0.321.

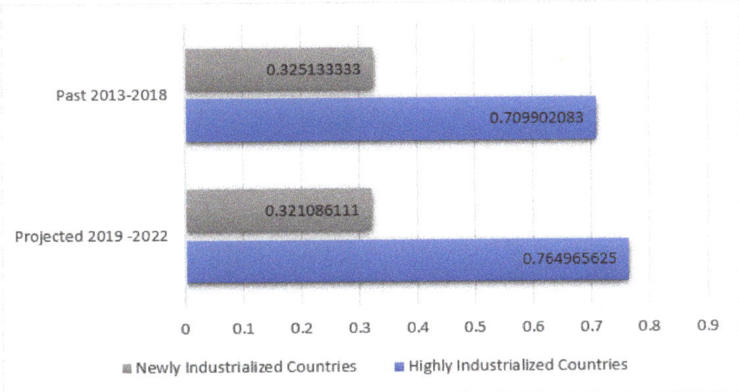


**Figure 7.** Comparative graph of HICs and NICs average efficiency scores for two different periods.

## 5. Conclusions

With the implementation of the Paris Agreement signed by the HICs and NICs, the results show different effects on each country. Some countries have efficiently utilized their renewable energy resources, but some are not doing so, while others have maintained their levels and it seems they are not moved by the Agreement terms. The use of the undesirable model has successfully calculated each country's position in terms of the utilization of their resources while considering greatly a very bad factor, which is the carbon emission.

Knowing the economic capabilities of the highly industrialized nations, the results show that Russia is the least efficient among them. A mean efficiency score of 0.1801 verifies the country's lack of attention to its renewable energy program. In fact, in 2015, renewable energy only comprised 20% of the total installed power capacity, which mostly came from bioenergy and hydropower. The country must focus on other sources such as geothermal, wind, and solar photovoltaic (PV) [42]. Canada, with a mean efficiency score of 0.4221 manages to place second to the last among the HICs. A report from 2015 reveals that Vancouver alone sourced its 69% of energy from fossil fuels and 31% from renewable sources [43]. The situation in Japan is a little different, as due to the nuclear disaster from a major earthquake years ago, the country aims their energy to be completely 100% supplied by renewable energy, especially in their local regions, by 2020 [44]. Additionally, a report states that there is a decrease in the use of fossil fuels in 2018, 78% from 81% share. However, a dependency of the country on nuclear energy still exists and is evident from the increase of its generation from 2.8% to 4.7% of the same year [45]. This can be one of the reasons for Japan's efficiency score drop from 2016 to 2018 and its acquisition of an average of 0.6671 throughout the studied period. The same goal is seen in Italy's energy program. With the introduction of the 20-20-20 EU goals, the Italian government aims to decrease the emission of greenhouse gases by 20%, improve energy savings to up to 20%, and attain a 20% generation of renewable energy by 2020 [46]. The result of this study shows that Italy has quite improved its renewable energy utilization with increased and maintained efficiency scores from 2015 to 2018. While reaching a perfect efficiency of 1.0 in 2018, the average score becomes 0.7108 due to lower scores acquired from previous years. These scores imply that the German government has effectively improved its renewable energy utilization throughout the study period. Gaining an efficiency score of 1.0 throughout the study period, France, the United Kingdom, and United States, have been consistently utilized their renewable energies even before or after the Paris agreement. Their energy and sustainability programs can be a benchmark for other HICs that aim to improve their use of own renewable resources.

In contrast with the efficiency scores of the HICs, none from the list of NICs is able to get a consistent perfect efficiency. In spite of this, Indonesia manages to acquire perfect efficiency during the period of 2015 to 2018 rising from low-efficiency scores. Getting the highest mean efficiency score of 0.7645 among the NICs, this can be due to the fact that the country is considered to have the biggest resources of geothermal energy in the world [47]. South Africa demonstrates an interesting efficiency score as it can be seen that during the period of 2013 to 2014, the country was able to gain a perfect efficiency. This performance is attributable to their noticeable increase in renewable energy production in previous years [48]. However, this performance by South Africa was not maintained since the country's efficiency dropped tremendously in succeeding years resulting in an average efficiency score of only 0.5185. Mexico, Turkey, Brazil, Malaysia, and Thailand are performing consistently at a low-level efficiency, with a minimum of 0.2278 and a maximum of 0.2734 mean scores. These results can be caused by the following factors: Mexico's dependency on fossil fuels wherein 85% of their total power was produced in 2012 [49]; Turkey's power sector industry as the biggest contributor to $CO_2$ emissions and a high reliance to coal-powered electricity generation accounting to 37.2% of their total electricity [50]; the old structure of Brazil's energy sector which limits their capacities in handling renewable energy demands, and other political factors that hinder the country's development to successfully execute their national energy programs [51]; and, Malaysia and Thailand's high proportional use of crude oil, coal, and natural gas, giving less importance to renewable energies [52,53].

The HIC group is expected to keep increasing the efficiency level by 7.76% from the past period to the projected period. This can be due to the expected development to the future renewable energy developments by Japan and Italy. However, the NIC group is expected a bit decline in the average efficiency with negative 1.23% due to the efficiency drop performance by Turkey and Mexico along with the low efficiency performances by other countries.

Future implications from the result of combined data gathered using grey prediction and the undesirable DEA model shows that the HICs and NICs will continuously follow the trend of the efficiency scores for all these countries with Germany joining the group of perfectly efficient ones together with Indonesia, France, United Kingdom, and the United States. All other HICs and NICs will perform otherwise, especially if they will not concede significant importance to renewable energy. Most of the developments in the use of the renewable energy begin with the improvement in the political and legal structure surrounding its use and regulation, providing incentives or tax exemptions to private power producers to shift in using the renewable energy. Countries can also build strong partnerships with non-government and international organizations that are focusing on providing assistance to countries that want to achieve sustainable and renewable energy production. Strengthening the support to the research and development sector to determine the suitable sites to deploy renewable energy sources such as solar, wind and wave energy. Long-term strategic plans for renewable energy development that will also consider the economic, social, and environmental impacts to the country.

This study contributes as a method to mathematically evaluate the energy utilization efficiency of HICs and NICs based on available public data. The DEA undesirable model treatment of the $CO_2$ emission factor as a less important factor made this study different from the others. Hence, this model is the most appropriate method to evaluate the energy sector that uses data with undesired factors. The results of this research may guide each country for improvement of its production and consumption towards sustainable renewable energy development. This can also help policymakers, government agencies and the energy sectors to address the problem in the existence of bad outputs such as $CO_2$ emissions. For future studies involving quantitative measurements, the authors recommend combining DEA with a qualitative evaluation approach such as the one described in Wang et al. [54], which uses the fuzzy analytical hierarchy process for analysis to improve studies of this kind.

**Author Contributions:** Conceptualization, C.-N.W. and H.T.; data curation, H.T. and D.H.D.; formal analysis, H.T. and D.H.D.; funding acquisition, C.-N.W.; investigation, H.T. and D.H.D.; methodology, C.-N.W., H.T., and D.H.D.; project administration, C.-N.W.; writing—original draft, H.T. and D.H.D.; writing—review and editing, C.-N.W. and H.T. All authors have read and agreed to the published version of the manuscript.

**Funding:** This research was partly supported by National Kaohsiung University of Science and Technology, and project number 108-2622-E-992-017-CC3 from the Ministry of Sciences and Technology in Taiwan.

**Acknowledgments:** The authors appreciate the support from Taiwan National Kaohsiung University of Science and Technology, Philippines Technological University of the Philippines Taguig, and Taiwan Ministry of Sciences and Technology.

**Conflicts of Interest:** The authors declare no conflict of interest.

## References


1. United Nations. Adoption of the Paris Agreement. In Proceedings of the Conference of the Parties, Paris, France, 30 November–11 December 2015.
2. Nejat, P.; Jomehzadeh, F.; Taheri, M.M.; Gohari, M.; Majid, M.Z.A. A global review of energy consumption, $CO_2$ emissions and policy in the residential sector (with an overview of the top ten $CO_2$ emitting countries). *Renew. Sustain. Energy Rev.* **2015**, *43*, 843–862. [CrossRef]
3. Ellabban, O.; Abu-Rub, H.; Blaabjerg, F. Renewable energy resources: Current status, future prospects and their enabling technology. *Renew. Sustain. Energy Rev.* **2014**, *39*, 748–764. [CrossRef]
4. The International Renewable Energy Agency (IRENA). Available online: https://www.irena.org/aboutirena (accessed on 4 December 2019).
5. Laub, Z. The Group of Eight (G8) Industrialized Nations. Counsil on Foreign Relations. 2014. Available online: http://www.cfr.org/backgrounder/group-eight-g8-industrialized-nations/ (accessed on 6 December 2019).
6. Szirmai, A.; Naudé, W.; Alcorta, L. *Pathways to Industrialization in the Twenty-First Century: New Challenges and Emerging Paradigms*; OUP Oxford: Oxford, UK, 2013.
7. Cleveland, C.J.; Kaufmann, R.K.; Stern, D.I. Aggregation and the role of energy in the economy. *Ecol. Econ.* **2000**, *32*, 301–317. [CrossRef]
8. Cîrstea, S.D.; Moldovan-Teselios, C.; Cîrstea, A.; Turcu, A.C.; Darab, C.P. Evaluating renewable energy sustainability by composite index. *Sustainability* **2018**, *10*, 811. [CrossRef]
9. Iddrisu, I.; Bhattacharyya, S.C. Sustainable Energy Development Index: A multi-dimensional indicator for measuring sustainable energy development. *Renew. Sustain. Energy Rev.* **2015**, *50*, 513–530. [CrossRef]
10. Lee, C.W.; Zhong, J. Construction of a responsible investment composite index for renewable energy industry. *Renew. Sustain. Energy Rev.* **2015**, *51*, 288–303. [CrossRef]
11. Schlör, H.; Fischer, W.; Hake, J.-F. Methods of measuring sustainable development of the German energy sector. *Appl. Energy* **2013**, *101*, 172–181. [CrossRef]
12. Liu, G. Development of a general sustainability indicator for renewable energy systems: A review. *Renew. Sustain. Energy Rev.* **2014**, *31*, 611–621. [CrossRef]
13. Doukas, H.; Papadopoulou, A.; Savvakis, N.; Tsoutsos, T.; Psarras, J. Assessing energy sustainability of rural communities using Principal Component Analysis. *Renew. Sustain. Energy Rev.* **2012**, *16*, 1949–1957. [CrossRef]
14. Štreimikienė, D.; Šliogerienė, J.; Turskis, Z. Multi-criteria analysis of electricity generation technologies in Lithuania. *Renew. Energy* **2016**, *85*, 148–156. [CrossRef]
15. Ren, J.; Sovacool, B.K. Prioritizing low-carbon energy sources to enhance China's energy security. *Energy Convers. Manag.* **2015**, *92*, 129–136. [CrossRef]
16. Troldborg, M.; Heslop, S.; Hough, R.L. Assessing the sustainability of renewable energy technologies using multi-criteria analysis: Suitability of approach for national-scale assessments and associated uncertainties. *Renew. Sustain. Energy Rev.* **2014**, *39*, 1173–1184. [CrossRef]
17. Yazdani-Chamzini, A.; Fouladgar, M.M.; Zavadskas, E.K.; Moini, S.H.H. Selecting the optimal renewable energy using multi criteria decision making. *J. Bus. Econ. Manag.* **2013**, *14*, 957–978. [CrossRef]
18. San Cristóbal, J. Multi-criteria decision-making in the selection of a renewable energy project in spain: The Vikor method. *Renew. Energy* **2011**, *36*, 498–502. [CrossRef]
19. Kabak, M.; Dağdeviren, M. Prioritization of renewable energy sources for Turkey by using a hybrid MCDM methodology. *Energy Convers. Manag.* **2014**, *79*, 25–33. [CrossRef]
20. Halkos, G.E.; Tzeremes, N.G. Measuring the effect of Kyoto protocol agreement on countries' environmental efficiency in $CO_2$ emissions: An application of conditional full frontiers. *J. Prod. Anal.* **2014**, *41*, 367–382. [CrossRef]
21. Wang, C.-N.; Ho, H.-X.T.; Luo, S.-H.; Lin, T.-F. An integrated approach to evaluating and selecting green logistics providers for sustainable development. *Sustainability* **2017**, *9*, 218. [CrossRef]

22. Oggioni, G.; Riccardi, R.; Toninelli, R. Eco-efficiency of the world cement industry: A data envelopment analysis. *Energy Policy* **2011**, *39*, 2842–2854. [CrossRef]
23. Woo, C.; Chung, Y.; Chun, D.; Seo, H.; Hong, S. The static and dynamic environmental efficiency of renewable energy: A Malmquist index analysis of OECD countries. *Renew. Sustain. Energy Rev.* **2015**, *47*, 367–376. [CrossRef]
24. Zhou, P.; Ang, B.; Han, J. Total factor carbon emission performance: A Malmquist index analysis. *Energy Econ.* **2010**, *32*, 194–201. [CrossRef]
25. Wang, C.-N.; Nguyen, T.-D.; Yu, M.-C. Energy Use Efficiency Past-to-Future Evaluation: An International Comparison. *Energies* **2019**, *12*, 3804. [CrossRef]
26. Zofío, J.L.; Prieto, A.M. Environmental efficiency and regulatory standards: The case of $CO_2$ emissions from OECD industries. *Resour. Energy Econ.* **2001**, *23*, 63–83. [CrossRef]
27. Xie, B.-C.; Shang, L.-F.; Yang, S.-B.; Yi, B.-W. Dynamic environmental efficiency evaluation of electric power industries: Evidence from OECD (Organization for Economic Cooperation and Development) and BRIC (Brazil, Russia, India and China) countries. *Energy* **2014**, *74*, 147–157. [CrossRef]
28. Cicea, C.; Marinescu, C.; Popa, I.; Dobrin, C. Environmental efficiency of investments in renewable energy: Comparative analysis at macroeconomic level. *Renew. Sustain. Energy Rev.* **2014**, *30*, 555–564. [CrossRef]
29. Wang, C.-N.; Luu, Q.-C.; Nguyen, T.-K.-L. Estimating Relative Efficiency of Electricity Consumption in 42 Countries during the Period of 2008–2017. *Energies* **2018**, *11*, 3037. [CrossRef]
30. Chien, T.; Hu, J.-L. Renewable energy and macroeconomic efficiency of OECD and non-OECD economies. *Energy Policy* **2007**, *35*, 3606–3615. [CrossRef]
31. Charnes, A.; Cooper, W.W.; Rhodes, E. Measuring the efficiency of decision making units. *Eur. J. Operat. Res.* **1978**, *2*, 429–444. [CrossRef]
32. Banker, R.D.; Charnes, A.; Cooper, W.W. Some models for estimating technical and scale inefficiencies in data envelopment analysis. *Manag. Sci.* **1984**, *30*, 1078–1092. [CrossRef]
33. Caves, D.W.; Christensen, L.R.; Diewert, W.E. The economic theory of index numbers and the measurement of input, output, and productivity. *Econ. J. Econ. Soc.* **1982**, *50*, 1393–1414. [CrossRef]
34. Färe, R.; Grosskopf, S.; Lindgren, B.; Roos, P. Productivity changes in Swedish pharamacies 1980–1989: A non-parametric Malmquist approach. *J. Prod. Anal.* **1992**, *3*, 85–101. [CrossRef]
35. Ahn, T.; Charnes, A.; Cooper, W.W. Some statistical and DEA evaluations of relative efficiencies of public and private institutions of higher learning. *Socio-Econ. Plan. Sci.* **1988**, *22*, 259–269. [CrossRef]
36. Tone, K. A slacks-based measure of efficiency in data envelopment analysis. *Eur. J. Operat. Res.* **2001**, *130*, 498–509. [CrossRef]
37. Cooper, W.W.; Seiford, L.M.; Tone, K. *Introduction to Data Envelopment Analysis and its Uses: With DEA-Solver Software and References*; Springer Science & Business Media: New York, NY, USA, 2006.
38. Wang, C.-N.; Nguyen, T.-D.; Le, M.-D. Assessing Performance Efficiency of Information and Communication Technology Industry-Forecasting and Evaluating: The Case in Vietnam. *Appl. Sci.* **2019**, *9*, 3996. [CrossRef]
39. Julong, D. Introduction to grey system theory. *J. Grey Syst.* **1989**, *1*, 1–24.
40. Tseng, F.-M.; Yu, H.-C.; Tzeng, G.-H. Applied hybrid grey model to forecast seasonal time series. *Technol. Forecast. Soc. Chang.* **2001**, *67*, 291–302. [CrossRef]
41. Ju-Long, D. Control problems of grey systems. *Syst. Control Lett.* **1982**, *1*, 288–294. [CrossRef]
42. Gielen, D.; Saygin, D. REmap 2030 Renewable Energy Prospects for Russian Federation. Abu Dhabi. 2017. Available online: http://www.irena.org/remap (accessed on 6 April 2020).
43. Ahmed, S.; Pourya, S.; Michael, W.; Doug, S.; Jinlei, F.; Ghislaine, K.; Verena, O.; Michael, R. *Adopting and Promoting a 100% Renewable Energy Target*; International Renewable Energy Agency: Vancouver, BC, Canada, 2018.
44. Esteban, M.; Portugal-Pereira, J. Post-disaster resilience of a 100% renewable energy system in Japan. *Energy* **2014**, *68*, 756–764. [CrossRef]
45. *Share of Renewable Energy Power in Japan, 2018 (Preliminary Report)*; Institute for Sustainable Energy Policies: Tokyo, Japan, 2019.
46. Antonelli, M.; Desideri, U.; Franco, A. Effects of large scale penetration of renewables: The Italian case in the years 2008–2015. *Renew. Sustain. Energy Rev.* **2018**, *81*, 3090–3100. [CrossRef]
47. Winarno, O.T.; Alwendra, Y.; Mujiyanto, S. Policies and strategies for renewable energy development in Indonesia. In Proceedings of the 2016 IEEE International Conference on Renewable Energy Research and Applications (ICRERA), Birmingham, UK, 20–23 November 2016; pp. 270–272.

48. Nakumuryango, A.; Inglesi-Lotz, R. South Africa's performance on renewable energy and its relative position against the OECD countries and the rest of Africa. *Renew. Sustain. Energy Rev.* **2016**, *56*, 999–1007. [CrossRef]

49. Vidal-Amaro, J.J.; Østergaard, P.A.; Sheinbaum-Pardo, C. Optimal energy mix for transitioning from fossil fuels to renewable energy sources—The case of the Mexican electricity system. *Appl. Energy* **2015**, *150*, 80–96. [CrossRef]

50. Karagöz, G.N. *Renewable Energy in Turkey: A Cleaner, Self-Sufficient Alternative to Coal*; Kadir Has Üniversitesi: Istanbul, Turkey, 2019.

51. de Melo, C.A.; de Martino Jannuzzi, G.; Bajay, S.V. Nonconventional renewable energy governance in Brazil: Lessons to learn from the German experience. *Renew. Sustain. Energy Rev.* **2016**, *61*, 222–234. [CrossRef]

52. Khor, C.S.; Lalchand, G. A review on sustainable power generation in Malaysia to 2030: Historical perspective, current assessment, and future strategies. *Renew. Sustain. Energy Rev.* **2014**, *29*, 952–960. [CrossRef]

53. Kumar, S. Assessment of renewables for energy security and carbon mitigation in Southeast Asia: The case of Indonesia and Thailand. *Appl. Energy* **2016**, *163*, 63–70. [CrossRef]

54. Wang, C.-N.; Nguyen, V.T.; Thai, H.T.N.; Tran, N.N.; Tran, T.L.A. Sustainable supplier selection process in edible oil production by a hybrid fuzzy analytical hierarchy process and green data envelopment analysis for the SMEs food processing industry. *Mathematics* **2018**, *6*, 302. [CrossRef]

*Article*

# Sustainable Solutions for Green Financing and Investment in Renewable Energy Projects

**Farhad Taghizadeh-Hesary** [1,*] **and Naoyuki Yoshino** [2,3]

1   Tokai University, Hiratsuka, Kanagawa 259-1292, Japan
2   Asian Development Bank Institute, Chiyoda-ku, Tokyo 100-6008, Japan
3   Faculty of Economics, Keio University, Minato-ku, Tokyo 108-8345, Japan; yoshino@econ.keio.ac.jp
*   Correspondence: farhad@tsc.u-tokai.ac.jp or farhadth@gmail.com

Received: 21 August 2019; Accepted: 6 February 2020; Published: 11 February 2020

**Abstract:** The lack of long-term financing, the low rate of return, the existence of various risks, and the lack of capacity of market players are major challenges for the development of green energy projects. This paper aimed to highlight the challenges of green financing and investment in renewable energy projects and to provide practical solutions for filling the green financing gap. Practical solutions include increasing the role of public financial institutions and non-banking financial institutions (pension funds and insurance companies) in long-term green investments, utilizing the spillover tax to increase the rate of return of green projects, developing green credit guarantee schemes to reduce the credit risk, establishing community-based trust funds, and addressing green investment risks via financial and policy de-risking. The paper also provides a practical example of the implementation of the proposed tools.

**Keywords:** green finance; green investment; green credit guarantee scheme; community-based trust funds; renewable energy

---

## 1. Introduction

In 2017 and 2018, the global investments (net capital flows) in renewable energy and energy efficiency projects reduced by 1% and 3%, respectively, and there is a risk that they will decrease even more [1]. This could threaten the achievements of the Sustainable Development Goals (SDGs) and the Paris Agreement for climate change. One reason behind the slowed development of green projects is difficulties in accessing private finance. The future of clean energy no longer concerns science and technology; it is all about access to finance [2]. People consider green energy projects, like other energy projects, as infrastructural projects. Infrastructural projects are capital-intensive and long-term projects. According to the Asian Development Bank [3], in developing Asia alone, there is a gap of $26 trillion investment from 2016 to 2030, or $1.7 trillion per year, if the region is to maintain its growth momentum, eradicate poverty, and respond to climate change (climate-adjusted estimate). Of the total climate-adjusted investment needs over the period 2016–2030, the largest share, or $14.7 trillion, will be for energy (power).

There are two major barriers associated with green energy projects: (a) a lower rate of return compared to fossil fuel projects and (b) a higher risk of investment compared to fossil fuel projects (see [4]). Due to the associated risk and due to the Basel capital requirements, many banks are reluctant to finance green energy projects. Another reason why debt finance is hard to secure for new green energy projects is that, historically, regulated utility rates spread risks in utilities across consumers. However, the lack of purchasing power agreements (PPAs), especially in less developed countries, generates uncertainty about the tariff and increases the risk of investment. The third reason behind the uncertainty that has shrunk the new investment in the renewable energy sector is rapid technological progress and cost reduction. Thanks to the technological progress, costs of renewable energy technologies reduced drastically. For example, solar PV module prices have reduced by around 80% since the end of 2009 and wind turbine prices have dropped by 30–40%. Although this is good news on one side, it prompts investors to pause to see how much the prices will drop on the other side.

All in all, it is important to take the necessary steps for mitigating the risks of green financing to unlock the participation of financial institutions in these projects [4]. One solution is to incentivize non-banking financial institutions (NBFIs), such as pension funds or insurance companies, to engage in green energy projects. The advantages of pension funds and insurance companies over banks are that these institutions pursue asset-liability matching and their resources are long-term (10, 20, or 40 years). Insurance companies or pension funds can finance infrastructure projects, including large green energy projects such as large hydropower, as they are long-term projects (10–20 years). Therefore, it is very important to develop pension funds and insurance companies in developing countries to fill the financing gap of infrastructure projects, including energy and green energy projects [4].

A good example of the role of non-banking financial institutions in unlocking green investment comes from Australia. The Powering Australian Renewables Fund (PARF) is a financing initiative that AGL Energy of Australia created and has $2-3 billion funds targeting to unlock the investments in large-scale renewable energy projects by diversifying the risks and reducing the financing costs. A partnership with the Queensland Investment Corporation (QIC) established the fund in 2016 on behalf of its clients, the Future Fund and those investing in the QIC Global Infrastructure Fund. The target is to accelerate Australia's transition to a low-carbon economy with the potential to meet 10% of the Federal Government's Renewable Energy Target (RET).

One point that is important to consider is that public financial institutions, including green banks [5], need to crowd in private investment. This means that the government needs to be the investor of last resort, which involves investing in green projects when there is an urgent need for them and the private sector is not showing eagerness to invest. As for the role of the government, one important issue is political incentive when there is a conflict in terms of the election horizon and the maturity of green projects, which are often long-term. To overcome this problem, [6] proposed the establishment of a regional network, namely the European Sustainable Banking Network (EU SBN) for Europe. The [6] proposal consisted of three major actions: (1) green certification of private and public financial institutions, (2) sustainability rating of project proposals, and (3) systematic monitoring of the performance of banks and the financed investment projects. This means that, despite changing governments, this network will continue to carry out the activities and its mission.

A method that this paper will present is the introduction of green credit guarantee schemes (GCGSs), which [7] developed and introduced, to reduce the risk of financing. It is also important for banks to have specific programs for a precautionary approach to green lending, as well as compliance and risk management, which this paper will highlight.

Our paper contributes to the literature by providing innovative solutions for unlocking green finance and investment from banks and NBFIs. These solutions will help financial institutions to minimize and manage the risk of green financing. They include developing green credit guarantee schemes for reducing the financial risk, utilizing community-based trust funds, introducing insurance mechanisms and de-risking to cover non-financial risks, and using the spillover tax to increase the rate of return on green projects.

The structure of the paper is as follows. In Section 2, we highlight the challenges of developing green projects. Section 3 focuses on introducing and analyzing the enabling conditions for green financing. Section 4 provides an example of the implementation of the proposed tools and instruments, and Section 5 delivers concluding remarks and outlines the policy implications.

## 2. Challenges for Development of Green Projects

This section highlights the challenges for the development of green projects.

## 2.1. Lack of Long-Term Financing

Large-scale green energy projects, such as hydro-power or large solar farms, are long-term projects; hence, they need long-term financing. A shortage of long-term financing hinders the progress of green development. Banks still dominate Asian economies, and the banking sector constrains long-term finance. The development of public financial institutions (PFIs) that provide long-term financing or the development of pension funds and insurance companies are major solutions for filling the long-term financing and investment gap. [8] by analyzing the state investment banks (SIBs) in Australia, the UK, and Germany, found that SIBs take a broader role than capital provision and de-risking to mobilize finance. SIBs have an educational role to enable financial sector learning. The authors showed that SIBs signal trust and produce track records to crowd in private finance.

Figure 1 shows the structure of financial markets in selected Asian countries. Banks dominate the financial systems in Asia. Banks' resources are deposits, and deposits are short to medium term. On the other hand, infrastructure projects and energy projects are long-term. A maturity mismatch arises if banks' resources are allocated to financing long-term projects.

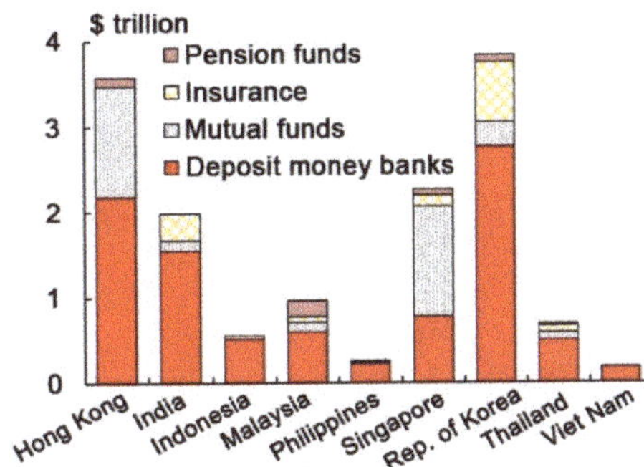


GDP = gross domestic product

Cap. = capital

Note: Emerging Asia is the aggregate of the People's Republic of China, India, Indonesia, the Republic of Korea, Malaysia, the Philippines, Thailand, and Vietnam.

**Figure 1.** Bank-dominated financial systems in Asia. Source: [9].

## 2.2. Existence of Various Risks

As most green energy technologies are new, there are several associated risks [10]. From mechanical breakdowns of wind power generator gearboxes to breakages of the panels of photovoltaic projects, the potential losses can reach millions of dollars, with major damage interrupting projects and businesses. This is not the whole story, however, as other risks accompany these projects—especially the weather. Most green energy projects depend on the climate and sunlight. The unpredictability of the weather, such as clouds that reduce the sun's irradiation or changes in wind strength, can have a significant negative impact on energy production and affect the feasibility of these projects. In addition, as many equipment for green projects are high-tech, it is expensive, creating risks regarding the feasibility. In order to identify and assess the diverse risks in green residential buildings that use green technologies, Ref. [11] conducted a survey of 30 construction companies in Singapore. Their survey results and study indicated that the top five critical risks in green residential building construction projects are:

"complex procedures to obtain approvals", "overlooked high initial cost", "unclear requirements of owners", "employment constraint", and "lack of availability of green materials and equipment".

In addition to the aforementioned risks, the manufacturing of green technologies depends on cross-country supply chains and trade. Economies that are net importers of final products may be major exporters of materials or subcomponents for the same technologies. Hence, the exchange rate is another risk for green technologies.

Figure 2 shows the balance of trade for the four major clean energy technologies. Crystalline silicon (c-Si) photovoltaic (PV) and LED packages are the most heavily traded, perhaps because they are easier to ship than the other end products. The balance of trade is not the full story, however. While major PV deployment markets, such as the United States and Germany, are net importers of PV modules, they are also the largest exporters of polysilicon to make those modules, which Japan and the People's Republic of China largely purchase.

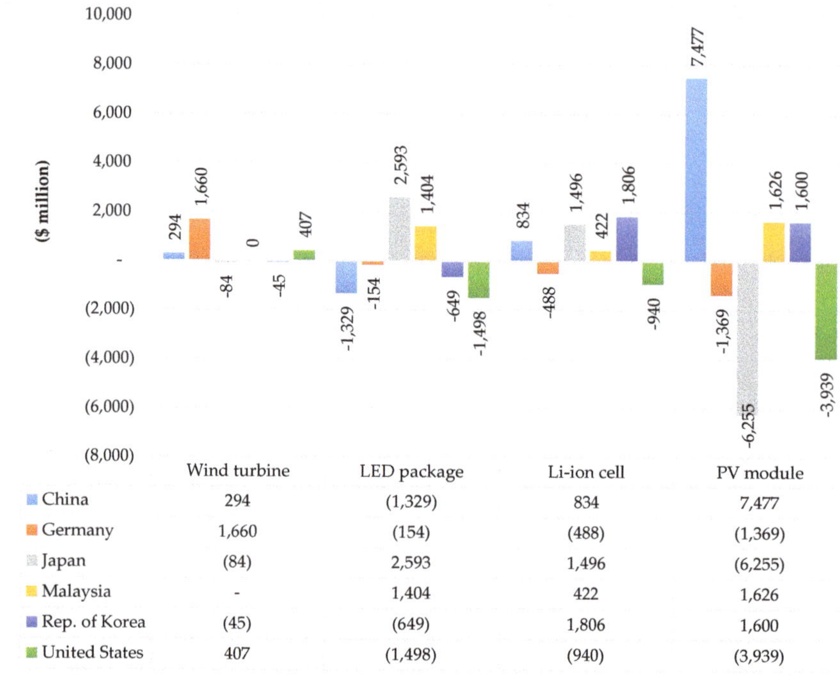

| | Wind turbine | LED package | Li-ion cell | PV module |
|---|---|---|---|---|
| China | 294 | (1,329) | 834 | 7,477 |
| Germany | 1,660 | (154) | (488) | (1,369) |
| Japan | (84) | 2,593 | 1,496 | (6,255) |
| Malaysia | - | 1,404 | 422 | 1,626 |
| Rep. of Korea | (45) | (649) | 1,806 | 1,600 |
| United States | 407 | (1,498) | (940) | (3,939) |

**Figure 2.** Balance of trade in select clean energy technology end products and across the C-Si PV module supply chain, 2014. ($ million). c-Si = crystalline silicon and PV = photovoltaic. Source: Authors' development based on data from the [12]. Note: The values in parentheses are negative.

Green projects also involve other associated risks. Some of them are project-specific (demand risk) and some are general (e.g., natural disaster and political risks).

### 2.3. Low Rate of Return and Lack of Capacity among Market Players

Green technologies are often earlier in the development stage and less commercially viable than technologies in the fossil fuel field, many of which date back 100 years. This makes green technologies more expensive and riskier. A lack of access to conventional financing sources increases the debt cost (borrowing interest rate). New and expensive green technologies and access to expensive debt markets reduce the rate of return in green projects compared with fossil fuel projects. On the other hand, the majority of energy subsidies globally supports fossil fuels rather than the green sector.

Longstanding subsidies for fossil fuels, combined with the lack of a price for carbon emissions, have favored fossil fuels, like oil and coal [13]. In 2015, both consumers and producers of fossil fuels received about $425 billion in subsidies globally—via direct payments, tax breaks, loan guarantees, cheap rental of public land, and research and development (R&D) grants [14]. According to the Organisation for Economic Co-operation and Development [15], almost 800 individual policies support the production or consumption of fossil fuels. According to [16], estimated fossil fuel subsidies were $4.9 trillion worldwide in 2013 and $5.3 trillion in 2015 (6.5% of global gross domestic product (GDP) in both years). Another form of subsidy, an indirect one, takes place when governments do not tax fossil fuel companies efficiently [16]. This means that the price that consumers pay for coal, gas, or oil does not consider the damage that these products cause, such as climate change or air pollution, making green projects less viable than fossil fuel energy projects. Targeting the subsidies and allocating them to the low-carbon sector will increase the rate of return of these projects and make them interesting to private investors.

On the other hand, another challenge is the lack of capacity for and information about green technologies among the market players (governments, investors, and financial institutions). Moreover, the lack of green data and green databases is an obstacle to the development of these projects. Lack of borrowers' green and environmental information limits banks' abilities to assess the environmental risks involved in project and corporate finance [17]. In this regard, establishment of a centralized database at the industry level would enable the lenders to assess the business and market risks related to the environment.

## 3. Enabling Conditions of Green Finance

To overcome the challenges mentioned in Section 2, this section provides practical solutions to create enabling conditions for green finance.

### 3.1. Increasing the Role of PFIs

The first and most important challenge for financing green projects is the lack of access to long-term finance and investment (Section 2.1). PFIs, or publicly created and/or mandated financial institutions, could be important entities for filling the financing gap in the green sector. For example, five PFIs in Europe—France's Caisse des Dépôts Group, Germany's KfW Bankengruppe, the United Kingdom's Green Investment Bank, the European Bank for Reconstruction and Development for transition economies, and the European Investment Bank—provided more than €100 billion in equity investment and financing for energy efficiency, renewable energy, and sustainable transport projects during the period 2010–2012 [18].

It is important for PFIs to open a separate file for green financing. They also need to integrate environmental considerations into conventional project financing. For example, the Japan Bank for International Cooperation (JBIC) launched GREEN operations in 2010, using measurement, reporting, and verification (J-MRV) as a method to evaluate the greenhouse gas emission reductions of the projects that it finances. Along with the basic concept and procedures for quantifying the reductions, the J-MRV guidelines include individual methodologies for each sector of the project and/or technology, varying from renewable energy to transport. When the GREEN operations began, J-MRV had only three methodologies for the projects that were most in demand, including renewable energy and the installation of energy-efficient industrial equipment. However, the JBIC gradually developed new methodologies, increasing to 10 in 2016 [19].

Although the role of PFIs could be very important, some important points need to be considered regarding the involvement of PFIs in green financing. The first point is that they need to focus more on long-term financing (long-term loans) than commercial private banks, the resources (deposits) of which are short-term (one, two, or three years). Private banks are not able to provide long-term loans, so the maturity of PFI loans has to be longer than that of private banks. The second point is to set a stable and fixed interest rate that is lower than that of private banks, the interest rates of which often fluctuate, since green projects need a stable and fixed interest rate for steady growth. Private banks have to pay taxes and set up branch offices, so they have more costs than governments, which translate into higher interest rates than those of PFIs. The third point is to mitigate the negative effects of government lending through PFIs by limiting the government's role as a lender. This implies offering PFI loans only when private banks cannot provide loans and avoiding the crowding-out effect on private banks. A successful case in this regard is the German KfW, in which the government funding with low interest rates passes through private banks to green projects, housing, small and medium-sized enterprises (SMEs), and so on rather than via direct lending from the government to projects. Accessing funds with lower interest rates is crucial for the development of the green sector. [20] examined 133 representative utility-scale photovoltaic and onshore wind projects in Germany over the last 18 years. Their empirical results revealed that the financing conditions have strongly improved. As drivers, they identified the macroeconomic conditions (the general interest rate) and experience effects within the renewable energy finance industry.

### 3.2. Increasing the Share of NBFIs in Long-Term Investments

Just in institutions in OECD countries, investors are managing more than $100 trillion of assets. An increasing number of institutional investors have adopted strategies to mitigate climate exposure. These include negative screening (the exclusion of non-green sectors/companies from portfolios), positive screening (the proactive identification of positive climate themes), active ownership (the exercising of statutory rights to promote green standards in portfolio companies), sustainability ratings (portfolio scoring based on green criteria), and the hedging of climate risks (through portfolio allocation or the use of derivatives). These strategies reflect specific fund manager mandates and the recognition that climate risks can have a tangible impact on corporate valuations and, as a result, institutional fund performance [21]. Most recently, it has become apparent that two major pressures from investors and regulators can also boost the participation of institutional investors in green projects. On the side of investors, environmentally friendly green concerns increasingly affect people's saving and investment decisions. This trend is happening especially among the younger generations. Savers with NBFIs are asking for stricter compliance with environmental, social, and governance criteria.

On the side of regulators, some jurisdictions are debating whether financial institutions should have a mandate to integrate environmental, social, and governance issues into their investment decision policies. One example is the Financial Stability Board's creation of the Task Force on Climate-Related Financial Disclosures, which has recommended that global organizations enhance their financial disclosures related to the potential effects of climate change [21].

However, when we look at the actual activities of institutional investors, their asset allocation to direct infrastructure investment in general remains small—less than 1% for OECD pension funds—and the "green" investment components are even more limited. These issues relate to the perception that green investments do not offer a sufficiently attractive risk-adjusted financial return and the fact that institutional investors still lack knowledge and expertise, as well as investment channels [22]. Using the spillover effect of the green energy supply and reducing the risk of their investment through GCGSs will increase their eagerness to engage in green projects (Sections 3.3 and 3.4).

### 3.3. Using Spillover Tax to Increase the Rate of Return

Governments often regulate electricity tariffs and usually do not determine them based on market mechanisms. Regulated tariffs make it difficult for private investors to invest in infrastructure projects because of the low rate of return on their investments. Increasing investment incentives requires the use of the spillover effects that energy supplies originally created and refunding the spillover tax revenues to investors in energy projects. The energy supply brings factories and businesses into the electrified region. The power supply enables the construction of new residences, and the property value increases. Corporate income taxes and sales taxes also rise in areas with a new energy supply. Local or central governments collect these spillover tax revenues, and they do not usually return them to investors in energy projects. Investors only receive the user charges/electricity tariffs accruing from the electricity supply. If part of the spillover tax revenues was returned to private investors, their rate of return would increase over a prolonged period, and this could support their maintenance costs [10].

Figure 3 shows that the total rate of return on a green energy project in the first year is almost zero because of the large initial investment. In addition, the spillover impact of the energy supply on the region is very low or almost zero, as it takes time for the spillover to affect the regional output and for the tax revenue of local and central governments to emerge.

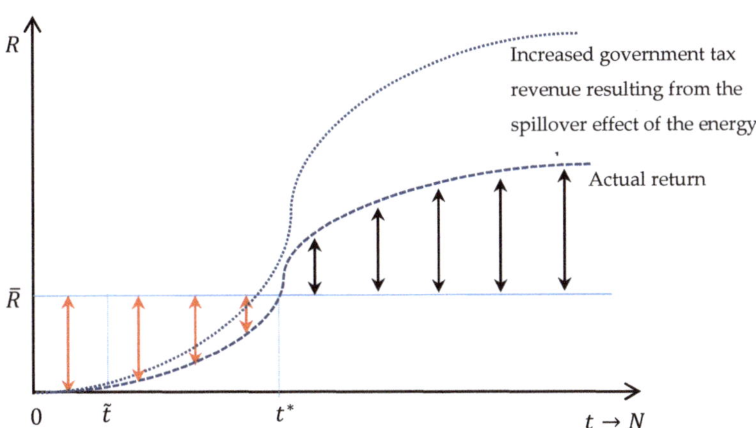


**Figure 3.** Using GCGSs and spillover tax in green projects. GCGS = green credit guarantee scheme. Source: Authors' compilation.

From $\tilde{t}$, the rate of return and the spillover tax start to increase. If the private investor relies only on user charges for the revenue of the project, the rate of return in the initial stages is very low and it takes time to increase, so the project is not viable. Hence, we suggest injecting the increase in the tax revenue generated from the spillover effect of the energy supply into the green project to secure at least the $\bar{R}$ rate of return (the benchmark rate). However, the spillover tax is not sufficient until $t^*$, so governments can issue long-term government bonds with $N$ years of maturity to support the private investors until then. In the absence of a bond market, a green credit guarantee fund/corporation could provide a supporting role for securing the $\bar{R}$ rate of return. We believe that, although the establishment of a GCGS will impose a cost and budget burden on the government in the initial years until $t^*$, the future increases in the tax revenue due to the spillover effect of the energy supply would compensate for it. In addition, thanks to the tax revenue due to the spillover effect, the government will obtain additional revenue, as Equation (1) shows:

$$\int_0^{t^*} \left(\bar{R} - Actual\ return\right) < \int_{t^*}^N \left(Actual\ return - \bar{R}\right) \tag{1}$$

### 3.4. Collecting Carbon Tax from Polluting Industries and Injecting It into Green Projects

One way to increase the rate of return on green projects is to inject the carbon tax collected from polluting industries and firms into green projects. This policy would make green projects attractive for private investors while forcing polluting firms to shift to greener technologies. Although this policy may increase the production costs and raise the price levels initially, it will increase the R&D expenditures on these technologies in the medium term because of the higher demand for green technologies—thus reducing the costs via technological progress. In this subsection, we show how imposing carbon taxation can change firms' behavior and induce them to use greener technologies.

Here, we assume that an economy with two firms (firms 1 and 2) has production functions, as in Equations (2) and (3):

$$y_t^1 = F_t^1\left(K_t^1, L_t^1\right) = \left(K^1\right)^{\alpha_1}\left(L^1\right)^{\beta_1} \tag{2}$$

$$y_t^2 = F_t^2\left(K_t^2, L_t^2\right) = \left(K^2\right)^{\alpha_2}\left(L^2\right)^{\beta_2} \tag{3}$$

where $y_t^1$ and $y_t^2$ are their total output, $K_t^1$ and $K_t^2$ denote their capital inputs, and $L_t^1$ and $L_t^2$ are their labor inputs. We consider the Cobb–Douglas production function for these firms, while $\alpha_i$ and $\beta_i$ are the elasticity of production of capital and labor, respectively. There is a constant return to scale, hence $\alpha_i + \beta_i = 1$.

Equations (4) and (5) show the profit equations for firms 1 and 2:

$$\pi_t^1 = P_t^1 y_t^1 - r_t^1 K_t^1 - w_t^1 L_t^1 \tag{4}$$

$$\pi_t^2 = P_t^2 y_t^2 - r_t^2 K_t^2 - w_t^2 L_t^2 \tag{5}$$

where $\pi_t^1$ and $\pi_t^2$ denote the profit of firms 1 and 2; $P_t^1$ and $P_t^2$ show the output prices for the products of firms 1 and 2, respectively; $r_t^1$ and $r_t^2$ denote the interest rate that firms 1 and 2 pay on their borrowed capital from the bank; and $w_t^1$ and $w_t^2$ denote the wage rates that firms 1 and 2 pay to their labor inputs.

Firms follow profit maximization behavior. To find the optimal level of $K_t^i$ that maximizes the profit of each firm, we obtain the first-order condition, as in Equations (6) and (7):

$$\frac{\partial \pi_t^1}{\partial K_t^1} = \frac{\alpha_1 y_t^1}{K_t^1} = r_t^1 \rightarrow K_t^1 = \frac{\alpha_1 y_t^1}{r_t^1} \tag{6}$$

$$\frac{\partial \pi_t^2}{\partial K_t^2} = \frac{\alpha_2 y_t^2}{K_t^2} = r_t^2 \rightarrow K_t^2 = \frac{\alpha_2 y_t^2}{r_t^2} \tag{7}$$

The allocation of the total funds in this economy is equal to the sum of the capital of both firms, as in Equation (8):

$$K_t^T = K_t^1 + K_t^2 \tag{8}$$

The objective of the government in this economy is to maximize the sum of the outputs of both firms, as in Figure 4.

In the previous case, we did not consider the level of emissions (carbon dioxide). However, in reality, each firm has carbon emissions. In the case below, we consider that each firm has not only a different level of output but also a different level of emissions, and the production functions are as in Equations (9) and (10):

$$g_1\left(y_t^1, CO_t^1\right) = f_t^1\left(K_t^1, L_t^1\right) \tag{9}$$

$$g_2\left(y_t^2, CO_t^2\right) = f_t^2\left(K_t^2, L_t^2\right) \tag{10}$$

where $CO_t^1$ and $CO_t^2$ show the carbon emissions of firm 1 and firm 2, respectively, at time t.

In addition to output (gross domestic product (GDP)) maximization, the second objective of the government is to minimize the carbon emissions of both firms (Equation (11)):

$$\text{Min } CO_t = CO_t^1 + CO_t^2 \tag{11}$$

Therefore, the ultimate objective of the government is to maximize the cumulative output (GDP) and minimize the cumulative carbon emissions, as in Equation (12):

$$W = W_1(y_t - y^*)^2 + W_2(CO_t - CO^*)^2 \tag{12}$$

where $y^*$ is the GDP in full employment (the desired GDP level) and $(y_t - y^*)$ is the GDP gap. $CO^*$ is the desired emission level, and $(CO_t - CO^*)$ is the gap between the current emission level and the desired emission level.

The chart on the left of Figure 5 shows two extreme cases. Point A reflects conventional economic theory—profit maximization without consideration of the environment. Point A is the maximization of the GDP (the sum of the outputs of firms 1 and 2). Point B just shows environmental concern and indicates the point that minimizes the sum of the emissions of firms 1 and 2. Since firm 1 has more carbon emissions, point B is where only firm 2, which has lower emissions, is producing and the output of firm 1 is zero. Points C and D show the optimal level of production for each firm, and the production function includes both the output and the carbon emission levels, as in Equations (9) and (10).

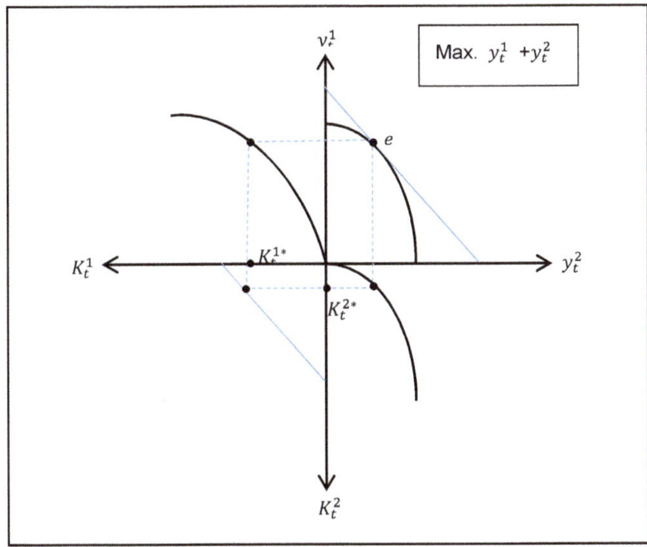


**Figure 4.** Allocation of funds to firms and output maximization. Notes: $e$ is the equilibrium point, $K_t^{1*}$ denotes the optimal capital for firm 1, and $K_t^{2*}$ denotes the optimal capital for firm 2. Source: Authors' compilation.

Next, the government charges a carbon tax, which affects the profits of polluting firms. Equations (13) and (14) show the carbon production, which is a function of the capital and labor inputs. A higher level of output will emit more carbon dioxide.

$$CO_t^1 = \varphi^1\left(K_t^1, L_t^1\right) = \left(K^1\right)^{\gamma_1}\left(L^1\right)^{\delta_1} \tag{13}$$

$$CO_t^2 = \varphi^2\left(K_t^2, L_t^2\right) = \left(K^2\right)^{\gamma_2}\left(L^2\right)^{\delta_2} \tag{14}$$

Equations (15) and (16) show the new profit equations of firms 1 and 2 after charging the carbon taxes, which will reduce their profits. We assume that the carbon tax rate is progressive, so a higher tax rate applies emission levels when industries pollute more. This is why firms 1 and 2 have different carbon tax rates, as in Equations (15) and (16) ($T_t^1$ and $T_t^2$):

$$\pi_t^1 = P_t^1 y_t^1 - r_t^1 K_t^1 - w_t^1 L_t^1 - T_t^1 CO_t^1 \tag{15}$$

$$\pi_t^2 = P_t^2 y_t^2 - r_t^2 K_t^2 - w_t^2 L_t^2 - T_t^2 CO_t^2 \tag{16}$$

As is clear from Figure 6, the optimal level of production of both firms in the case of the charging of a carbon tax is not A or B but E, which lies between them; its position depends on many factors, including the level of emissions and the tax ratio.

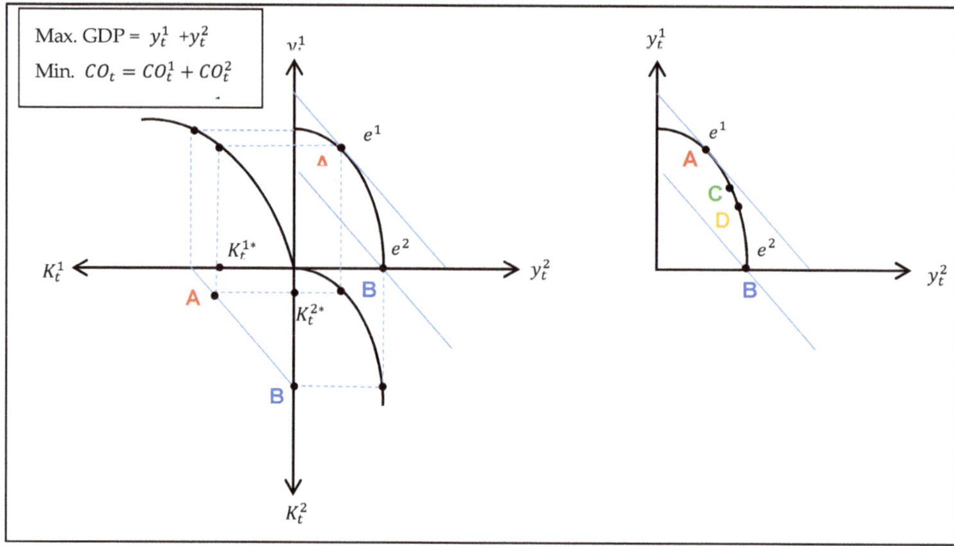


**Figure 5.** Equilibrium levels of output with different objectives. GDP = gross domestic product. Note: *e* is the equilibrium point considering different objectives, $K_t^{1*}$ denotes the optimal capital for firm 1, and $K_t^{2*}$ denotes the optimal capital for firm 2. Source: Authors' compilation.

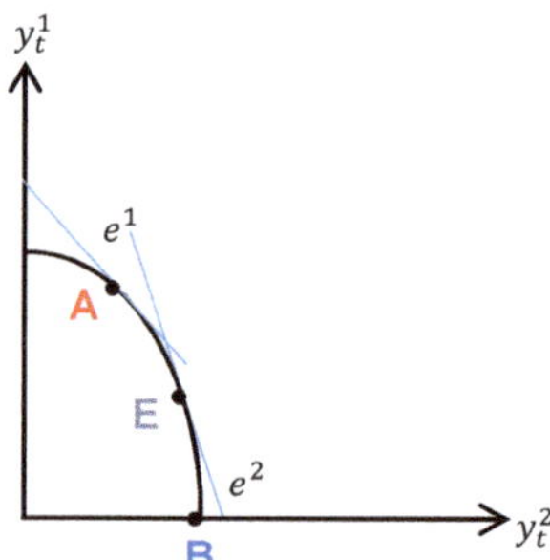


Note: Point E is the new equilibrium point after charging the carbon tax. Source: Authors' compilation.

**Figure 6.** Optimal level of output when charging carbon tax.

This carbon tax system will induce new firms to start investing in green technologies and establish their industries with green technologies. This will create a spillover effect of green industry/infrastructure in that region, as Section 3.3 explains.

We can express the spillover effects of green technologies as follows:

$$y_t^{Gi} = h\left(K_t^{Gi}, L_t^{Gi}, E_t^{Gi}\right) \tag{17}$$

where $y_t^G$ is the output of the firm that has green production (e.g., green or green energy) and $K_t^G, L_t^G,$ and $E_t^G$ are the capital, labor, and green energy production inputs, respectively.

Equation (18) depicts the spillover effect of the green energy supply. The first part in the parenthesis shows the spillover effect and the second part shows the direct effect.:

$$dy_t^{Gi} = \left( \frac{\partial h}{\partial K_t^{Gi}} \frac{\partial K_t^{Gi}}{\partial E_t^{Gi}} + \frac{\partial h}{\partial L_t^{Gi}} \frac{\partial L_t^{Gi}}{\partial E_t^{Gi}} + \frac{\partial h}{\partial E_t^{Gi}} \right) dE_t^{Gi} \tag{18}$$

The spillover effect of the green (green) energy supply will increase the tax revenue of the government from the region that has the green energy supply. In Equation (19), we assume that the government will inject 50% of the increase in government tax revenue into green projects and retain the other 50% as the government's ultimate revenue:

$$Tdy_t^{Gi} = 50\% \, (goverment) + 50\% \, (low - carbon \, projects) \tag{19}$$

As Equations (20) and (21) show, the injection of tax revenue originally generated from the spillover effect of the green energy supply, and the carbon taxes collected from polluting industries will increase the rate of return on green projects and induce private sector investment in the green sector:

$$r_t^{G1} = \frac{\alpha^{G1} y_t^{G1}}{K_t^{G1}} + T_t^1 CO_t^1 + a\left( Tdy_t^G \right) \tag{20}$$

$$r_t^{G2} = \frac{\alpha^{G2} y_t^{G2}}{K_t^{G2}} + T_t^2 CO_t^2 + b\left( Tdy_t^G \right) \tag{21}$$

where $r_t^{Gi}$ is the rate of return on the green (green) project and $\frac{\alpha^{Gi} y_t^{Gi}}{K_t^{Gi}}$ is the initial rate of return on the green project, which is very low. Relying on this alone, the project will not be feasible or attractive to private investors. $T_t^i CO_t^i$ is the carbon tax that the government charges on $i$ polluting projects (firms) and then injects into green projects, while $Tdy_t^G$ is the tax revenue of the government from the spillover effect of the green energy supply. Percentage $a$ of this increase in tax revenue will return to project 1 and percentage $b$ will return to project 2 to increase their rates of return. We assumed earlier that $a + b = 0.5$, which means that the government will inject 50% of the increase in tax revenue that the spillover effect of the green energy supply caused into green projects and take the other half as its final tax revenue. As the implementation of this scheme shows, the rate of return on green projects will increase; on the other hand, the carbon taxation will force the polluting industries to shift to cleaner industries and green technologies.

### 3.5. Development of Green Credit Guarantee Schemes to Reduce the Credit Risk

Credit guarantee corporations (CGCs) are public institutions supporting sectors that lack access to finance (SMEs and start-ups). CGCs serve as a guarantor and cover the risk of lending to risky sectors, such as SMEs. Japan initiated CGCs in the 1930s, and presently, many developing and developed economies are using CGCs, especially in the SME sector. The green CGCs that [7] initially proposed improve the creditworthiness of green (green) projects that lack physical collateral and have a weak credit rating.

Figure 7 shows three participants in GCGs—banks, green projects, and green CGCs. Green CGCs will increase the loan supply to green projects. The green credit guarantee for green projects will reduce the asymmetry of information and decrease the expected default losses, because the CGC (government) guarantees a portion of the loan default, so banks will want to lend money to guaranteed green projects. An investor in a green project applies for a green credit guarantee when submitting the loan application. Then, a green CGC performs a creditworthiness evaluation of the project and the project owner (individual or corporate). Not all green projects are eligible to receive a guarantee—this depends on the borrower's credit score and an evaluation of the probability of success of the green project. Depending on the results of the creditworthiness assessment of a green credit guarantee corporation (GCGC), the investor pays a guarantee fee or premium. The fee depends on the project risk rating and the borrower's credit score. Using the same guarantee fee for all borrowers would create a moral hazard [23]. Banks also need to apply to a GCGC for a green credit guarantee, which issues the relevant certificate. Next, the bank disburses the loan to the green project (borrower), and the borrower starts to pay the loan instalments. In the case of default, the GCGC compensates a portion of the loan amount—the credit guarantee ratio—and subrogates the bank. An adjustment of the optimal credit guarantee ratio is necessary to avoid moral hazard [23]. This means that healthy banks that manage their nonperforming loans and have higher creditworthiness should receive a higher credit guarantee ratio from the government, while unsound banks need a lower guarantee and very risky banks do not obtain a guarantee. The regulator for a green finance's selection of an optimal guarantee ratio also creates an incentive for financial institutions to improve their creditworthiness to receive a higher guarantee ratio.

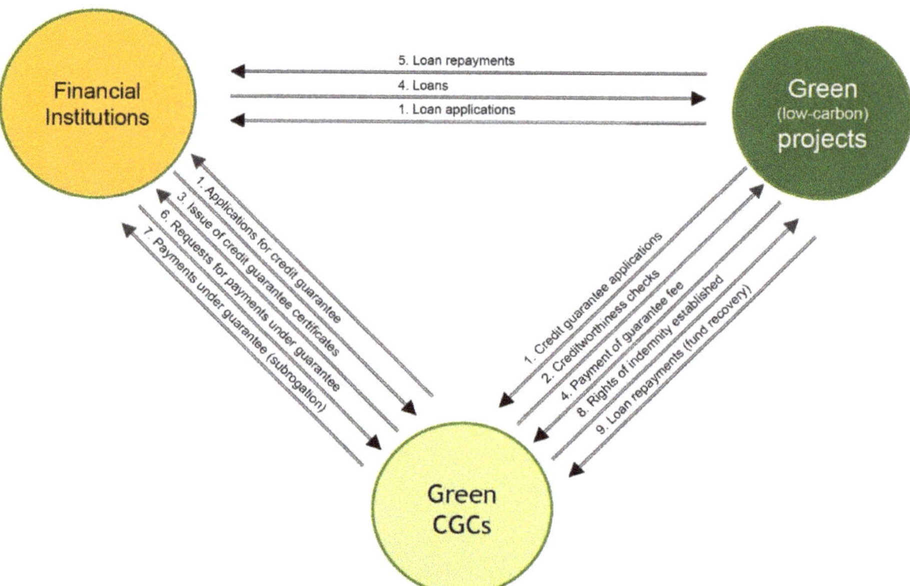


**Figure 7.** GCGS for the management of green credit risk. CGC = credit guarantee corporation and GCGS = green credit guarantee scheme. Source: The authors following [7]. Note: The numbering shows the sequence of the workflow.

### 3.6. Addressing Green Investment Risks via De-Risking

Various financial and nonfinancial risks are associated with green projects, as Section 2.2 explained. Since risks have an impact on access to credit, it is very important to mitigate them. De-risking is a potentially powerful policy option to redirect financial flows to green investments in two ways: financial and policy. Financial de-risking lowers the perceived risks and requires returns and, thus, reduces the investment costs. Financial de-risking can occur as a result of transferring a large portion of the risk to another party; for example, insurance risks of governments or development banks, loan guarantees, or support policies like feed-in-tariffs, subsidies, or low-carbon promotion tools, such as carbon pricing. These measures are not necessarily limited to countries receiving investments, and multilateral organizations, such as development banks, employ them equally. The UNFCCC Green

Climate Fund (GCF) with a target value of US $100 billion/yr by 2020 is expected to play an important role in financial de-risking and delivering the required investment levels for the large-scale renewable energy projects in developing countries in line with the Paris Agreement [24].

Unfortunately, in many developing countries, financial de-risking to date has mainly related to coal, while only a small fraction has focused on green alternatives [25]. The green credit guarantee scheme that this paper proposes is a tool for reducing the financial risk for private investors, as the GCGS (government) covers a portion of the risk. [26] discussed the possibility of financial de-risking by increasing the transparency and improving the laws and regulations. De-risking can increase the effectiveness and efficiency of policies aiming to attract green investments [27]. Policy de-risking instruments are including programs, policies, and regulations that reduce the risks of private sector investments in green and low-carbon projects that are typically implemented by the governments. In utility-scale renewable energy, these policies and instruments include auction processes and reforms to ensure financially sound utilities (cost-recovery). In energy efficiency, the de-risking instruments include the design, implementation, and enforcement of various minimum energy efficient standards, such as green building codes or in lighting and appliances [28].

*3.7. Summary of Tools and Instruments for Green Investments*

Table 1 presents the tools and instruments that this section outlined, as well as suggestions for reducing the risk of green projects, raising the rate of return, increasing the capacity of the investors and other stakeholders, and facilitating access to finance and investment.

**Table 1.** Tools and instruments for green investments.

| Goal | Functions | Tools and Instruments |
| --- | --- | --- |
| Facilitate access to finance/investment | Providing long-term finance/capital Facilitating access to private finance/capital | Equity investment International climate funds Public–private partnerships Institutional investors (pension funds, insurance companies, etc.) |
| Reduce risk | Risk-sharing Credit enhancement mechanism | Green credit guarantee scheme Financial de-risking Policy de-risking Structured finance Public–private partnership |
| Raise the rate of return | Making green projects feasible | Utilizing the spillover effect in the form of tax refunds to private investors |
| Increase capacity | Aiding project development Reducing project risks | Technical assistance Capacity building Information tools (e.g., energy certificate tracking, etc.) |

Sources: Authors based on: [7,10,18,27,29].

## 4. Example of Green Finance Management

In this section, we provide an example of the development of an environmental project. The objective is to show how the implementation of a GCGS, which this paper proposes, can reduce the risk of investment in projects. One of the major challenges that cities, mainly mega cities, in developing countries face is the environmental impact of generating huge amounts of solid waste. This issue is severe in Asia, as the Asian urban population increases by 44 million people every year. By 2050, half of the world's population will live in Asia and the Pacific countries [30]. The People's Republic of China generates 150 million tons of waste annually, and globally, it is the largest generator of municipal solid waste; India ranks second in the world [31]. One of the major obstacles to the development of solid waste management projects is the lack of a municipal budget and the low interest of private investors in this sector because of the low rate of return. Many municipalities in large Asian cities allocate more than 20% of the municipal budget to solid waste management.

In this section, we suggest practical funding schemes for fixed capital and working capital to incentivize solid waste management for private investors. These schemes are applicable to green and environmental projects, including waste-to-energy projects, which face similar barriers.

### 4.1. CGSs for Providing Fixed Capital

As mentioned earlier, many countries have used CGSs in various forms over the decades to increase the flow of funds to targeted sectors and segments of the economy that have difficulties accessing finance, including SMEs. A CGS absorbs the risk, and the guarantee that it provides acts as collateral. Therefore, by reducing the level of risk, banks are more willing to lend to borrowers. In addition, as the CGS acts as a guarantor, it needs to assess the creditworthiness of the borrower by monitoring the status of the project or borrower to improve the quality of lending.

Figure 8 shows that a CGS has three players. The first is the borrower, which can be a green project. In this example, it is a waste management project seeking finance. When borrowers approach a bank, they often refuse to lend because of the asymmetry of information and a lack of collateral. The second player is the lender, which is a financial institution (bank). The third player is the guarantor, which is the CGC, which usually the government runs, providing a full or partial guarantee. The CGS has a cost, so the borrower needs to pay a credit premium to the CGC. However, in the early stages of development of a CGC, it needs enough capital to cover the risks, so it requires government support. After some years, it can become financially sustainable. For this example, this scheme is especially applicable to sectors that require large fixed capital, waste-to-energy projects, or green projects. In this example, as is clear from Figure 8, the central government or the municipality funds the CGC. After assessing the credit history of the borrower (individual or corporate) and evaluating the feasibility of the project, the CGC agrees to provide this project with a guarantee. Then, it guarantees a certain amount (e.g., 80%), and the borrower proceeds to approach a bank to apply for a loan. When the bank sees that about 80% of the loan amount has a guarantee, it is eager to lend to this project. For the remaining 20%, the bank may ask for collateral. As this is a small amount, it is easier for the project owner to provide it.

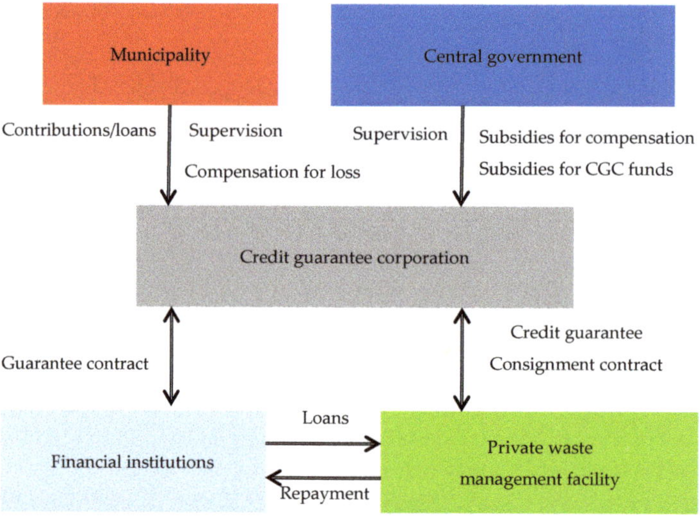


**Figure 8.** Establishment of a CGS for reducing the supply–demand gap of finance in waste management projects. CGC = credit guarantee corporation. Source: Authors.

### 4.2. Establishment of Community-Based Funds for Providing Working Capital

In addition to the fixed capital, the second major challenge facing waste management and many green projects (e.g., waste-to-energy projects) is the difficulty in funding their working capital. Therefore, it is important to design a scheme that can adapt to the socioeconomic environment in Asia to help the private sector fund the working capital of these projects.

In many large cities, landfills occupy large tracts of land, and the space is limited. By establishing sorting, recycling, composting, and waste-to-energy facilities, the freed landfills could work better to support other, more beneficial purposes for generating user charges, rent, and making revenue from the sale of electricity generated from waste for municipalities or private investors, which could be a sustainable source of funding for working capital.

Figure 9 illustrates a waste management trust fund, a type of community-based funding or hometown investment trust fund for providing working capital for these projects.

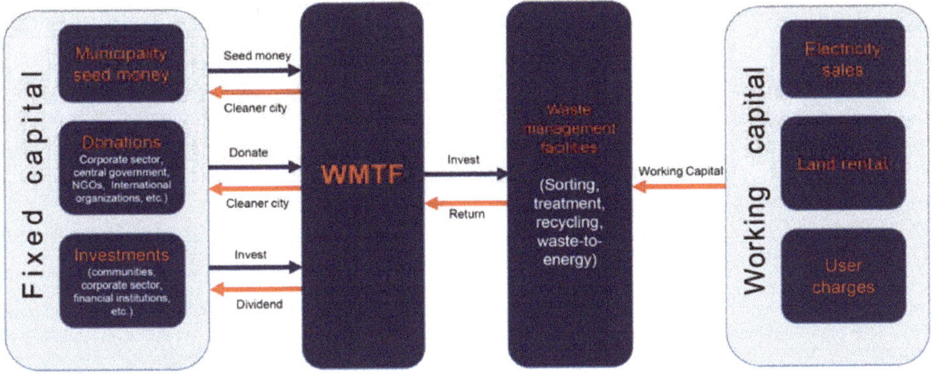


**Figure 9.** Establishment of community-based funds for waste management projects. NGO = non-governmental organization and WMTF = waste management trust fund. Source: Authors.

In Japan, the development of hometown investment trust funds occurred mainly after the Fukushima nuclear power disaster in March 2011, when the government shut down the nuclear power plant as it was unsafe. Many people, especially in the affected region of Fukushima, showed an interest in renewable energy, such as solar and wind power, instead of nuclear power [32]. However, green projects carry a high risk, and most banks are reluctant to lend to them. Therefore, local people started to collect small amounts of money ($100–$5000) from the region through a local fund to build a solar power plant and wind power generator. They planned to establish a green energy plant, generate electricity, use the power that they generated, sell the excess to the power company, and make some profit. This was the reason for the establishment of the hometown investment trust funds, with the basic idea of connecting the investors to the projects in their hometown. Individual investors can invest a small amount of money through the internet in the projects toward which they are sympathetic. The investors not only seek a profit but also are well-disposed toward the project or the region. If these projects are successful, the banks will also be interested in lending to them. We believe that hometown investment trust funds are suitable for waste management projects, and the waste management trust fund is a new type of hometown investment trust fund targeting waste management projects. Hometown investment trust funds have expanded from Japan to Cambodia, Malaysia, Vietnam, and Mongolia [4]. Similar funds apply to green projects in developing countries, especially in regions where communities are integrated and trust exists among their members. These funds will help high-risk sectors, including the green sector, to grow.

As Figure 9 shows, waste management trust funds are project-oriented and designed for running waste management projects. The working capital comes from three sources: (1) rents from the freed landfills, (2) user charges from waste generators (the facility can burn the waste of other regions or other countries and receive the charges and fees—as occurs in many European cities), and (3) the sale of the electricity generated by the waste. In addition, the waste management trust funds receive donations from locals, as well as central governments from their municipal and urban management budgets in the form of seed money for increasing the rate of return of the projects.

## 5. Conclusions and Policy Recommendations

The lack of long-term financing, the existence of various risks, the low rate of return, and market actors' lack of capacity are major challenges for developing green projects. Using the spillover effects on green energy projects would increase the rate of return of these projects. PFIs can use both traditional and innovative approaches to link green projects with finance by enhancing their access to capital, facilitating risk reduction and sharing, improving the capacity of market actors, and shaping broader market practices and conditions. PFIs should avoid the negative effects of government lending (crowding out of private investment) by engaging in long-term lending at stable rates and only lending where private banks cannot lend. The green R&D sector is among the sectors for which PFI lending is suitable. GCGSs will reduce the asymmetry of information and decrease the expected default losses, thereby covering part of the risk and unlocking financial institutions' private investments and lending to green projects. GCGCs and PFIs can play important roles in credit enhancement and reduce the risks and improve the capacity for the adoption of a green economy. To prevent moral hazard, the guarantee ratio of GCGCs needs to be variable, depending on the creditworthiness of borrowers and financial institutions, and not fixed. To achieve a sustainable financing scheme of banks and NBFIs, it is important for the government to consider leverage, transparency, and specific results in the financial scheme (the case of a waste management project).

**Author Contributions:** F.T.-H. contributed to this research article in the formal analysis, investigation, resources, data curation, and writing of the original draft. N.Y. contributed to this research article for conceptualization and methodology. All authors have read and agreed to the published version of the manuscript.

**Funding:** This research was supported by JSPS Kakenhi (2019-2020) Grant-in-Aid for Young Scientists No. 19K13742 and Grant-in-Aid for Excellent Young Researcher of the Ministry of Education of Japan (MEXT).

**Acknowledgments:** This research paper is output of the research project titled: "Unlocking the potentials of Private Sector Financing for Accelerated Low-Carbon Energy Transition (Phase II)" of the Economic Research Institute for ASEAN and East Asia (ERIA) which done by the authors. Authors are grateful to ERIA and the Asian Development Bank Institute (ADBI) for supporting this research. The paper benefited from the precious comments received from the anonymous referees of *Energies* and authors are grateful to them.

**Conflicts of Interest:** The authors declare no conflicts of interest.

## References

1. Sachs, J.; Woo, W.T.; Yoshino, N.; Taghizadeh-Hesary, F. Importance of green finance for achieving sustainable development goals and energy security. In *Handbook of Green Finance: Energy Security and Sustainable Development*; Sachs, J., Woo, W.T., Yoshino, N., Taghizadeh-Hesary, F., Eds.; Springer: Singapore, 2019; pp. 3–12.

2. Donovan, C.W. *Renewable Energy Finance: Powering the Future*; World Scientific Books; World Scientific Publishing Co. Pte. Ltd.: Singapore, 2015; p. 1030.

3. Asian Development Bank (ADB). *Meeting Asia's Infrastructure Needs*; Asian Development Bank: Manila, Philippines, 2017.

4. Yoshino, N.; Taghizadeh-Hesary, F. Alternatives to private finance: Role of fiscal policy reforms and energy taxation in development of renewable energy projects. In *Financing for Low-Carbon Energy Transition: Unlocking the Potential of Private Capital*; Anbumozhi, V., Kalirajan, K., Kimura, F., Eds.; Springer: Tokyo, Japan, 2018; pp. 335–337.

5. Schub, J. Green banks: Growing clean energy markets by leveraging private investment with public financing. *J. Struct. Financ.* **2015**, *21*, 26–35. [CrossRef]

6.	Gabbi, G.; Ticci, E.; Vercelli, A.; Hall, C. Financialization, economy, society and sustainable development: A European Union sustainable banking network. *European Policy Brief.* 2016. Available online: https://ec.europa.eu/research/social-sciences/pdf/policy_briefs/fessud_policy_brief_10.pdf (accessed on 10 October 2019).

7.	Taghizadeh-Hesary, F.; Yoshino, N. The way to induce private participation in green finance and investment. *Financ. Res. Lett.* **2019**, *31*, 98–103. [CrossRef]

8.	Geddes, A.; Schmidt, T.S.; Steffen, B. The multiple roles of state investment banks in low-carbon energy finance: An analysis of Australia, the UK and Germany. *Energy Policy* **2018**, *115*, 158–170. [CrossRef]

9.	Bank of Japan. Asian Financial Markets—20 Years since the Asian Financial Crisis, and Prospects for the Next 20 Years. In Proceedings of the Keynote Speech of Governor Haruhiko Kuroda at the 2017 Annual General Meeting of the Asia Securities Forum, Tokyo, Japan, 28 November 2017.

10.	Yoshino, N.; Taghizadeh-Hesary, F.; Nakahigashi, M. Modelling the social funding and spill-over tax for addressing the green energy financing gap. *Econ. Model.* **2019**, *77*, 34–41. [CrossRef]

11.	Hwang, B.-G.; Shan, M.; Phua, H.; Chi, S. An exploratory analysis of risks in green residential building construction projects: The case of Singapore. *Sustainability* **2017**, *9*, 1116. [CrossRef]

12.	Clean Energy Manufacturing Analysis Center. *Benchmarks of Global Clean Energy Manufacturing*; Clean Energy Manufacturing Analysis Center: Washington, DC, USA, 2017.

13.	Leonard, W.A. Clean is the new green: Clean energy finance and deployment through green banks. *Yale Law Policy Rev.* **2014**, *3*, 197–229.

14.	Merrill, L.; Bridle, R.; Klimscheffskij, M.; Tommila, P.; Lontoh, L.; Sharma, S.; Gerasimchuk, I. *Making the Switch: From Fossil Fuel Subsidies to Sustainable Energy*; Nordic Council of Ministers: Copenhagen, Denmark, 2017.

15.	Organisation for Economic Co-operation and Development (OECD). *OECD Companion to the Inventory of Support Measures for Fossil Fuels 2015*; OECD: Paris, France, 2015.

16.	Coady, D.; Parry, I.; Sears, L.; Shang, B. how large are global fossil fuel subsidies? *World Dev.* **2017**, *91*, 11–27. [CrossRef]

17.	United Nations Environment Programme (UNEP). 2016. Available online: http://unepinquiry.org/wp-content/uploads/2016/09/Synthesis_Report_Full_EN.pdf (accessed on 28 December 2019).

18.	Cochran, I.; Hubert, R.; Marchal, V.; Youngman, R. *Public Financial Institutions and the Low-Carbon Transition: Five Case Studies on Low-Carbon Infrastructure and Project Investment*; OECD Environment Working Papers; OECD: Paris, France, 2014.

19.	Japan Bank for International Cooperation (JBIC). *Japan's Role towards Global Climate Change Resolution*; JBIC Today, November; Japan Bank for International Cooperation: Tokyo, Japan, 2016.

20.	Egli, F.; Steffen, B.; Schmidt, T. A dynamic analysis of financing conditions for renewable energy technologies. *Nat. Energy* **2018**, *3*, 1084–1092. [CrossRef]

21.	Gianfrate, G.; Lorenzato, G. Stimulating non-bank financial institutions' participation in green investments. In *Handbook of Green Finance: Energy Security and Sustainable Development*; Sachs, J., Woo, W.T., Yoshino, N., Taghizadeh-Hesary, F., Eds.; Springer: Singapore, 2019; pp. 213–236.

22.	Kaminker, C.; Kawanishi, O.; Stewart, F.; Caldecott, B.; Howarth, N. *Institutional Investors and Green Infrastructure Investments: Selected Case Studies*; OECD Working Papers on Finance, Insurance and Private Pensions; OECD: Paris, France, 2013.

23.	Yoshino, N. Taghizadeh-Hesary, F. Optimal credit guarantee ratio for small and medium-sized enterprises' financing: Evidence from Asia. *Econ. Anal. Policy* **2019**, *62*, 342–356. [CrossRef]

24.	Sweerts, B.; Dalla Longa, F.; van der Zwaan, B. Financial de-risking to unlock Africa's renewable energy potential. *Renew. Sustain. Energy Rev.* **2019**, *102*, 75–82. [CrossRef]

25.	Chen, H. Why Are G20 Governments Financing Coal over Renewables? NRDC Expert Blog (Blog). 2017. Available online: https://www.nrdc.org/experts/han-chen/why-are-g20-governments-financing-coal-over-renewables (accessed on 12 October 2019).

26.	Steckel, C.J.; Jakob, M. The role of financing cost and de-risking strategies for clean energy investment. *Int. Econ.* **2018**, *155*, 19–28. [CrossRef]

27.	Schmidt, T.S. Low-carbon investment risks and de-risking. *Nat. Clim. Chang.* **2014**, *4*, 237–239. [CrossRef]

28.	United Nations. 2018. Available online: https://sustainabledevelopment.un.org/content/documents/17549PB_5_Draft.pdf (accessed on 27 December 2019).

29. UNCTAD. *The Continuing Relevance of Development Banks*; UNCTAD Policy Brief, No. 4; United Nations Conference on Trade and Development: Geneva, Switzerland, 2012.
30. ADB (Asian Development Bank). *Key Indicators for Asia and the Pacific 2011*; Asian Development Bank: Manila, Philippines, 2011.
31. ADB (Asian Development Bank). ADB Supports Clean Waste-to-Energy Project in the PRC. 2009. Available online: https://www.adb.org/news/adb-supports-clean-waste-energy-project-prc (accessed on 10 January 2019).
32. Taghizadeh-Hesary, F.; Yoshino, N.; Rasoulinezhad, E. Impact of Fukushima nuclear disaster on the oil-consuming sectors of Japan. *J. Comp. Asian Dev.* **2017**, *16*, 113–134. [CrossRef]

*Article*

# Linking Environmental Policy Integration and the Water-Energy-Land-(Food-)Nexus: A Review of the European Union's Energy, Water, and Agricultural Policies

Sandra Venghaus *, Carolin Märker, Sophia Dieken and Florian Siekmann

Institute of Energy and Climate Research: Systems Analysis and Technology Evaluation (IEK-STE),
Forschungszentrum Jülich, Wilhelm-Johnen-Straße, 52428 Jülich, Germany; c.maerker@fz-juelich.de (C.M.);
s.dieken@fz-juelich.de (S.D.); f.siekmann@fz-juelich.de (F.S.)
* Correspondence: s.venghaus@fz-juelich.de; Tel.: +49-2461-61-6541

Received: 17 October 2019; Accepted: 21 November 2019; Published: 22 November 2019

**Abstract:** Against the backdrop of climate and environmental pressures, as well as limited resource availability and trade conflicts, devising policies for energy and the use of natural resources in general becomes exceedingly complex. Moreover, policies are required to account for interrelations between individual resources and between different sectors and policy fields, but implementation often lacks. To evaluate the current state of integrated policy design in the EU, a review of European energy, water, and agricultural policies was conducted. Using a qualitative comparative research approach, the objective was to identify and explain the differing degrees and variations in policy integration among them. To this aim, the concepts "Environmental Policy Integration" and "Water-Energy-Land Nexus" were jointly applied as analytical frameworks. The analysis revealed that currently, different authorities are endowed with largely sectoral mandates. Accordingly, the respective sectoral policy sets are historically grown based on differing sets of formal and informal rules and processes, thus making policy integration among the sectors, let alone within the nexus, a highly challenging task.

**Keywords:** water-energy-land nexus; environmental policy integration; policy analysis; European Union

---

## 1. Introduction

A multitude of environmental, economic, political, and social challenges demonstrate that current practices of natural resource management are unsustainable. In this context, current research shows that sectoral interdependencies among resources are increasingly important [1–6] and they influence each other through complex feedback [7]. With Agenda 2030, the international community has confirmed its commitment to adopting measures supporting sustainable development. Reflecting on sectoral interdependencies, the 17 Sustainable Development Goals (SDGs) are interrelated, and associated actions and policies require integrated and holistic approaches. Already since the 1990s, integrative policy concepts such as Environmental Policy Integration (EPI) and, more recently, the nexus around the conflicting interdependencies among water, energy, and land (WEL Nexus) have entered public and political agendas. The UN Global Sustainability Panel in its 2012 report highlighted the importance of the water, energy, and food nexus [8], and reinforced this focus in the 2014 UN Global Sustainability Development Report by attributing a special section to 'the climate-land-energy-water-development nexus' [8]. In 2011, the World Economic Forum in its 'Global Risks Report' identified the water-energy-food (WEF) nexus as one of its three cross-cutting global risks [9], and the World Business Council for Sustainable Development (WBCSD) under its 'Vision 2050' has centered several activities around the water-energy-land-food nexus [10]. The 'World Energy Outlook 2012' explicitly addresses the nexus among water and energy [11], and the U.S. National

Intelligence Council (US NIC) in its report 'Global Trends 2030' identified the nexus among water, energy, and food as one of four global megatrends [12]. With a focus on developing countries, the European Commission in 2012 published a report entitled 'Confronting Scarcity: Managing Water, Energy and Land for Inclusive and Sustainable Growth' [13]. On a global scale, providing all people sufficient access to clean water, adequate nutrition and a clean, reliable source of energy has been the primary objective. Within Europe, however, sufficient access to these resources has long been achieved and the focus instead has shifted to their effective and sustainable management:

"The effective management of water, energy and land can and should contribute to economic growth. We draw on water, energy and land for a broad range of productive activities, from the production of food and fiber to the generation of power that moves our society. Water, energy and land are key inputs to the economic system, and thus play a crucial role in creating wealth [13]."

Effective management first and foremost builds on political processes and governance structures. This holds especially true for the EU with its large territory and population, and its complex multi-level governance system. However, to this date, only limited research addresses integrated natural resources governance on the EU level. With regard to EPI, there is considerable literature on EU governance, but the concept is almost exclusively applied to individual sectors and policy fields such as agriculture [14] or climate and energy policies [15]. Also, according to the literature, the concept lacks implementation, since EU efforts on including environmental concerns have been limited primarily to a strategic, discursive level [14]. In turn, with regard to the WEF Nexus concept, existing literature reveals that in the EU, the concept is often used to address links between only two sectors, or to analyze impacts of one sector on another. Examples include the nexus between energy consumption and economic growth [16,17], the electricity-fuel nexus [18], or the climate-energy security nexus [19,20]. More closely in line with the nexus approach as it was internationally promoted, Karabulut et al. [21] and Ziv, Watson, Young, Howard, Larcom, and Tanentzap [3] apply the water-energy-food nexus approach to case studies in the Danube river basin and the UK, whereas Siciliano et al. [22] analyze the relations between European large-scale farmland investments and the nexus. However, such comprehensive applications of the WEF Nexus concept seldom include governance aspects, e.g., [23]. This brief overview of studies demonstrates that, so far, little research has been conducted on the degree of integration of natural resource governance in the EU. This research gap needs to be addressed in order to overcome the prevailing silo-thinking, especially on higher governance levels such as the EU.

The aim of this paper is thus to review, in how far the European Union and its policy strategies are currently designed to pursue the objective of an integrated perspective on natural resources governance. It further investigates the challenges and barriers to a more coherent policy design. Therefore, current EU policies of high relevance for the water, energy, and land sectors are identified. The analysis of these policies' degree of integration is based on the literature on EPI differentiating between vertical and horizontal policy integration and the WEL Nexus literature on integrated governance. The paper provides a review of the current state of policy integration, which is assessed against the background of the two conceptual frameworks, EPI and WEL Nexus. The remainder of this paper is structured as follows. In Section 2, we further introduce the conceptual frameworks. In Section 3, the results of the qualitative in-depth analysis of implemented policy integration in current, nexus-relevant EU policies are presented. Section 4 includes a discussion of research outcomes and offers respective policy implications. The paper then concludes with a section on research implications, limitations, as well as further research needs (Section 5).

## 2. Conceptual Framework and Methodology

This paper offers a review of EU policies relevant to energy, water, and land with regard to their degree of integration. Already in form of the EU's predecessor, the European Coal and Steel Community, governing natural resources has been a crucial field of common action. However, how individual resources are regulated and managed differs widely. Despite its historical role in European integration, energy policy has long been a strictly national prerogative. Article 194 of the Treaty on the Functioning of the European Union affirms the 'Member State's right to determine the conditions for exploiting its energy resources, its choice between different energy sources and the general structure of its energy supply' [24]. However, with the Energy Union in 2015, the EU has moved towards a more communitized energy policy, especially with regard to supply security, the internal energy market, and decarbonization efforts. Similarly, the provision of water is a national task, but against the background of environmental pressures, the EU has enforced several regulatory frameworks on water quality and the treatment of wastewater as part of its environmental policy and strives for integrating water concerns into related policy fields. Issues of land management are primarily regulated by either EU environmental policies or agricultural policies. The Common Agricultural Policy (CAP) is a core concern of the EU, managing regulations and support for farmers and rural areas since 1962. It is among the most contested and financially significant EU policy fields. In sum, EU management of natural resources, specifically energy, water, and land is very complex, and providing more integrated policies constitutes a significant governance challenge.

As mentioned in the introduction, there are primarily two concepts of relevance for integrating natural resources policies in the EU: EPI and the WEL Nexus. As analytical frameworks they provide the foil against which to reflect EU policies in order to identify and explain variations in integration. In this section, these concepts and the review approach will be introduced.

The review primarily draws on policy documents that can directly contribute to or are of relevance for achieving policy integration among the energy, water, and land policy sectors in Europe. For this purpose, the major political frameworks from the energy, water, environmental, and agricultural policy fields, as well as overarching policy documents dealing with sectoral interdependencies and general sustainability issues, were considered and qualitatively analyzed. For the identification of the relevant documents, keywords were deductively derived based on the available scientific literature [25,26] and searched for on the homepage of the European Union [27]. A list of the identified policy documents (N = 41) is provided as Supplementary Materials Table S1.

The policy documents were supplemented by relevant secondary literature. To this aim, a general web search was conducted for all specific policy measures of relevance (e.g., the Water Framework Directive (WFD) & CAP). This search produced both non-governmental and scientific reviews of different policies. The identified documents thus included original policy documents as well as policy reviews, totaling over 100 documents of relevance. These documents were then reviewed with regard to the assumptions and requirements about policy integration provided by the concepts EPI and WEL Nexus.

Before introducing the concepts, the terminology of policy integration is specified. The term policy integration, both in policy documents as well as in much of the scientific literature, is often used interchangeably with policy coherence. To carefully frame our research approach, in the remainder of this paper, we will differentiate between the two concepts based on Nilsson et al.'s [28] definition, according to which policy coherence refers to policy outputs whereas policy integration refers to policy processes and the institutional setting. Analytically, policy coherence thus results from an integrated policy framing. In actual policy-making, however, they can hardly be considered separately.

## 2.1. EPI Concept

Both the EPI and the WEL Nexus concepts originate in the sustainable development discourse. The concept of EPI emerged in the 1990s, following the 1987 Brundtland Report, as a means to harmonize economic, social, and environmental policies [29]. The primary objective of EPI is to integrate environmental concerns into non-environmental sectors. EPI has been politically backed internationally, but especially by the EU [29]. However, this principle has seen little implementation since, and thus, research has turned towards investigating how governance structures are formed [29].

With respect to governance structures, EPI is often divided into a horizontal dimension (defined as the 'extent to which a central authority has developed a comprehensive cross-sectoral strategy') and a vertical dimension (defined as the 'degree to which sectoral governance structures have been "greened"') [30]. The vertical perspective primarily addresses procedures and interactions that have been established to promote environmental objectives within policy domains, such as sector-specific environmental targets [31]. Research, however, has shown that such sectoral strategies most likely will not automatically converge in order to best meet the overall objectives set in a comprehensive strategy [30]. Instead, successful EPI will require an even balance between both dimensions as well as the political will to achieve them [31–33]. Thus, the EPI concept accounts for environmental objectives within sectoral policies and a cross-sectoral strategy.

EPI has a long history in European policy design. The first legal basis was implemented as part of the Single European Act in 1986, demanding the integration of environmental objectives into other sector policies. This objective was further reinforced by the Maastricht and Lisbon treaties [34]. Although these treaties lay out a legal basis for the formal consideration of environmental objectives, their relevance has remained far inferior to sectoral objectives. Merely providing environmental concerns a legal basis, hence, does not guarantee an effective implementation [32]. Nonetheless, increasing consideration of environmental objectives becomes apparent, e.g., when looking at the development of the EU's Environment Action Programs [34].

A significant stream of EPI research on energy, water, and land policies has focused on its implementation in the CAP, e.g., [14,35]. In the literature, the food and agricultural policy is described as 'an extreme case of the ensuing compartmentalized and "exceptionalist" policy-making, where sector-specific policy ideas and institutions provide privileged access for sectoral interest groups and generate policies that benefit their members' [35]. Correlate finding of the research is that, despite a noticeable increase in the environmental discourse or rhetoric, the comparatively low priority of environmental policy and the 'closed agricultural policy network' make it difficult to 'move from political commitment to genuine EPI' [14]. Instead, environmental concerns seem to fulfill mainly a strategic role in legitimizing existing practices [36]. Concluding this research, Alons [14] thus recently stated: 'Environmental objectives have become a variable in the agricultural policy-making equation, but its coefficient remains small.' The example of EPI in agricultural policy is of direct relevance to the WEL Nexus approach, since environmental concerns in the agricultural sector very often directly relate to water concerns. Generally, however, EPI is actively promoted by actors in the broader environmental sector, whereas its implementation in other relevant policy sectors so far has had only limited impact [29]. So, while there is considerable literature on EPI, the concept has not been translated into the integration of natural resources governance in the EU.

### 2.2. WEL Nexus Concept

The WEL Nexus is a comparatively recent concept, which was formulated in the context of the Bonn 2011 Nexus Conference as a means to optimize the management of the interdependent resources with the twin objectives of achieving a sustainable, fair resource allocation and economic growth [2,37]. As such, it was introduced into EU thinking on sustainable development with the 2012 report "Confronting Scarcity: Managing water, energy and land for inclusive and sustainable growth" [13]. Although 'the nexus' has been extensively discussed on many levels during the past years, a consistent definition, or even terminology of the concept, does not exist. Most publications refer to the 'water, energy, food' (or 'food, energy, water')-security nexus (WEF/FEW Nexus) [2,5,10,38–43]. Others instead refer to the nexus among 'water, energy and land' (WEL Nexus) [13,44,45], or a combination of the two [46,47]. Also, some publications specifically investigate certain relations or sectors, such as the nexus between seafood and hydropower [48]. Others again explicitly include a fourth sector—namely climate—to the nexus [11,45,49,50]. Depending on specific perspectives, several publications address merely bilateral relationships among any two of the nexus sectors—most notably, the water-energy relationship, e.g., [1,11,51–64]. In this paper, the WEL Nexus perspective was chosen, since it reflects a holistic approach, including all three sectors and a focus on natural resources instead of exclusively human needs (i.e., food), while subsuming overarching climate aspects under the three sectors, respectively (e.g., emissions originating from agricultural practices or power generation).

Despite these variations in nexus terminologies and perspectives, their common ground is the realization that the sectoral fragmentation among water, energy, and land resources constitutes a crucial challenge in achieving sustainable development: 'silo-thinking' in managing these sectors leads to unintended side-effects and conflicts [65]. The aim is therefore to identify potential trade-offs as well as synergies, enabling the design of optimized, cross-sectoral resource management strategies [2,39]. In contrast to different approaches of integrated resource management (such as, e.g., integrated water resources management or the integrated landscape approach), the idea of the nexus approach is to not prioritize an individual resource, but regard the water, energy, and land dimensions of the nexus at once, in their complex interrelations [39].

'The nexus' has been addressed by a number of different methods and disciplines in the past years. While current research suggests that quantifying the WEL Nexus could offer possibilities for an enhanced understanding of interconnections, related approaches and applications face several methodological obstacles [66,67]. Even though research on socio-economic issues increases the nexus, academic literature, so far, has been dominated by techno-economic approaches [68–71]. However, especially in industrialized countries, problems or inefficiencies in the use of natural resources are often caused by fragmented management and governance of these resources. Common barriers are a high level of bureaucracy, historically grown sector policies and institutions, and differentiated responsibilities [33]. Thus, in this paper, the WEL Nexus approach is specifically applied to governance issues.

In this paper, we define the WEL Nexus as an analytical approach for optimized solutions of natural resources management based on a holistic assessment of challenges and opportunities [13,39,72]. Applying the EPI terminology, we understand the nexus as a concept aiming at both vertical and horizontal policy integration across related policy fields, such as energy, water, or agricultural policy. Given these objectives, a nexus-enabled integrated policy-design for managing those resources should systematically account for the cross-sectoral effects of sectoral policies. Such an approach may 'support a transition to sustainability, by reducing trade-offs and generating additional benefits that outweigh the transaction costs associated with stronger integration across sectors' [2]. However, it will also likely alter the costs and benefits of existing policies and actions [13]. So although a nexus-enabled policy approach is expected to lead to better solutions, it may confront initial resistance in the different sectoral domains. Research has further shown that currently societies are only vaguely organized as to effectively implement and enact integrated (i.e., coherent) planning and action [1,38], and that even within the sectors, different policy packages often conflict with each other [73]. This holds true especially in the case of the EU, where sectoral fragmentation is strongly prominent [15]. Thus, the WEL Nexus offers a new, alternative approach for EU policy integration to advance.

With respect to managing the WEL Nexus, the EU is confronted with diverse conflicts of interests and unintended interrelations. Two prominent examples are the nitrogen pollution of ground water by agriculture [74] and the link between energy and water infrastructures in industry [75]. The EU emphasizes that internal policies affecting sustainable consumption and production patterns in the EU are important to help prevent resource scarcities [13]. Although a significant stream of research on the WEL Nexus has developed, its vast application and inconsistent definitions impose challenges on operationalizing the nexus for EU policies [76]. Attempts to achieve better coherence in the governance of the nexus resources inevitably lead to questions about the most adequate level of integration among the resource sectors. By emphasizing the need for considering interlinkages among the sectors, the nexus concept frames a new objective of policy integration among the related resource sectors. The nexus concept, in contrast to the main purpose of EPI, does not aim at implementing general environmental objectives into other policy sectors, but instead sets focus on the interlinkages between the water, energy, and land sectors [2]. Given this difference, existing, but often single-sector findings from the EPI literature can provide valuable insight, but will not be sufficient. Accordingly, a multi-sector analysis of the specific cross-sectoral interdependencies among the nexus sector policies shall provide the fundament for assessing the current state of policy integration in the European Union.

Reviewing EU policy integration based on the WEL Nexus concept faces methodological challenges. While the EPI concept provides specific analytical categories, the WEL Nexus concept has been heavily criticized for its lack of operationalization [76]. The EPI concept also entails differing interpretations and implementation options, but to a comparatively smaller degree [29]. As the above discussion shows, there is considerable overlap in the challenges for policy integration between EPI and the WEL Nexus concepts, given that water, energy, and land as natural resources are inevitably also part of the environmental sector. Though nexus aspects might fall under the concept of EPI, the WEL Nexus concept offers a more systemic perspective than the EPI concept. While the WEL Nexus approach looks at the overall integration of different policy sectors, the EPI approach entails the integration of specific sectoral policy goals. Thus, both concepts complement each other well for reviewing the state of integration in EU policy. Based on the WEL Nexus' systemic perspective and EPI's analytical categories, in the subsequent review we will investigate EU policy integration with regard to cross-sectoral and nexus-informed strategies and policies, and the mutual consideration of related objectives within sectoral policies.

## 3. Review

### 3.1. Cross-Sectoral Considerations in EU Policy Documents by Policy Sector

As point of departure for the in-depth qualitative analysis of policy integration, we examined policy documents for considerations of policy coherence. In a first step, explicit references to policy coherence were considered by policy sector. Policy coherence, in this context, was defined in line with its understanding in the Sustainable Development Goals as 'synergies and trade-offs among [ . . . ] targets, between different sectoral policies, and between diverse actions at the local, regional, national, and international levels' [77]. This definition served as the basis for analysis, where the specific focus was set on the cross-effects among the sectors. The analysis revealed that, within the policy sectors of direct relevance to water, energy and land, considerations of policy coherence are most prominent in the sector of general environmental policy. Of the directly resource-relevant sectors—water, energy, and agriculture—policy coherence is most strongly considered in water policy documents (cf. Table S1: WFD, WFD_DecMak, WatScarcImpact, WFD_IntegrWat, NitrDir_Backgr, SoilWaterStudy, ResEcEffWatDistr, WFD_Leaflet). Given that environmental considerations and sustainability objectives are of direct and spanning relevance for all resource sectors, this result is not unexpected. Furthermore, as discussed (Section 2.1), the EPI focus of integrating environmental concerns into other sectors has long been part of EU policies, which accordingly is reflected in the documents.

In a second step, the documents were reviewed with regard to implicit considerations of policy coherence for sector-specific policy considerations. The results indicate that of the relevant policy sectors, in the energy sector, cross-sectoral considerations are least integrated into policy design. In the considered energy policy documents, no reference to water policy was found. Agricultural policy was singularly referred to with regard to the combined emission of $CO_2$ in the case of energy crop production for bioenergy (cf. Table S1: En2030Strt). The same trend shows in the analysis of the agricultural and water policy sectors with none or only few energy policy considerations (cf. Table S1: WatScarcImpact, SoilWaterStudy, ResEcEffWatDistr). According to the results, it can further be assumed that policy coherence, at least to some degree, has been achieved among agricultural and water policies with frequent cross-sectoral considerations, respectively.

To further differentiate these first, general results, a more detailed analysis of policy coherence was conducted. For this purpose, the documents were analyzed for reflections of cross-sectoral considerations. Specifically, the implementation of policy coherence as a policy objective was searched for and found primarily in general environmental policy with few instances in water and agricultural policy (cf. Table S1: IndicIntegrEnvCAP, ReviewEIA, CAP_2013RefOverv, WatScarcImpact, ResEcEffWatDistr). These observations again confirm that the implementation of cross-sectoral policy considerations is least advanced in the energy policy sector.

Furthermore, references to the respective other resources were searched for. One central result was that the agricultural policy sector seems to play a crucial role when analyzing the integrated policy perspective, since the respective cross-sectoral considerations with energy and water resources were repeatedly and specifically addressed (cf. Table S1 CAP_Markets, CAP_RuralDev, CAP_Tow2020, CAP_2003CrossComp, CAP_Payments, IndicIntegrEnvCAP). Also, these considerations were found in different documents and address a broad number of different objectives. The analysis of energy policy documents, in contrast, revealed references to both agriculture and water policies; however, all originate from the same document, namely the Renewable Energy Directive (2009/28/EC) (cf. Table S1: RenEnDir). Within this document, general sustainability considerations play an important role, so that merely the very general objective of 'measures taken for soil, water and air protection, the restoration of degraded land, the avoidance of excessive water consumption in areas where water is scarce' is referred to [78]. The review of policy coherence finds that the agricultural sector plays a central role for integrated resources governance. In order to look beyond these indicative results, they serve as the starting point for the subsequent comparative analysis of policy integration among the sectors.

### 3.2. In-Depth Analysis of Vertical and Horizontal Policy Integration in Nexus-Relevant European Policies

To consolidate the above analysis, this section provides an in-depth qualitative comparative analysis and discussion of those EU energy, water, and agricultural policies that were identified to show a significant degree of implemented vertical and horizontal policy integration. Firstly, we evaluated the vertical dimension. This allows us to investigate to what extent EU sector policies are nexus-informed.

### 3.2.1. Vertically Integrated EU Water and Agricultural Policies

In case of water and agricultural policies, a high degree of vertical policy integration can be found. In Europe, the management of water resources is primarily regulated via three binding directives: (a) the Water Framework Directive (Directive 2000/60/EC), (b) the Bathing Water Directive (Directive 2006/7/EC), and (c) the Marine Strategy Framework Directive (Directive 2008/56/EC). The WFD, as the EU's main instrument for water protection, requires its member states to achieve 'good status for surface and groundwater' by 2015 (Article 4 WFD). It was adopted in 2000 and in this early version highlights the need to integrate water protection and sustainable management into other policy areas such as 'energy, transport, agriculture, fisheries, regional policy and tourism' [79].

One integral part of the WFD is the Nitrates Directive, which came into effect in 1991, and specifically addresses the prevention of ground and surface water pollution from agriculture by promoting the use of good farming practice [80]. Within the WFD, several measures have been defined with a direct link to agriculture, including, e.g., the management of water demand, fertilizer emissions, as well as efficiency and reuse measures [81]. It thus shows that the negative side-effects of intensive agriculture production were considered in water management. To date, however, specific analyses to determine the actual effects of such policy frameworks on other resource sectors are rare [2]. However, the link between water and agriculture policies is bilateral, as the above results indicate.

The CAP is currently the most important EU policy framework to mandate the formal consideration of externalities among the nexus resources, especially low water quality resulting from agriculture. With a total spending of over 58 billion Euro, the CAP accounts for almost 40% of the EU budget [81] and pursues three main long-term objectives: (a) viable food production, (b) sustainable management of natural resources, and (c) climate action and balanced territorial development [82]. In March 2013, the cornerstones of the post-2013 CAP were defined. 'Cross-compliance' was reinforced, directing payments for farmers to comply with rules on farming practices that account for the environment, food safety, animal and plant health, and animal welfare while maintaining 'agricultural land in good agricultural and environmental condition' [81]. The post-2013 reform of the CAP was motivated by three main challenges within the agricultural sector, which—although not explicitly referring to the WEL Nexus—very closely correlate to the nexus challenges (Table 1):

**Table 1.** Challenges in the agricultural sector [82].

| | |
|---|---|
| Economic challenges | • food security and globalization<br>• a declining rate of productivity growth<br>• price volatility<br>• pressures on production costs due to high input prices and the deteriorating position of farmers in the food supply chain |
| Environmental challenges | • resource efficiency<br>• soil and water quality<br>• threats to habitats and biodiversity |
| Territorial challenges | • rural development<br>• demographic, economic and social developments including depopulation and relocation of businesses |

In order to address these challenges, a new policy instrument has been added to the new CAP: the 'Green Direct Payment' [82]. It rewards farmers for complying with three agricultural practices: (a) maintenance of permanent grassland, (b) ecological focus areas, and (c) crop diversification [82]. Furthermore, as part of the second pillar, rural development, at least '30% of the budget of each rural development program must be reserved for voluntary measures that are beneficial for the environment and climate change' and will be implemented as part of the national (or regional) rural development programs [82]. Given the financial endowments of almost 100 billion Euro for the period from 2007 to 2013, rural development could thus contribute substantially to funding the protection of water resources [81].

The main challenge to do so consists in that EU water policy objectives are anchored in different policy areas sub-ordinate to different authorities with partially contradictory interests [81]. In order to overcome these challenges, the re-established mechanisms of cross-compliance and the European Agricultural Fund for Rural Development provide the means to more strongly encourage good farming practice in compliance with environmental legislations [82].

Integrating policies across the water protection and agriculture policy realms in the EU constitutes a valuable contribution towards an integrated policy framework, which comprehensively considers both direct and secondary policy effects on each of the nexus resources. Notwithstanding the progress made, the European Court of Auditors' analysis pointed out that the integration of EU water policy goals as part of the CAP cannot be regarded as completed yet, and that several major challenges remain, often related, e.g., to weaknesses in the definition of standards or inconsistencies among member states [81].

### 3.2.2. Approaches for Horizontal Policy Integration in Nexus-Relevant European Policies

Next, horizontal policy integration was investigated, i.e., the degree to which the EU has developed cross-sectoral strategies. Specifically, this analysis evaluates to what extent the EU pursuits an overall integrative approach. In addition to the sectoral approaches of policy integration, further instruments of horizontal EPI exist in the EU in the form of political initiatives, overarching strategies, plans and assessment tools that are of great relevance for achieving a nexus-enabled policy approach. Prominent examples of such merely horizontal approaches are EU roadmaps. As part of the Energy Roadmap 2050, the EU reaffirmed its objective to reduce greenhouse gas emission to 80–95% below 1990 levels by 2050 [83]. Furthermore, the share of renewable energies in final energy consumption is to increase to at least 20% by 2020. More specifically, in terms of the energy sector, the 'Roadmap for Moving to a Competitive Low-Carbon Economy in 2050' broadly refers to potential negative side-effects of renewable energies, e.g., biofuels, concluding that 'any negative impacts on other resources (e.g., water, soil and biodiversity) will need careful management' [84].

Thus, for the energy sector, horizontal policy approaches exist within these roadmaps, formulating overarching objectives that are to be reached by vertical processes. Similar approaches exist for water management. The 'Roadmap to a Resource Efficient Europe' defines a number of non-binding water management objectives including, for example, to keep water abstraction below 20% of available renewable water resources, while minimizing the impacts of droughts and floods through adapted crops, increased water retention in soils, and efficient irrigation mechanisms [85]. In accordance with the nexus understanding, the roadmap further calls for a comprehensive and integrative management approach to achieve these objectives [85]. Such overarching strategies and plans—despite their reference to the main nexus idea—currently lack an operational implementation in day-to-day policy making [48].

In addition to those long-term strategies, which propose concrete and, in some cases, binding quantitative targets, other horizontal approaches also have a long-term background. One such approach to putting environmental objectives onto a stronger legal basis was the Cardiff-Process in the early 2000s. It moved forward the integration of environmental and sustainability considerations into the different policy sectors and stipulated learning that has translated into further nexus-relevant policy measures. It followed upon the addition of Article 6 to the EC Treaty (COM (1998) 333) in 1998, according to which 'environmental protection requirements must be integrated into the definition and implementation of the Community policies [...] in particular with a view to promoting sustainable development'. Accordingly, it requested a number of Councils to develop strategies for the integration of environmental and sustainability considerations into their respective policy fields for the European Council meeting in Goeteborg in 2001. Post-Goeteborg analyses, however, revealed huge differences in the quality, scope, and ambition of objectives among the submitted documents [86]. Unfried [87] identified the 'lack of support from political leaders together with—maybe also as a result—the lack of commitment among the councils of ministers [ ... ] as reasons the Cardiff process had not achieved the intended level of integration', so that the Cardiff process was eventually 'singled out for its unrealistic potential to lead to policy learning for EPI' [15]. Nonetheless, the Cardiff process has instigated some important progress in the policy integration of environmental considerations. It was an innovative policy measure that stipulated, e.g., cross-sectoral councils and working groups as well as significant progress in sectoral environmental integration indicators [88]. It further revealed important challenges inherent in cross-sectoral policy integration that, in a similar form, must be overcome in order to implement an effective nexus policy framework. These challenges include (a) promoting the implementation of the policy measures in the member states, (b) the development of national processes for integration, (c) establishing harmonized reporting mechanism for information exchange, and (d) promoting a process of trans-national policy learning and the development of networks of experts [86].

Other measures implemented to environmentally assess operational public policies include, for example, Environmental Impact Assessment (EIA) (Directive 2011/92/EU) and Strategic Environmental Assessment (SEA) (Directive 2001/42/EC) [89] as procedures to ensure that 'the environmental implications of decisions are taken into account before the decisions are made' [90]. Within the EIA, the direct and indirect effects of a project on (a) human beings, fauna and flora, (b) soil, water, air, climate and landscape, (c) material assets and cultural heritage, and (d) interaction between the above listed are to be identified [91]. In contrast to the EIA, SEA applies to public plans and programs—not, however, policies [90]. In addition to identified implementation gaps, the absence of obligatory environmental standards was identified as a major drawback [92]. This gives rise also to challenges related to the fact that many projects are under the jurisdiction of different member states, which in turn inheres the risk of duplications, inconsistencies, burdens (e.g., administrative) and thus potential conflicts [92]. Such inefficiencies often result from differences in the national EIA procedures, e.g., with respect to different stages of the project proposal process or differing timeframes, but could likely be reduced, if more formal consultation on transboundary impacts among neighboring countries was implemented, e.g., in the form of specified timeframes for consultation with neighboring countries or joint procedures for international projects [92].

The effect of these European environmental assessment approaches on actual policy making, however, has been limited [89]. According to Jordan and Lenschow [29], this hints at a weak impact of environmental policy in general, but can partly also be explained by the lack of clarity as to what these measures are to achieve: Is it to 'strengthen environmental policies, integrate the environment into other sectors, promote public participation or deliver sustainable development' [29]. Furthermore, the attempt to regulate and better inform decision makers by using these assessment tools is often diminished by their improper application or even abuse [28].

## 4. Discussion and Policy Implications

Given the EU objective to develop a more integrated policy framework that explicitly accounts for the dynamic interplay between different natural resources, the above analysis of current, relevant policy measures can serve as a first foundation. The analysis showed that including the objectives of the WFD into the CAP is currently the most prominent attempt of especially vertical policy integration on a sectoral level. With regard to the energy sector, vertical integration does not seem to play a role at all. Despite the fact that numerous studies have affirmed the direct links between energy production and both water resources e.g., [51,53,55,56,93,94] as well as land resources and food production [47,95], the explicit consideration of these cross-resource effects is not yet formally integrated into current energy directives or action plans. Thus, overall, the degree of vertical integration among energy and water, as well as energy and land sectors, has been identified as rather weak.

One reason can be seen in the fact that, currently, authority is spread across different actors with mandates for specific sectors, such as energy, agriculture, or water. Within the EU Commission, water policy falls within the responsibilities of the Directorate General (DG) Environment, which officially aims at 'greening' other policy areas. The energy and agriculture sectors instead form their own DGs. Such different sub-units usually act based on historically grown—and thus different—sets of formal and informal rules and processes, which can be inconsistent or conflicting, thus making policy integration hard to reach [96]. This fragmented responsibility is in line with our finding that these two sectors show a relatively weak level of policy coherence. Additionally, their uneven legal 'weight' impedes the development of a common and integrated 'nexus-enabled' policy understanding [32]. Whereas the CAP is completely communitized, the energy sector is a policy field of shared responsibilities between the EU and its member states. The EU is thus limited to setting overarching policy objectives in the form of directives, which leave the actual implementation in the responsibility of the member states. The CAP, however, is based on regulations that are transferred directly into national law [97]. The different legal character, therefore, makes it hard to achieve a common level of integration among the nexus sectors. Furthermore, not only the different legal character but also differing, partly contesting, underlying values and objectives impede a more integrated policy design. For example, a CAP that aims at economically efficient and intensive production, which inevitably comes with a high demand of fertilizers as well as a high quantity of manure, undermines or outright contradicts environmental objectives such as low nitrate levels [14].

However, with certain challenges remaining, selected aspects of water management have already been successfully integrated into the CAP by binding payments to farmers to conformance with the standards of cross-compliance, which now include standards explicitly related to water management. Given the often trans-national character of externalities related to the provision of water, energy or food, the WFD can serve as a reference for a policy measure that explicitly recognizes the relevance of geographical—rather than national—system boundaries (e.g., river basins) and is thus conceptualized as to provide a mandatory policy framework that leaves the development of specific, trans-national implementation strategies to local authorities. Nevertheless, as the example of Germany shows, the success of these instruments highly depends on the national implementation. In 2018, the European Court of Justice found Germany guilty of failing to reach required maximum levels of nitrate pollution of groundwater [98].

In terms of horizontal policy integration, the impact of existing European roadmaps, as well as impact assessment tools, is strongly limited by lacking enforcement measures. Nevertheless, the cross-cutting procedures of environmental impact assessment, for example, may serve as the conceptual foundation for mandating the explicit analysis of cross-sectoral effects among the nexus resources when proposing plans or programs within the EU. The existing variety of approaches—if systematically brought together—can provide a solid foundation for developing a coherent European nexus policy framework.

In order to address the identified remaining challenges of such a policy framework, a combined top-down and bottom-up policy approach is needed that can stipulate cross-sectoral cooperation and the consolidation of political system boundaries with natural-geographic, often trans-national requirements. Important assessment tools in this regard are cross-compliance mechanisms and support for regional development, in line with the approach taken by CAP. Additionally, however, the policy framework must carefully address the challenges of determining a reasonable degree of regulation, which—while framing the path towards its policy objectives—leaves sufficient room for competition-based market structures or locally adapted implementation strategies. Providing universal access to the limited natural resources essential for human life is of indispensable importance. Furthermore, our analysis highlighted that tools and measures are only useful if integrated and used in day-to-day policy making. Specifically, implementation could be improved by streamlining standards and procedures, and formulating clear and specific policy goals. Therefore, a revised and updated EU sustainable development strategy seems to be urgently necessary in order to coordinate policy objectives towards more sustainable pathways. This demand is backed by e.g., the German Federal Government since the last EU strategy was published in 2006 [99]. In this case, the role of member states also needs to be considered. For example, national implementation of EU regulations as well as existing national sustainable development strategies play an essential role in achieving integrated policy making [32]. A central aspect of this is also the promotion of policy learning, e.g., by cross-sectoral working groups as introduced by the Cardiff-Process, to increase the political will to follow through on integration efforts.

These policy implications are especially relevant for EU energy policy. The analysis revealed that the energy sector lags behind most in terms of integration. Though horizontal integration approaches exist in the form of overarching strategies, vertical approaches are almost non-existent. Cross-sectoral considerations are lacking beyond very general sustainability considerations. Here, EU energy policy has to clearly and specifically include concerns of and consequences for other sectors as well in order to enable a generally more integrated natural resources policy. Energy policy could especially profit from policy learning, since EU agricultural and water policy have considerable experience in comparison.

With regard to the applied analytical frameworks, each has its limits, but there is also notable overlap. Analysis along the frames of EPI and WEL Nexus revealed that including environmental aspects into other sector policies often remains on a rhetorical level, and that instead, changes to the governance structures are required that transcend the historically grown sectoral boundaries. Here, the Nexus concept enables a broad, comprehensive approach to integrating natural resource policies, while the EPI concept is capable of identifying and evaluating specific governance challenges, although in a sectorally defined context. However, both the EPI and the WEL Nexus concept are considered to allow for various interpretations and implementation options [29,76]. This conceptual ambiguity is difficult to translate into jurisdiction since it does not define the degree of integration required [29], and thus, is mirrored in the incomplete realization of more integrated EU natural resource policies. Here, synthesizing policy integration and nexus research can offer valuable insights, such as considering cognitive factors in the analysis of nexus governance [65]. In this paper, synthesizing both integration concepts provided valuable insight into the different dimensions of integrating natural resource policies. The paper thus offers a new method for analyzing integrated natural resources governance and contributes to the ongoing discussion on finding 'the right level' of policy integration in the European Union.

## 5. Conclusions

At present, policy design for natural resources, especially in the energy sector, only rudimentarily accounts for cross-sectoral effects among the different resources, despite the clearly visible interdependencies among the resource systems. Given the discussed sustainability objectives, adequate policy design, however, must encompass and account for the full spectrum of social, ecological, and economic preconditions. Traditionally, policy design has been framed and analyzed per sector. Only more recently, cross-sectoral effects among the nexus resources have been accounted for, and mostly rudimentarily. Against the background of the advanced EPI literature, the approach used in this paper allowed the assessment of the state of policy integration across the sectors energy, water, and land. The review of EU resource governance indicates that the adaptation or consolidation of singular measures under the paradigm of EPI does not suffice to meet the needs of cross-sectoral and transnational challenges, such as the integrated management of natural resources. Instead, it will require the institutionalization of a comprehensive system perspective as conceptualized by the WEL Nexus in order to overcome the current particular interests. Thus, further research is required on the implementation of a nexus policy perspective in the EU, specifically, whether fundamentally new institutional settings are required, or integrating a coherent, cross-sectoral perspective into the existing structures provides a more efficient approach to policy coherence. Specifically, the question of what role the member states play needs to always be taken into account.

Furthermore, comprehensive scientific research is needed to systematically analyze the systemic interrelations between natural resources (i.e., the true feedbacks and interactions among the resources and their drivers) for deriving scenarios that can serve as a foundation for the development of effective nexus governance options.

In order to analyze progress towards the objective of more integrated policy making, research into the operationalization of relevant concepts is necessary. As this review has demonstrated, synthesizing EPI and WEL Nexus provides a useful analytical framework for estimating the state of policy integration. At the same time, the already existing policy approaches should be further refined as to successively obviate the currently remaining policy inefficiencies. As the above analysis shows, a policy framework in line with a comprehensive WEL Nexus approach will likely confront a number of challenges very similar to those encountered in previous measures of policy integration, so that the handling of these challenges may provide significant insights for the effective implementation of a WEL Nexus approach. Answering these questions will require careful analysis, given the current debates around the 'right' degree of EU integration and the balance between fully integrated and intergovernmental politics.

**Supplementary Materials:** The following are available online at http://www.mdpi.com/1996-1073/12/23/4446/s1, Table S1: List of policy documents investigated, according to document type and year of publication.

**Author Contributions:** Conceptualization, S.V.; Methodology, S.V. and C.M.; Investigation, all; Formal Analysis, all; Writing—Original Draft Preparation, S.V. and C.M.; Writing—Review and Editing, S.D. and F.S.; Supervision, S.V.

**Funding:** This research received no external funding.

**Conflicts of Interest:** The authors declare no conflict of interest.

## References

1. Howells, M.; Rogner, H.H. Water-energy nexus: Assessing integrated systems. *Nat. Clim. Chang.* **2014**, *4*, 246–247. [CrossRef]
2. Hoff, H. Understanding the Nexus: Background Paper for the Bonn2011 Nexus Conference. In Proceedings of the Bonn 2011 Conference The Water, Energy and Food Security Nexus-Solutions for the Green Economy, Bonn, Germany, 16–18 November 2011.

3.   Ziv, G.; Watson, E.; Young, D.; Howard, D.C.; Larcom, S.T.; Tanentzap, A.J. The potential impact of Brexit on the energy, water and food nexus in the UK: A fuzzy cognitive mapping approach. *Appl. Energy* **2018**, *210*, 487–498. [CrossRef]

4.   Jeswani, H.K.; Burkinshaw, R.; Azapagic, A. Environmental sustainability issues in the food–energy–water nexus: Breakfast cereals and snacks. *Sustain. Prod. Consum.* **2015**, *2*, 17–28. [CrossRef]

5.   Azapagic, A. Special issue: Sustainability issues in the food–energy–water nexus. *Sustain. Prod. Consum.* **2015**, *2*, 1–2. [CrossRef]

6.   Scott, C.; Sugg, Z. Global Energy Development and Climate-Induced Water Scarcity—Physical Limits, Sectoral Constraints, and Policy Imperatives. *Energies* **2015**, *8*, 8211–8225. [CrossRef]

7.   IPCC. *Climate Change and Land-An IPCC Special Report on Climate Change, Desertification, Land Degradation, Sustainable Land Management, Food Security, and Greenhouse Gas Fluxes in Terrestrial Ecosystems*; Intergovernmental Panes on Climate Change: Geneva, Switzerland, 2019.

8.   United Nations Secretary-General's High-level Panel on Global Sustainability. *Resilient People, Resilient Planet: A Future Worth Choosing*; United Nations: New York, NY, USA, 2012.

9.   World Economic Forum. *Global Risks 2011*; World Economic Forum: Geneva, Switzerland, 2011.

10.   WBCSD. *Water, Food and Energy Nexus Challenges*; WBCSD: Geneve, Switzerland, 2014.

11.   IEA. *World Energy Outlook 2012*; IEA: Paris, France, 2012.

12.   US NIC. *Global Trends 2030: Alternative Worlds*; United States National Intelligence Council: Washington, DC, USA, 2012.

13.   ODI; ECDPM; GDI/DIE. *Confronting Scarcity: Managing Water, Energy and Land for Inclusive and Sustainable Growth*; Overseas Development Institute (ODI), European Centre for Development Policy Management (ECDPM), German Development Institute/Deutsches Institut für Entwicklungspolitik (GDI/DIE): Brussels, Belgium, 2012.

14.   Alons, G. Environmental policy integration in the EU's common agricultural policy: Greening or greenwashing? *J. Eur. Public Policy* **2017**, *24*, 1604–1622. [CrossRef]

15.   Adelle, C.; Russel, D. Climate Policy Integration: A Case of Déjà Vu? *Environ. Policy Gov.* **2013**, *23*, 1–12. [CrossRef]

16.   Pirlogea, C.; Cicea, C. Econometric perspective of the energy consumption and economic growth relation in European Union. *Renew. Sustain. Energy Rev.* **2012**, *16*, 5718–5726. [CrossRef]

17.   Śmiech, c.; Papież, M. Energy consumption and economic growth in the light of meeting the targets of energy policy in the EU: The boot strap panel Granger causality approach. *Energy Policy* **2014**, *71*, 118–129. [CrossRef]

18.   Gianfreda, A.; Parisio, L.; Pelagatti, M. Revisiting long-run relations in power markets with high RES penetration. *Energy Policy* **2016**, *94*, 432–445. [CrossRef]

19.   Criqui, P.; Mima, S. European climate—Energy security nexus: A model based scenario analysis. *Energy Policy* **2012**, *41*, 827–842. [CrossRef]

20.   Lechón, Y.; De La Rúa, C.; Cabal, H. Impacts of Decarbonisation on the Water-Energy-Land (WEL) Nexus: A Case Study of the Spanish Electricity Sector. *Energies* **2018**, *11*. [CrossRef]

21.   Karabulut, A.; Egoh, B.N.; Lanzanova, D.; Grizzetti, B.; Bidoglio, G.; Pagliero, L.; Bouraoui, F.; Aloe, A.; Reynaud, A.; Maes, J.; et al. Mapping water provisioning services to support the ecosystem–water–food–energy nexus in the Danube river basin. *Ecosyst. Serv.* **2016**, *17*, 278–292. [CrossRef]

22.   Siciliano, G.; Rulli, M.C.; D'Odorico, P. European large-scale farmland investments and the land-water-energy-food nexus. *Adv. Water Resour.* **2017**, *110*, 579–590. [CrossRef]

23.   Venghaus, S.; Hake, J.F. Nexus thinking in current EU policies–The interdependencies among food, energy and water resources. *Environ. Sci. Policy* **2018**. [CrossRef]

24.   European Union. Consolidated version of the Treaty on the Functioning of the European Union. *Off. J. Eur. Union* **2012**, *2012/C 326/01*, 0001–0390.

25.   Maxwell, J.A.; Chmiel, M. Notes Toward a Theory of Qualitative Data Analysis. In *The SAGE Handbook of Qualitative Data Analysis*; Flick, U., Ed.; SAGE Publications: London, UK, 2014; pp. 21–34.

26.   Gibbs, G.R. Using Software in Qualitative Analysis. In *The SAGE Handbook of Qualitative Data Analysis*; Flick, U., Ed.; SAGE Publications: London, UK, 2014; pp. 277–294.

27.   European Union. Official Website of the European Union. Available online: http://europa.eu/european-union/index_en (accessed on 30 September 2019).

28.   Nilsson, M.; Zamparutti, T.; Petersen, J.E.; Nykvist, B.; Rudberg, P.; McGuinn, J. Understanding Policy Coherence: Analytical Framework and Examples of Sector-Environment Policy Interactions in the EU. *Environ. Policy Gov.* **2012**, *22*, 395–423. [CrossRef]

29. Jordan, A.; Lenschow, A. Environmental policy integration: A state of the art review. *Environ. Policy Gov.* **2010**, *20*, 147–158. [CrossRef]

30. Lafferty, W.M.; Hovden, E. Environmental Policy Integration: Towards an Analytical Framework. *Environ. Politics* **2003**, *12*, 1–22. [CrossRef]

31. Di Gregorio, M.; Nurrochmat, D.R.; Paavola, J.; Sari, I.M.; Fatorelli, L.; Pramova, E.; Locatelli, B.; Brockhaus, M.; Kusumadewi, S.D. Climate policy integration in the land use sector: Mitigation, adaptation and sustainable development linkages. *Environ. Sci. Policy* **2017**, *67*, 35–43. [CrossRef]

32. Herodes, M.; Adelle, C.; Pallemaerts, M. *Environmental Policy Integration at the EU Level–A Literature Review*; Ecologic–Institute for International and European Environmental Policy: Berlin, Germany, 2007.

33. Märker, C.; Venghaus, S.; Hake, J.-F. Integrated governance for the food–energy–water nexus–The scope of action for institutional change. *Renew. Sustain. Energy Rev.* **2018**, *97*, 290–300. [CrossRef]

34. Lenschow, A. New Regulatory Approaches in 'Greening' EU Policies. *Eur. Law J.* **2002**, *8*, 19–37. [CrossRef]

35. Daugbjerg, C.; Feindt, P.H. Post-exceptionalism in public policy: Transforming food and agricultural policy. *J. Eur. Public Policy* **2017**, *24*, 1565–1584. [CrossRef]

36. Daugbjerg, C.; Swinbank, A. Three Decades of Policy Layering and Politically Sustainable Reform in the European Union's Agricultural Policy. *Governance* **2016**, *29*, 265–280. [CrossRef]

37. Weitz, N.; Huber-Lee, A.; Nilsson, M.; Davis, M.; Hoff, H. *Cross-Sectoral Integration in the Sustainable Development Goals-A Nexus Approach*; Stockholm Environment Institute: Stockholm, Sweden, 2014.

38. Bazilian, M.; Rogner, H.; Howells, M.; Hermann, S.; Arent, D.; Gielen, D.; Steduto, P.; Mueller, A.; Komor, P.; Tol, R.S.J.; et al. Considering the energy, water and food nexus: Towards an integrated modelling approach. *Energy Policy* **2011**, *39*, 7896–7906. [CrossRef]

39. FAO. *The Water-Energy-Food Nexus-A New Approach in Support of Food Security and Sustainable Agriculture*; FAO: Rome, Italy, 2014.

40. Federal Government of Germany. *National Sustainability Development Strategy-2012 Progress Report*; Federal Government of Germany: Berlin, Germany, 2012; p. 245.

41. Lawford, R.; Bogardi, J.; Marx, S.; Jain, S.; Wostl, C.P.; Knüppe, K.; Ringler, C.; Lansigan, F.; Meza, F. Basin perspectives on the Water–Energy–Food Security Nexus. *Curr. Opin. Environ. Sustain.* **2013**, *5*, 607–616. [CrossRef]

42. Lele, U.; Klousia-Marquis, M.; Goswami, S. Good Governance for Food, Water and Energy Security. *Aquat. Procedia* **2013**, *1*, 44–63. [CrossRef]

43. Ozturk, I. Sustainability in the food-energy-water nexus: Evidence from BRICS (Brazil, the Russian Federation, India, China, and South Africa) countries. *Energy* **2015**, *93*, 999–1010. [CrossRef]

44. Stockholm Environment Institute. Water-Land-Energy Nexus. Available online: http://www.sei-international. org/rio20/water-land-energy-nexus (accessed on 22 October 2014).

45. US DOE. *Climate and Energy-Water-Land System Interactions. Technical Report to the U.S. Department of Energy in Support of the National Climate Assessment*; Pacific Northwest National Laboratory: Richland, WA, USA, 2012.

46. Tranatlantic Academy. *The Global Resource Nexus. The Struggles for Land, Energy, Food, Water, and Minerals*; Tranatlantic Academy: Washington, DC, USA, 2012.

47. Ringler, C.; Bhaduri, A.; Lawford, R. The nexus across water, energy, land and food (WELF): Potential for improved resource use efficiency? *Curr. Opin. Environ. Sustain.* **2013**, *5*, 617–624. [CrossRef]

48. Golden, C.; Shapero, A.; Vaitla, B.; Smith, M.; Myers, S.; Stebbins, E.; Gephart, J. Impacts of Mainstream Hydropower Development on Fisheries and Human Nutrition in the Lower Mekong. *Front. Sustain. Food Syst.* **2019**, *3*. [CrossRef]

49. UNC Water Institute. *Building Integrated Approaches into the Sustainable Development Goals-A Declaration from the Nexus 2014: Water, Food, Climate and Energy Conference in the name of the Co-directors held at the University of North Carolina at Chapel Hill, 5–8 March 2014*; Submitted to the UN Secretary-General on the 27th of March, 2014; UNC Water Institute: Chapel Hill, NC, USA, 2014.

50. World Economic Forum. *Water Security-The Water-Food-Energy-Climate Nexus*; World Economic Forum Water Initiative: Washington, DC, USA; Covelo, CA, USA; London, UK, 2011.

51. Rio Carrillo, A.M.; Frei, C. Water: A key resource in energy production. *Energy Policy* **2009**, *37*, 4303–4312. [CrossRef]

52. Siddiqi, A.; Anadon, L.D. The water–energy nexus in Middle East and North Africa. *Energy Policy* **2011**, *39*, 4529–4540. [CrossRef]

53. Glassman, D.; Wucker, M.; Isaacman, T.; Champilou, C. *The Water-Energy Nexus: Adding Water to the Energy Agenda*; World Policy Institute: New York, NY, USA, 2011.

54. Ackerman, F.; Fisher, J. Is there a water–energy nexus in electricity generation? Long-term scenarios for the western United States. *Energy Policy* **2013**, *59*, 235–241. [CrossRef]

55. Dubreuil, A.; Assoumou, E.; Bouckaert, S.; Selosse, S.; Maizi, N. Water modeling in an energy optimization framework–The water-scarce middle east context. *Appl. Energy* **2013**, *101*, 268–279. [CrossRef]

56. Bartos, M.D.; Chester, M.V. The Conservation Nexus: Valuing Interdependent Water and Energy Savings in Arizona. *Environ. Sci. Technol.* **2014**, *48*, 2139–2149. [CrossRef] [PubMed]

57. UN Water. *The United Nations World Water Development Report 5: Water and Energy*; UNESCO: Paris, France, 2014.

58. Nogueira Vilanova, M.R.; Perrella Balestieri, J.A. Exploring the water-energy nexus in Brazil: The electricity use for water supply. *Energy* **2015**, *85*, 415–432. [CrossRef]

59. Venkatesh, G.; Chan, A.; Brattebø, H. Understanding the water-energy-carbon nexus in urban water utilities: Comparison of four city case studies and the relevant influencing factors. *Energy* **2014**, *75*, 153–166. [CrossRef]

60. Abegaz, B.W.; Datta, T.; Mahajan, S.M. Sensor technologies for the energy-water nexus–A review. *Appl. Energy* **2018**, *210*, 451–466. [CrossRef]

61. Chen, P.-C.; Alvarado, V.; Hsu, S.-C. Water energy nexus in city and hinterlands: Multi-regional physical input-output analysis for Hong Kong and South China. *Appl. Energy* **2018**, *225*, 986–997. [CrossRef]

62. Logan, L.H.; Stillwell, A.S. Probabilistic assessment of aquatic species risk from thermoelectric power plant effluent: Incorporating biology into the energy-water nexus. *Appl. Energy* **2018**, *210*, 434–450. [CrossRef]

63. Wu, X.D.; Chen, G.Q. Energy and water nexus in power generation: The surprisingly high amount of industrial water use induced by solar power infrastructure in China. *Appl. Energy* **2017**, *195*, 125–136. [CrossRef]

64. Martín, M.; Grossmann, I.E. Water–energy nexus in biofuels production and renewable based power. *Sustain. Prod. Consum.* **2015**, *2*, 96–108. [CrossRef]

65. Weitz, N.; Strambo, C.; Kemp-Benedict, E.; Nilsson, M. Closing the governance gaps in the water-energy-food nexus: Insights from integrative governance. *Glob. Environ. Chang.* **2017**, *45*, 165–173. [CrossRef]

66. Chang, Y.; Li, G.; Yao, Y.; Zhang, L.; Yu, C. Quantifying the Water-Energy-Food Nexus: Current Status and Trends. *Energies* **2016**, *9*. [CrossRef]

67. Venghaus, S.; Dieken, S. From a few security indices to the FEW Security Index: Consistency in global food, energy and water security assessment. *Sustain. Prod. Consum.* **2019**, *20*, 342–355. [CrossRef]

68. Dai, J.; Wu, S.; Han, G.; Weinberg, J.; Xie, X.; Wu, X.; Song, X.; Jia, B.; Xue, W.; Yang, Q. Water-energy nexus: A review of methods and tools for macro-assessment. *Appl. Energy* **2018**, *210*, 393–408. [CrossRef]

69. Endo, A.; Tsurita, I.; Burnett, K.; Orencio, P.M. A review of the current state of research on the water, energy, and food nexus. *J. Hydrol. Reg. Stud.* **2017**, *11*, 20–30. [CrossRef]

70. Al-Saidi, M.; Elagib, N.A. Towards understanding the integrative approach of the water, energy and food nexus. *Sci. Total Environ.* **2016**, *574*, 1131–1139. [CrossRef]

71. Allouche, J.; Middleton, C.; Gyawali, D. *Nexus Nirvana or Nexus Nullity? A Dynamic Approach to Security and Sustainability in the Water-Energy-Food Nexus: Water and the Nexus*; STEPS Centre: Brighton, UK, 2014.

72. Schlör, H.; Hake, J.F.; Venghaus, S. An Integrated Assessment Model for the German Food-Energy-Water Nexus. *J. Sustain. Dev. Energy Water Environ. Syst.* **2018**, *6*, 1–12. [CrossRef]

73. Sovacool, B.K.; Saunders, H. Competing policy packages and the complexity of energy security. *Energy* **2014**, *67*, 641–651. [CrossRef]

74. Cruz, S.; Cordovil, C.M.; Pinto, R.; Brito, A.G.; Cameira, M.R.; Gonçalves, G.; Poulsen, J.R.; Thodsen, H.; Kronvang, B.; May, L. Nitrogen in water-Portugal and Denmark: Two contrasting realities. *Water* **2019**, *11*, 1114. [CrossRef]

75. Oliveira, M.C.; Iten, M.; Matos, H.A.; Michels, J. Water–Energy Nexus in Typical Industrial Water Circuits. *Water* **2019**, *11*, 699. [CrossRef]

76. Albrecht, T.R.; Crootof, A.; Scott, C.A. The Water-Energy-Food Nexus: A systematic review of methods for nexus assessment. *Environ. Res. Lett.* **2018**, *13*, 043002. [CrossRef]

77. OECD. *Policy Coherence for Sustainable Development 2017: Eradicating Poverty and Promoting Prosperity*; OECD Publishing: Paris, France, 2017. [CrossRef]

78. European Commission. Directive 2009/28/EC of the European Parliament and of the Council on the promotion of the use of energy from renewable sources and amending and subsequently repealing Directives 2001/77/EC and 2003/30/EC (Renewable Energy Directive). *Off. J. Eur. Union* **2009**, *L 140*, 16–62.

79. European Commission. Directive 2000/60/EC of the European Parliament and of the Council (Water Framework Directive). *Off. J. Eur.* **2000**, *L 327*, 1–72.

80. European Commission. The EU Nitrates Directive Factsheet. In *Environment Fact Sheets*; European Commission: Brussels, Belgium, 2010.

81. European Court of Auditors. *Integration of EU Water Policy Objectives with the CAP: A Partial Success*; European Court of Auditors: Luxembourg, Luxembourg, 2014.

82. European Commission. Overview of CAP Reform 2014–2020. In *Agricultural Policy Perspectives Brief*; DG Agriculture and Rural Development, Unit for Agricultural Policy Analysis and Perspectives: Brussels, Belgium, 2013.

83. European Commission. *COM (2011) 885: Energy Roadmap 2050*; European Comission: Brussels, Belgium, 2012.

84. European Commission. *A Roadmap for Moving to A Competitive Low Carbon Economy in 2050*; European Commission: Brussels, Belgium, 2011.

85. European Commission. *COM (2011) 571: Roadmap to A Resource Efficient Europe*; Office for Official Publications of the European Communities: Luxembourg, Luxembourg, 2011.

86. Kraemer, A. *Results of the "Cardiff-Processes"–Assessing the State of Development And Charting the Way Ahead*; Ecologic Institute for International and European Environmental Policy: Berlin, Germany, 2001.

87. Unfried, M. The Cardiff Process: The Institutional and Political Challenges of Environmental Integration in the EU. *Rev. Eur. Community Int. Environ. Law* **2000**, *9*, 112–119. [CrossRef]

88. Wilkinson, D.; Skinner, L.; Ferguson, M. *The Future of the Cardiff Process–A Report for the Danish Ministry of the Environment*; Institute for European Environmental Policy: London, UK, 2002.

89. Runhaar, H. Tools for integrating environmental objectives into policy and practice: What works where? *Environ. Impact Assess. Rev.* **2016**, *59*, 1–9. [CrossRef]

90. European Commission. Environmental Assessment. Available online: http://ec.europa.eu/environment/eia/home.htm (accessed on 30 September 2019).

91. European Commission. Council Directive of 27 June 1985 on the assessment of the effects of certain public and private projects on the environment (85/337/EEC). *Off. J. Eur. Union L* 175, 40–48.

92. European Commission. *On the Application and Effectiveness of the EIA Directive (Directive 85/337/EEC, as Amended by Directives 97/11/EC and 2003/35/EC)*; Commission of the European Communities: Brussels, Belgium, 23 July 2009.

93. Gerbens-Leenes, P.W.; Lienden, A.R.v.; Hoekstra, A.Y.; van der Meer, T.H. Biofuel scenarios in a water perspective: The global blue and green water footprint of road transport in 2030. *Glob. Environ. Chang.* **2012**, *22*, 764–775. [CrossRef]

94. Scott, C.A.; Pierce, S.A.; Pasqualetti, M.J.; Jones, A.L.; Montz, B.E.; Hoover, J.H. Policy and institutional dimensions of the water–energy nexus. *Energy Policy* **2011**, *39*, 6622–6630. [CrossRef]

95. Canning, P.; Charles, A.; Huang, S.; Polenske, K.R.; Water, A. *Energy Use in the U.S. Food System*; United States Department of Agriculture: Washington, DC, USA, 2010.

96. Russel, D.J.; den Uyl, R.M.; de Vito, L. Understanding policy integration in the EU—Insights from a multi-level lens on climate adaptation and the EU's coastal and marine policy. *Environ. Sci. Policy* **2018**, *82*, 44–51. [CrossRef]

97. Feindt, P.H. Policy-learning and environmental policy integration in the Common Agricultural Policy, 1973–2003. *Public Adm.* **2010**, *88*, 296–314. [CrossRef]

98. European Court of Justice. European Commission v Federal Republic of Germany. In *Failure of a Member State to Fulfil Obligations—Directive 91/676/EEC—Article 5(5) and (7)—Annex II, A, Points 1 to 3 and 5—Annex III(1), Points 1 to 3, and (2)—Protection of Waters Against Pollution Caused by Nitrates from Agricultural Sources—Inadequacy of Measures in Force—Supplemental or Enhanced Measures—Review of the Action Programme—Limitation of Land Application—Balanced Fertilisation—Periods of Land Application—Capacity of Tanks for Storing Manure—Land Application on Steeply-Sloping Surfaces and on Frozen or Snow-Covered Land*; European Court of Justice: Kirchberg, Luxemburg, 2018; Volume Case C-543/16.

99. Bundesregierung. *Deutsche Nachhaltigkeitsstrategie*; Bundesregierung: Berlin, Germany, 2016.

 *energies*

*Article*

# Oil Price and Stock Prices of EU Financial Companies: Evidence from Panel Data Modeling

**Alexandra Horobet [1,*], Georgiana Vrinceanu [1], Consuela Popescu [1] and Lucian Belascu [2]**

[1]  Department of Business and Economics, Faculty of International Business and Economics, The Bucharest University of Economic Studies, 010374 Bucharest, Romania; vrinceanugeorgiana16@stud.ase.ro (G.V.); consuela.popescu@rei.ase.ro (C.P.)

[2]  Department of Management, Marketing and Business Administration, Faculty of Economic Sciences, "Lucian Blaga" University of Sibiu, 550024 Sibiu, Romania; lucian.belascu@ulbsibiu.ro

*  Correspondence: alexandra.horobet@rei.ase.ro

Received: 25 September 2019; Accepted: 23 October 2019; Published: 25 October 2019

**Abstract:** Crude oil is an indispensable resource for the world economy and European Union (EU) countries are strongly dependent on oil imports. In a framework defined by generally positive correlations between oil and stock prices, the paper investigates the relationship between financial companies' stock prices and crude oil price using a sample of major financial companies headquartered in the EU. The link between stock prices and oil price risk is modelled using a set of macroeconomic variables that includes local stock market indices, the EUR/USD exchange rate, the oil imports dependency, inflation rate, and global volatility indices. We employ panel data as the base econometric model and an ARDL extension that is more appropriated for our research objectives. Our findings show that the EU financial sector is pervasively exposed to oil price changes over the long-run and this exposure is a component of financial companies' exposure to real economy risk factors, which points towards the key role of the financial sector in the EU economy in transmitting systemic shocks. At the same time, we detect signs of a different behavior of market investors over the short-versus the long-run concerning the valuation of financial companies' stock prices in relation to oil price and other macroeconomic variables, which raises distressing challenges for financial authorities.

**Keywords:** oil prices; stock prices; panel data analysis; ARDL; financial sector

---

## 1. Introduction

Crude oil is considered nowadays the most influential natural resource for the entire world economy. Moreover, it is a critical input for oil consumer countries and a substantial source of revenue for oil supplier countries; thus, any change or shock in the price of oil has the potential to impact the global economy. At the same time, given that oil price is determined at an international level, it is the price of reference for many sector industries, oil-, or non-oil related, which generates strong connections between them.

In this paper, we build on the idea that financial companies' exposure to oil price is a consequence of the countries' exposure to systemic risk at domestic, regional, and global level. In addition, based on the assumption that the stocks' market value represents the sum of the discounted expected future cash flows provided to investors by the issuer of the stock, oil prices influence economic growth, inflation, and overall market expectations of near-term up to long-term volatility, all of them having an indirect effect on the interest rate, which is critical for the discounting of future cash flows. Provided that interest rates increase, bonds will be more attractive to investors and the stock market prices are expected to fall. Exchange rate volatility is also expected to distress stock returns; for example, when EUR depreciates and USD appreciates, taking into account that American investors own shares in Europe, this might lead to a run-off from these shares in Europe. Additionally, stock indices are generally regarded as "barometers" of the markets hence we would expect a positive relationship between stock returns and local indices.

Changes in oil prices and, on the extreme side, shocks in prices, raise significant challenges to all economic actors that see themselves highly interconnected by a variable that, at first sight, impacts just a few of them. As such, although movements in the price of oil have undoubtedly the most significant impact on industries that are oil consumers, financial institutions might be exposed to oil price risk through their links with all economic sectors and industries either through traditional banking operations but also through the design and offer of hedging instruments and risk diversification tools. When we consider banking institutions, their performance is influenced, on the one hand, by changes in the probability of loan default of their customers that may originate from oil or non-related sectors, but also by the propensity towards investments of all market participants. Certainly, both of them are affected by the level of risk in the economy and, as long as oil price shocks increase the economy-wide risk level, this will have negative consequences on financial companies' performance. Furthermore, as long as the amount of money lent out from financial companies to the companies whose business relies heavily on oil is high, their exposure to changes in the oil price will be significant, but one should not forget that financial institutions are also well diversified and hedged with respect to the loans extended to the oil-and other non-oil related sectors. Moreover, the difficulties concerning oil price forecasting and the reality of open economies in a globalized world makes the study of the financial sector exposure to oil price changes interesting and thought-provoking. In this framework, an investigation of the largest financial institutions in the European Union, as the one proposed in this paper, offers a better understanding of the economy-wide role of these institutions in the transmission of risk.

Figure 1 illustrates the evolution of crude oil imports prices (USD per barrel of oil) in the European Union (EU) between 2000 and 2016. It is easily observable the surge in oil prices after 2000 until 2008, as well as between 2009 and 2011, while crude oil and petroleum imports in the European Union remained rather stable. Oil import prices dropped significantly after 2013 to reach approximately 40 US dollars per barrel in 2016, but the recent attacks on Saudi Arabia oil fields might trigger significant rises in the recent future.

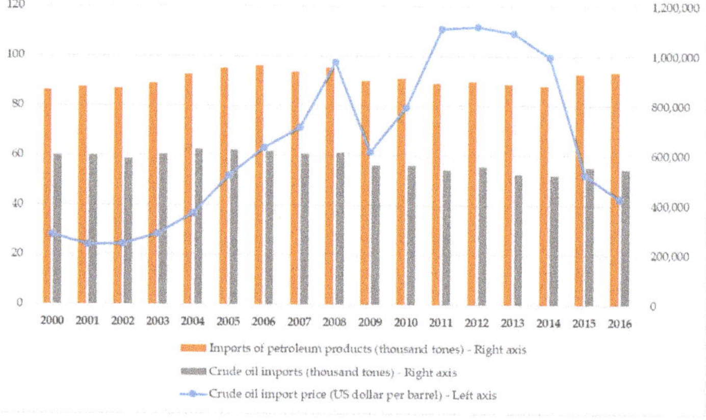


**Figure 1.** Crude oil import prices and imports of crude oil and petroleum products in the European Union, 2000–2016. Note: The crude oil import prices at the European Union Level (EU) level include oil import prices from Austria, Belgium, Czech Republic, Denmark, Finland, France, Germany, Greece, Ireland, Italy, Netherlands, Portugal, Spain, Sweden, and United Kingdom. Data source: Organization for Economic Cooperation and Development (OECD) and Eurostat.

Given that EU member countries are mainly net oil importers, expectation is that when oil price rises the imports would generally decrease; however, considering that oil is such an important and not easily replaceable resource, it could happen that no effect is seen in the short-term. This, in turn, would be incorporated into higher risk-adjusted returns of financial companies. Incidentally, the European Union relies heavily on imports for oil consumption—in 2017, the ratio between net oil imports and gross inland energy consumption, known as the oil import dependency ratio, was 87% at the overall EU level, only slightly lower than the peak value of 89% recorded in 2015. Figure 2 shows the values of this ratio for EU member countries in 2017. In 18 out of 28 EU countries, net oil imports levels were close to energy consumption levels, with dependency ratio values between 96% and 115%. Only Denmark, with a rate of −4% indicates that a net exporting country is an outlier in this EU landscape of heavy reliance on oil imports.

**Figure 2.** Oil import dependency rate in EU countries, 2017 (%). Note: Oil import dependency is defined as the ratio between net imports and gross inland energy consumption. Data source: Eurostat-https://ec.europa.eu/eurostat/web/products-eurostat-news/-/DDN-20190828-1.

One of the best documented and consistent studies on the European Union oil dependency reveals other concerning issues for business and authorities alike [1]. Thus, the study shows that a high proportion of EU oil imports comes from regions that are politically unstable and have been prone to terrorism, conflicts, and wars (such as Russia, Nigeria and sub-Saharan Africa, Middle East, and North Africa). This generates a higher exposure of EU consumers and businesses to the risk of oil supply interruptions and shortages. Moreover, this exposure varies substantially across EU member states; of all countries, the Eastern EU countries (Poland, Slovakia, and Hungary) rely on Russia for over 90% of their crude oil supply, which makes this region highly vulnerable to both geopolitically and price-related oil shocks.

To the best of our knowledge, it is the first time in the literature when the short and long-term relationships between stock returns of financial companies, oil price, and several macroeconomic variables are studied using a panel ARDL model. Thus, our paper genuinely contributes both to the literature in the field and the more thorough understanding of the economic sectors' specific exposure to sources of systemic risk. Furthermore, we contribute to the debate on the particular role played by the financial sector in any open economy, even beyond its traditional functions.

The paper is structured as follows: Section 2 offers insights into the research directions and results in the existing literature, Section 3 presents the data used and the specifications of the model employed—an ARDL panel that is able to cope with the specific features of the data set and provides a useful understanding of the short-versus long-term relationship between oil and stock prices—, Section 4 outlines the most important and relevant results and Section 5 concludes and explores possibilities for future research. By far, the most important conclusion of our endeavor resides in confirming the critical role of the financial sector for the EU financial stability, given its position of intermediary and transmitter of economic shocks.

## 2. Literature Review

The empirical literature on this subject is currently growing and there are several studies that focus on the impact of oil price fluctuations on stock market returns. The current empirical evidence suggests that oil price changes lead to fluctuations in stock prices despite rather mixed results. As such, there are studies that found no relationship between oil price fluctuations and stock market returns, while others provided completely different findings.

One of the first tests of the impact of oil price changes on stock returns has been implemented for the Japanese economy [2]. The authors use an Arbitrage Pricing model to investigate the influence of Japanese macroeconomic factors—oil price among them—over the Japanese equity prices and find no risk premiums for oil price changes and unexpected changes in the currency rate allocated by market investors; they consider their results surprising given the high importance of international trade for the Japanese economy. Later, another study contradicts these findings and discovers that Japanese stock market returns are exposed to changes in oil prices [3]. The authors explain the difference between the two studies as stemming from the sample period and the research methodology—the latter paper uses a VAR model in the investigation. Moving to the US market, an investigation of the link between returns of oil futures contracts and stock returns during the 1980s using the same VAR methodology finds no correlation between the two variables, except for the returns of oil companies [4]. As a result, the authors propose the use of oil futures contracts in stock portfolios as good diversification tools. Enlarging the examination of the oil–stock price relationship to other developed economies, Jones and Kaul [5] find that the United States and Canadian stock prices react to changes in oil prices and this reaction is explained by the impact of oil price shocks on real cash flows, but the same pattern cannot be found for United Kingdom and Japanese stock prices; in the latter case, oil price shocks' impact on stock prices are larger than the changes in real cash flows or expected returns.

In a more recent research conducted on the Chinese market, oil price shocks are not found to significantly impact real stock returns of the most important Chinese equity indices, but the impact is significant in the case of an index of stocks from the manufacturing sector and of some oil companies [6]. Somehow in the same vein, an analysis of the impact of crude oil price shocks on stock market returns from United Kingdom, France, and Japan between 1989 and 2007, which uses a more sophisticated methodology–wavelet analysis coupled with a Markov Switching Vector Autoregressive (MS-VAR) approach, shows that shocks in oil prices do not negatively affect the recession phases in the stock market (with the exception of Japan) [7]. Contrarily, the same shocks are able to reduce the moderate and/or expansion phases of the stock markets, although this is only a temporary effect.

One of the most influential papers indicating an important role of oil price volatility in negatively influencing real stock returns concludes that changes in oil prices are able to explain a larger part of the forecast error variance in real stock returns in a VAR model compared to interest rates [8]. Moreover, the same author finds consistent evidence that shocks in oil price volatility impact asymmetrically the economy. On the same line, but this time using non-linear model specifications, Ciner [9] points toward a non-linear impact of oil price shocks on the S&P 500 stock index returns. Moreover, the author documents a feedback relation from the stock index to oil futures markets, particularly in the 1990s, which highlights the significant interdependence between financial and real markets in the United States.

An interesting study examines the Granger-causal relationships between oil price fluctuations and global stock returns using time-varying causality tests in mean and in variance by splitting the sample into emerging and developed countries [10]. Its main finding consists in revealing that the relationship between oil prices and stock markets depends on the economic cycles' phases. Thus, during economic downturns there seems to be a volatility spillover from G7 stock markets to crude oil prices (due to increased overall uncertainty). The same conclusion is obtained in the case of emerging markets, even though of a different magnitude. The causal link from oil price volatility to G7 stock returns volatility also shows up during turbulent times (1994, 1999, and 2007–2009), but this finding is not supported in the case of emerging countries. Nevertheless, during late 2014 and late 2015, crude oil prices caused increased volatility in emerging stock markets' returns, a phenomenon which disappeared afterwards. What is also interesting is the volatility spillover experienced from G7 stock markets to emerging stock markets in the period before the global financial crisis, a relationship that ended in the post-crisis period and became very weak afterwards.

In addition to research that examines the general impact of oil price volatility and shocks on various economies, there are important results obtained as a result of investigations concerning the effects of oil price changes on stock returns from various economic sectors and industries. For example, Dominioni et al. [11] investigated the relationship between oil and renewable energy stock prices between 2006 and 2016 using an integrable non-autonomous Lotka–Volterra model. They discovered that the relationship between oil price and renewable energy sources stock prices is characterized by major structural breaks taking place in 2008 and around 2013. Reboredo et al. [12] studied the relationship between oil price and stock market returns using daily data based on the aggregate S&P 500 and Dow Jones Stoxx Europe 600 indexes and US and European industrial sectors (banking, automobile and parts, chemical, oil and gas, telecommunications, industrial goods, utilities, technologies) from 2000 to 2011. Using wavelet multi-resolution analysis authors discovered that oil price changes have little effect on stock market returns in the pre-crisis period at both the aggregate and sectoral level. During the financial crisis, their findings show the positive interdependence between oil price shocks and stock returns at both aggregate and sector level.

At the same time, oil price fluctuation can be influenced by political and economic events that would lead to structural change, which on their side might generate biased estimations [13]. Thus, in a period of six months in 2017, several structural changes occurred in the Brent oil price. The results indicate a cointegrating relationship (positive or negative) between oil price and VIX (a global volatility index) or OVX (CBOE crude oil volatility index) depending on the period. All in all, OVX is considered to be a better predictor of oil price (as a measure of fear in the market) and authors advise investors to observe not only macroeconomic and political events when trying to predict oil price (for hedging or speculating purposes) but also the relationship between the latter and market volatility.

Considering that the study of oil price fluctuations impact on the stock market industrial sectors was rather scarce until now, it is meaningful to explore each industry's specific response to oil price changes, as each industry has its own particularities [14]. When it comes to the financial sector, results showed no clear pattern until now. Positive effects of oil price fluctuations were experienced on oil related and oil substitute products industries and negative effects on sectors that use oil but also inconsiderable effects on the financial sector. These findings may be explained by the supply chain dependency to oil markets as the key to explore the impact of oil price fluctuations in any sector [15].

A study on the relationship between oil price fluctuations and European stock returns using the Dow Jones Stoxx 600 index and 12 industrial sector indices from European countries shows that Financial, Oil and Gas, Industrials, Basic Materials, and Personal and Household Goods sectors present positive responses to oil price movements [16]. Additionally, in a subsequent study on the same relationship, it is revealed that the Financial and Consumer Goods sectors exhibit a negative response to oil price fluctuations [17]. Moreover, a study on the relationship between oil prices and the stock returns of US companies listed on the NYSE using a GARCH model also finds that oil price movements have asymmetric effects on stock returns taking into consideration the sector they originate from; for the Electricity, Engineering, and Financial sectors, the results were inconclusive [18]. Another study proves that most industries from European countries would benefit from the negative oil price movements. The authors found that the impact of oil prices shocks on 38 stock markets at the industry level notably varies along the European industries over the period 1983–2007 [19]. These results are also confirmed by Bagirov et al. [20]. They also demonstrate that the performance of listed oil and gas firms is significantly and positively impacted by crude oil prices; however, other factors impact the performance of unlisted firms. During the global financial crisis only the performance of the listed companies was negatively influenced by the situation.

Another recent study finds that stock market return and oil price are negatively correlated and oil price changes cause more volatility in stock returns. The relationship is explained by the fact that oil price represents an input for many industries, thus an increase in oil price leads to economic turmoil by creating inflation and unemployment [21]. Jinghzen et al. [22] analyzed the influence of oil price changes on stock returns of UK oil and gas companies at different time scales using the wavelet analysis. By decomposing the original return series into multiscale orthogonal components the changes can be better monitored in the short, medium, and long-term. Authors find that there is a positive relationship between the two variables which increases in the medium and long-term while in the short-run there is low interdependence.

Overall, the existing results revealed in the literature do not indicate a clear-cut and well-defined relationship between oil price changes and stock prices and returns, but suggests that this link has sectoral particularities that deserve to be better explored. In this framework, our research adds to the literature on the exposure of economic sectors to sources of systematic risk and highlights the (sometimes) neglected or less understood impact of a macroeconomic or even global variable—i.e., the oil price—on an economic sector that is not directly related to oil price trading and consumption, but may be a catalyst of risk transmission at the macroeconomic level.

## 3. Materials and Methods

We investigate the relationship between financial companies' stock prices and oil prices using a sample of 76 financial companies headquartered in EU included in the Forbes 2000 Ranking of the World's Largest Public Companies [23]. Table A1 in Appendix A shows the financial companies selected for the analysis, originating from six industries (Consumer financial services, Diversified insurance, Investment services, Major Banks, Regional Banks, Thrifts, and Mortgage Finance) and from 16 EU member countries (Austria, Belgium, Denmark, Finland, France, Germany, Greece, Hungary, Ireland, Italy, Netherlands, Poland, Portugal, Spain, Sweden, and the United Kingdom).

The macroeconomic set of explanatory variables is presented in Tables 1 and 2. It includes Brent crude oil price, local stock market indices, EUR/USD exchange rate (EUR as base currency), the oil imports dependency ratio—defined as the ratio of imports of oil and oil products to the total imports —and the harmonized index of consumer prices (HICP)—as a measure of consumer price inflation—for each headquarter country, VIX—a real-time market index that represents the market's expectation of 30-day forward-looking volatility (it measures market risk, fear, and stress) and VSTOXX-based on EURO STOXX 50 real-time options prices, is designed to reflect the market expectations of near-term up to long-term volatility by measuring the square root of the implied variance across all options of a given time to expiration. While VIX is a measure of global volatility and widely used for such purpose, VSTOXX incorporates the perception of market investors regarding the volatility of European equity markets.

**Table 1.** Variable description.

| Variable | Notation | Measurement Unit | Data Source |
|---|---|---|---|
| Stock prices | P | EUR | Forbes, Bloomberg |
| Oil prices | OIL | USD per | Bloomberg |
| Exchange rate | FX | EUR/USD (EUR as base currency) | Bloomberg |
| Consumer price index/Inflation rate | HICP | Units | Eurostat |
| Oil imports dependency | IMP | Percentage | UN Comtrade Database |
| Market volatility | VIX<br>VSTOXX | Units | Bloomberg |

Source: Authors' work.

**Table 2.** Local stock market indices.

| Country of Financial Company's Headquarters | Local Stock Market Index |
| --- | --- |
| Austria | ATX |
| Belgium | BEL20 |
| Denmark | OMX Copenhagen all shares |
| Finland | OMX Helsinki |
| France | CAC Large 60 |
| Germany | DAX |
| Greece | Athens General Composite (ATG) |
| Hungary | Budapest SE (BUX) |
| Ireland | ISEQ Overall |
| Italy | FTSE Italia All Share |
| Netherlands | AEX All Share |
| Poland | WIG |
| Portugal | PSI 20 |
| Spain | IBEX 35 |
| Sweden | OMX Stockholm |
| United Kingdom | FTSE–All Share |

Source: Authors' work.

All data has a monthly frequency and was included in our analysis in logarithmic form in order to mitigate the different measurement units across variables. The period under investigation is January 2010 to December 2018. Table 3 presents the common descriptive statistics of the data series.

**Table 3.** Descriptive statistics.

| Variable | Mean | Median | Maximum | Minimum | Standard Deviation | Skewness | Kurtosis |
| --- | --- | --- | --- | --- | --- | --- | --- |
| Stock price | 266.068 | 18.319 | 59921.060 | 0.144 | 1909.682 | 20.960 | 527.265 |
| Oil price | 64.185 | 63.595 | 94.012 | 28.329 | 17.058 | −0.138 | 1.751 |
| EUR/USD | 1.243 | 1.249 | 1.480 | 1.052 | 0.114 | 0.000 | 1.760 |
| HICP | 99.193 | 99.800 | 107.800 | 87.780 | 3.510 | −0.618 | 3.529 |
| Oil imports dependency | 0.1053 | 0.095 | 0.414 | 0.022 | 0.055 | 2.047 | 8.876 |
| Local indices | 6889.808 | 4547.812 | 26,255.000 | 29.439 | 6708.770 | 1.238 | 3.607 |
| VIX | 13.924 | 12.956 | 32.093 | 8.050 | 4.346 | 1.512 | 5.865 |
| VSTOXX | 21.960 | 21.050 | 46.680 | 11.990 | 6.195 | 1.032 | 4.433 |

Source: Authors' calculations.

Taking each variable separately, expectation is that oil price will influence in a positive way companies' stock returns as an increase in the former might create leverage investments in stocks leading to a higher demand and thus an increase in overall stock price for oil companies. Consequently, for companies in the same business network, i.e., financial companies, this might build up their stock price as they are directly influenced by cash flows of oil companies. When general economic prospects, as reflected in local stock market indices, are good, we expect financial companies' stock prices to go up and vice-versa, given their general exposure to market risk. As EU economies are generally net oil importers, a depreciation of the EUR against the USD means bad news for oil importers (sign is expected to be negative in this case). The majority of these countries being net importers of oil an increase in imports of oil would have as an effect a higher market risk and a lower stock return. At the same time, expectation may be that when oil price rises the imports would generally decrease; however, considering that oil is an important and not easily replaceable resource, it could happen that no effect is seen in the short-term.

As HICP measures inflation, expectation is that it can either positively or negatively impact stocks, depending on the ability of the investor to hedge inflation risk and the government's fiscal policy. At the same time, during economic contractions, high inflation leads to low stock returns (spillover effect), which shows the negative relationship between the consumer price index—as a measure of inflation and stock returns [24]. VIX and VSTOXX would normally impact in a negative way stock returns as both of them designate market volatility and empirical evidences show that higher market volatility is typically associated to lower returns [25,26].

We employ panel data as the base econometric model as it helps control for heterogeneity of cross-section units (companies' stock prices in our case) over time but also helps in obtaining more unbiased estimations. For this paper, an autoregressive distributed lag (ARDL) specification is used, which is a standard least square estimation that includes lags of both the dependent and independent variables. The ARDL framework examines cointegrating relationships between variables and allows for a long-run versus short-run view over the link between variables; thus, the ARDL specification results in a dynamic model where the effect of regressors on the dependent variable occurs over time and not immediately [27,28].

The equation of the ARDL ($p$, $q_1$, $q_2$ ... $q_n$) model, where $p \geq 1$ and $q \geq 0$—assuming that the lag order $q$ is identical for all variables in the vector X, is:

$$Y_{it} = \sum_{j=1}^{p}\left(\alpha_{ij}Y_{i,t-j}\right) + \sum_{j=0}^{q}\left(\delta'ijX_{i,t-j}\right) + \mu_i + \varepsilon_{it} \tag{1}$$

where $Y$ is the stocks' price and the dependent variable, while $X$ is the vector of explanatory variables: Brent crude oil price (OIL), local stock market indexes (IND), EUR/USD exchange rate (FX), the oil imports dependency (IMP), the harmonized index of consumer prices (HICP), VIX, and VSTOXX. $\alpha_{i,j}$ and $\delta'ij$ are the coefficients associated with a linear trend, lags of $Y$, and lags of the $q$ regressors X for $j = 0, \dots q$, $\mu_i$ is the short-run coefficient and $\varepsilon_{it}$ is the error term.

Equation (1) allows for the parameters to vary between units and they can be estimated using the mean group estimator per company and then the average for the group [28]. The pooled mean group (PMG) allows for the short-run parameters to vary across companies but makes the long-run parameters homogeneous [29]. The PMG estimation is also consistent from the point of view of variable endogeneity, as an alternative to the more traditional Arellano-Bond GMM dynamic panel estimation [30,31]. If we reparametrize (as a way to demonstrate the short-run dynamics but also the long-run relationship of the underlying variables) the model we obtain is:

$$\Delta Y_{it} = \varphi_i(Y_{i,t-1} - \beta'_iX_{i,t}) + \sum_{j=1}^{p-1}\left(\alpha_{ij}^*\Delta Y_{i,t-j}\right) + \sum_{j=0}^{q-1}\left(\delta_{ij}^{*\,'}\Delta X_{i,t-j}\right) + \mu_i + \varepsilon_{it} \tag{2}$$

where $\beta_i$ is a vector of interest which is used to measure the long-run impact of the regressors on the returns of stocks and $\varphi_i$ is the error corrector mechanism impact on the dependent variable, $\varepsilon_{it}$ is the error term, and the rest of the parameters are short-run coefficients ($\mu_i$, $\delta'ij$).

Various authors have used in their estimations endogeneity tests for ARDL models, (see [32,33], for example), but although panel ARDL modelling does not intrinsically embed a test of variable endogeneity, this problem is unlikely to arise as long as the errors are serially uncorrelated because the regressors are at their lagged levels. Additionally, as long as cointegration is present, the OLS regression is consistent. On the same vein, the ARDL model produces consistent coefficients despite the possible presence of endogeneity because it includes lags of dependent and independent variables [29,34].

ARDL models have been used for decades to investigate the relationships between economic variables given their ability to decipher short-versus long-term connections between variables [35–37]. The oil–stock price link has not been an exception. Thus, Donggyu et al. [38] use the non-linear autoregressive distributed lag (ARDL) approach to test whether there is any relationship between oil price and stock prices of renewable energy firms. They find that oil price changes have asymmetrical effects on renewable energy stock prices in the short-run but not in the long-run. Moving further, another study confirmed both the existence of a long-and short-term relationship between the Romanian energy market and the capital market using, among others, an ARDL model, on the basis of a mixture between stationary and non-stationary time series [39]. In addition to this, another research found that in the long-run oil price, similar to interest rates and real effective exchange rates, has a negative impact on Malaysia's stock returns, while industrial production has a positive impact [40]. Results of another study using ARDL confirm that both in the long-and short-run there is a negative relationship between Shanghai stock returns and oil price [41].

In order to choose the most appropriate model of the long-run underlying equation, it is important to determine the optimum lag length by selecting the model with the smallest Akaike information criterion (AIC), Schwartz Bayesian criterion (SIC), or Hannah-Quinn criterion (HQ) or small standard errors and highest Adjusted $R^2$ [42]. These tests also account for robustness of the model and results.

As the regression equation gives only the short-run relationship between the variables and we are also interested in the long-run relationship between the variables, the concept of cointegration and reparametrization of the ARDL model into an error correction model becomes imperative [43]. Thus, with the help of ECM, Equation (2) provides both short-term and long-term information about the relationship between the dependent variable and the set of independent variables.

## 4. Results and Discussion

In order to see if the variables can be good predictors, tests for unit root were performed. Thus, Levin, Lin, Chu test (LLC), Breitung t-stat, Im, Pesaran and Shin W-Stat (IPS), ADF–Fisher Chi-square, and PP Fisher Chi-square tests were used. The advantage of using these tests is that the null hypothesis does not change for any of them, meaning the null hypothesis assumes common/individual unit root process.

The results in Table 4 show that the series of oil imports dependency, harmonized index of consumer prices, VIX and VSTOXX were stationary at level, while oil price, local indexes, EUR/USD exchange rate, and stock prices proved to be stationary at first difference. Considering that not all of the series are stationary at level, the use of ARDL model is more than justified as this technique is preferable when dealing with variables that are integrated at different orders [34,44].

**Table 4.** Unit root test of the variables.

| Variables | Order of Integration | LLC | Breitung | IPS | ADF | PP |
|---|---|---|---|---|---|---|
| Stock return | 1 | −67.964 (0.000) | −28.608 (0.000) | −65.278 (0.000) | 30.196 (0.000) | 45.113 (0.000) |
| Oil price | 1 | −57.337 (0.000) | −6.59729 (0.000) | −51.8059 (0.000) | 2272.75 (0.000) | 3493.71 (0.000) |
| Local indices | 1 | −72.942 (0.000) | −23.0325 (0.000) | −62.5537 (0.000) | 2893.06 (0.000) | 4327.37 (0.000) |
| EUR/USD | 1 | −61.924 (0.000) | −50.2184 (0.000) | −58.9631 (0.000) | 2686.40 (0.000) | 4874.68 (0.000) |
| HICP | 1 | −74.759 (0.000) | −40.7939 (0.000) | −76.6122 (0.000) | 3601.05 (0.000) | 4882.43 (0.000) |
| Oil imports dependency | 0 | −6.913 (0.000) | −11.2161 (0.000) | −9.52687 (0.000) | 336.120 (0.000) | 889.111 (0.000) |
| VIX | 0 | −27.210 (0.000) | −3.61176 (0.000) | −18.8560 (0.000) | 618.617 (0.000) | 1081.68 (0.000) |
| VSTOXX | 0 | −33.108 (0.000) | −17.3583 (0.000) | −16.8271 (0.000) | 543.504 (0.000) | 1103.24 (0.000) |

Note: The null hypothesis is that the series is a unit-root process; $p$-values are reported in parentheses. Source: Authors' calculations.

The results of our panel ARDL estimations are summarized in Table 5. The table shows the estimation of the long-run and short-run coefficients by using only AIC as the lag length criterion, as results obtained using SC or HQ are very similar. Nevertheless, we discuss differences in our estimations obtained by using the three lag length criteria whenever they are relevant. It is important to mention that the long-run coefficients are constrained to be the same across financial companies, while the short-run coefficients are allowed to vary. The estimations whose results are presented in Table 5 are performed by varying the independent variables as a test for results' robustness. Thus, we vary regressors in order to examine the presence of oil price exposure when the panel regression specification is changed; the results are organized in the table from Model 1 to Model 8.

**Table 5.** Results of panel ARDL estimations.

| Model | Lag length | Independent Variables | | | | | | | | | SE | LL |
|---|---|---|---|---|---|---|---|---|---|---|---|---|
| | | CQ1 | CO (−1) | FX | HICP | IMP | IND | OIL | VIX | VSTOXX | | |
| 1 | ARDL (2, 1, 1, 1, 1, 1, 1) | *Long-run equation* | | | | | | | | | 0.159 | 10,938.63 |
| | | na | na | −1.280 * | −2.807 * | 0.243 ** | 1.288 * | 0.230 ** | −0.247 * | – | | |
| | | *Short-run equation* | | | | | | | | | | |
| | | −0.047 * | −0.016 | 0.517 * | −0.197 | −0.030 ** | 1.315 * | 0.055 | 0.094 * | – | | |
| 2 | ARDL (2, 1, 1, 1, 1, 1, 1) | *Long-run equation* | | | | | | | | | 0.156 | 10,931.51 |
| | | na | na | −1.339 * | −3.653 * | 0.366 * | 1.162 * | 0.132 | – | −0.383 * | | |
| | | *Short-run equation* | | | | | | | | | | |
| | | −0.047 * | −0.016 | 0.517 * | −0.197 | −0.030 ** | 1.315 * | 0.055 | 0.094 * | – | | |
| 3 | ARDL (1, 1, 1, 1, 1, 1, 1) | *Long-run equation* | | | | | | | | | 0.180 | 8854.49 |
| | | na | na | −4.952 * | −5.821 * | 1.328 * | – | −0.396 ** | −1.476 * | – | | |
| | | *Short-run equation* | | | | | | | | | | |
| | | −0.042 * | na | 0.890 * | −0.548 ** | −0.024 * | – | 0.117 * | −0.008 | – | | |
| 4 | ARDL (2, 1, 1, 1, 1, 1, 1) | *Long-run equation* | | | | | | | | | 0.168 | 9158.56 |
| | | na | na | −4.591 * | −9.452 * | 1.361 * | – | −0.437 * | – | −1.782 * | | |
| | | *Short-run equation* | | | | | | | | | | |
| | | −0.046 * | −0.008 | 0.715 * | −0.154 | −0.030 * | – | 0.144 * | – | −0.066 ** | | |
| 5 | ARDL (2, 1, 1, 1, 1, 1, 1) | *Long-run equation* | | | | | | | | | 0.160 | 10,804.28 |
| | | na | na | – | −0.110 * | 0.219 * | 1.300 * | −0.126 * | −0.161 * | – | | |
| | | *Short-run equation* | | | | | | | | | | |
| | | −0.052 * | −0.008 | – | −0.050 * | −0.022 * | 1.330 * | 0.028 * | 0.069 * | – | | |
| 6 | ARDL (2, 1, 1, 1, 1, 1, 1) | *Long-run equation* | | | | | | | | | 0.160 | 10,804.28 |
| | | na | na | – | −0.311 * | 0.238 * | 1.254 * | −0.152 * | – | −0.161 * | | |
| | | *Short-run equation* | | | | | | | | | | |
| | | −0.006 | −0.100 * | – | −0.020 * | 1.338 * | 0.031 * | 0.091 * | – | −0.006 | | |
| 7 | ARDL (2, 1, 1, 1, 1, 1, 1) | *Long-run equation* | | | | | | | | | 0.158 | 10,886.99 |
| | | na | na | −0.574 * | – | 0.267 * | 1.242 * | 0.002 * | −0.112 * | – | | |
| | | *Short-run equation* | | | | | | | | | | |
| | | −0.050 * | −0.011 | 0.504 * | – | −0.025 * | 1.324 * | 0.056 * | 0.092 * | – | | |
| 8 | ARDL (2, 1, 1, 1, 1, 1, 1) | *Long-run equation* | | | | | | | | | 0.156 | 10,873.16 |
| | | na | na | −0.506 * | – | 0.348 * | 1.238 * | −0.078 | – | −0.128 * | | |
| | | *Short-run equation* | | | | | | | | | | |
| | | −0.051 * | −0.009 | 0.376 * | – | −0.023 ** | 1.328 * | 0.051 | – | 0.104 ** | | |

* and ** denote statistical significance at least at 1% and 5% level, respectively. SE and LL indicate the standard error of the regression and log likelihood, respectively. na—not available. "—"[M1]—the variable is not included in the model. All results are available from authors. Source: Authors' calculations.

In the long-run, we notice statistically significant coefficients for all independent variables with signs that are consistent across the models. At the same time, we find statistically significant coefficients in the short-run equations as well, but not for all variables in all models and, more important, with different signs compared to the long-run equation. We further present and discuss the results for each of the eight models used in our estimations.

In Model 1, which uses VIX as a volatility measure, the long-run coefficient for oil price is positive, which shows that an increase in oil prices leads to an increase in stock prices. Regarding the EUR/USD exchange rate, we notice the opposite situation; for this variable, the coefficient is negative, which means that in the long-run an appreciation of the EUR would hurt financial companies' stock prices. Contrarily, in the short-run the oil price does not show a significant coefficient, but the cross-section coefficients indicate that 61 out of 76 cross-sections' coefficients are statistically significant (28 positive and 33 negative)—see Table 6. In other words, even if the short-run coefficient for oil price for the entire group of financial companies does not show statistical significance, there are enough cross-sectional features that induce specific exposures to oil price changes. By using SIC and HQ, in the long-run, oil price does not have a significant coefficient, but for the other variables we notice significant coefficients; moreover, the value of the coefficients and their signs are the same for all variables.

**Table 6.** Short-run statistically significant cross-section coefficients—number and signs.

| Model | Independent Variables | | | | | | | | | | | | | | | | | |
|---|---|---|---|---|---|---|---|---|---|---|---|---|---|---|---|---|---|---|
| | CQ1 | | CO (−1) | | FX | | HICP | | IMP | | IND | | OIL | | VIX | | VSTOXX | |
| | + | − | + | − | + | − | + | − | + | − | + | − | + | − | + | − | + | − |
| 1 | 2 | 74 | 34 | 33 | 45 | 10 | 0 | 3 | 31 | 35 | 74 | 1 | 28 | 33 | 55 | 15 | — | |
| 2 | 3 | 73 | 36 | 36 | 43 | 11 | 0 | 4 | 31 | 37 | 75 | 1 | 29 | 31 | — | | 52 | 21 |
| 3 | 0 | 76 | na | | 48 | 9 | 0 | 1 | 15 | 49 | — | | 48 | 11 | 8 | 68 | — | |
| 4 | 0 | 76 | 35 | 31 | 45 | 5 | 0 | 1 | 21 | 50 | | | 48 | 14 | | | 7 | 67 |
| 5 | 1 | 75 | 43 | 33 | — | | 27 | 49 | 40 | 36 | 75 | 1 | 35 | 41 | 51 | 25 | — | |
| 6 | 1 | 75 | 42 | 34 | — | | 27 | 49 | 39 | 37 | 75 | 1 | 36 | 40 | — | | 43 | 25 |
| 7 | 2 | 74 | 40 | 36 | 56 | 20 | — | | 36 | 40 | 75 | 1 | 37 | 39 | 56 | 20 | | |
| 8 | 2 | 74 | 38 | 33 | 39 | 13 | — | | 32 | 39 | 75 | 1 | 27 | 34 | — | | 47 | 22 |

Note: "—" indicates a variable that is not included in the respective model. "+"—positive coefficients; "-"—negative coefficients. na—not available. Source: Authors' calculations.

Model 2 is similar to Model 1 but VIX is replaced by VSTOXX as a variable incorporating market volatility. This model shows no significant coefficient for oil price, but the EUR/USD exchange rate, local stock market index, and VSTOXX show statistically significant coefficients in both long-run and short-run equations. The VSTOXX results confirm the previously obtained result in the case of VIX, i.e., financial market volatility depresses stock prices. The same is true for the long-run coefficients identified for the currency rate and the market index. We also notice, as in the case of Model 1 estimations, that while in the long-run the signs for the EUR/USD exchange rate and VSTOXX are negative, in the short-run the signs are positive. This suggests that market investors might perceive differently the exposure of financial companies to various risk factors over the short versus the long-run, which is reflected in their assessment of prices and returns. Regarding the signs of the coefficients for the local stock market index, they are positive both in the long- and short-run, which confirms our hypothesis that good economic conditions, mirrored by stock market indexes increases, generate higher stock returns, regardless of the industry or economic sector. Again, as for Model 1, there is no significant difference between SIC and HQ in results.

Model 3 excludes the local stock market index and uses VIX as a volatility measure. In the short-run, as well in the long-run, we identify statistically significant coefficients for oil price, but with different signs. The fact that all long- and short-run coefficients for oil price become statistically significant when removing the local stock market index from the equation represents somehow a surprising result because it suggests that market investors do take into account an exposure of financial companies to oil price movements that is beyond the overall exposure to market risk. As such, this result might demonstrate the exposure of the financial sector to real economy risk factors. All criteria show significant coefficients for all variables in the long-run, as well as in the short-run, but except VIX. It is important to mention that the signs of the coefficients for the EUR/USD exchange rate are the same as in Models 1 and 2, i.e., negative in the long-run and positive in the short-run.

In Model 4 we removed the local stock market index and VSTOXX is now the measure of volatility this time. As in Model 3, we confirm the oil price exposure of financial companies in the absence of the local stock market index. The long- and short-run coefficients for oil price are statistically significant, but this time the statistical significance is found at least at 1% in all cases, not just in the short-run equation. The long-run coefficients are negative and the short-run coefficients are positive, similar to Model 3. All long-run coefficients are statistically significant for all criteria, except for the short-run coefficients for HICP and IMP; actually, the IMP coefficient is significant only by using AIC. We interpret this result as indicating a weak and residual exposure of financial companies' stock returns to both inflation risk and oil imports dependency of the country where the company is headquartered.

Model 5 excludes the EUR/USD exchange rate and uses VIX as volatility measure. In this case, all long-run coefficients for oil price are statistically significant and the signs are negative, as in Models 3 and 4. In the short-run, all coefficient for oil price are significant, but only by using AIC the sign is positive; when SIC or HQ are used, the sign becomes negative. We interpret this result as a weak exposure of financial companies' stock prices to oil price in the absence of the exchange rate and as an intermediate role of the currency rate for the oil exposure. Moreover, under this specification, the coefficients for HICP became significant, as indicated by all criteria; this also points toward the intermediary role for the inflation rate in the relationship between stock and oil prices.

In Model 6 we eliminate the EUR/USD exchange rate and VIX from our panel equations, while using VSTOXX as volatility measure. The estimations are very similar to the ones for Model 5. All long-run coefficients are statistically significant and the signs are negative, while all short coefficients are significant, but only by using AIC, and with a positive sign. When SIC and HQ are used, the short-run coefficients for oil price became negative. Again, all the coefficients for HICP are significant as a result of the removal of the exchange rate from the equations. We interpret this result as a confirmation of the results obtained with Model 5.

Model 7 removes HICP and VSTOXX from the panel equations. All long-and short-run coefficients for oil price are significant. Due to the presence of the local stock market index in the equations, we did not expect significant coefficients for oil price. Looking back at Models 1 and 2 results, where both HICP and local stock market index were included in the equations, the coefficients for oil were not significant (except using AIC in Model 1). Again, this result suggests the existence of strong exposure of financial companies' stock prices to changes in the price of oil. Thus, the exposure of the financial sector to real economy risk factors is reinforced by our findings.

The last model used, Model 8, differs from Model 7 by employing VSTOXX as volatility measure. Here, the coefficients for oil price are statistically significant in the long-run only by using SIC and HQ. As in Models 1 and 2, the short-run coefficients for oil price are not significant, but if we look at the cross-section coefficients we notice that more than half of our companies show statistically significant coefficients (see Table 6). For example, by using AIC, for 61 out of 76 financial companies the cross-section coefficient for oil price is significant; of them, for 27 companies the sign is positive and for 34 companies the sign is negative. In other words, even if the short-run coefficient for oil price for the entire group of financial companies does not show statistical significance, there are enough cross-sections features that induce exposure to oil price risk. Except oil price, all other variables show significant coefficients regardless of the criterion, which indicates a ubiquitous exposure of financial companies' stock prices to macroeconomic risk factors.

It is important to mention that we notice the existence of a statistically significant long-term cointegration relationship between variables, regardless of model. In addition, overall we find a connection between each independent variable and the dependent one, but particularities appear depending on each panel specifications but also on the long-versus the short-run perspective. The exposure of financial companies' stock prices to market risk is evident, as indicated by positive coefficients for the local market indices in both long-run and short-run equations. At the same time, inflation risk impacts the stock prices of financial companies, but the link between them is negative; this is in line with the negative relationship between interest rates and the price of financial securities, given the positive relationship between inflation and nominal interest rates. Another pervasive exposure of financial companies' stock prices is identified in the case of market volatility—either through VIX and VSTOXX—, but while volatility surges negatively influence stock prices over the long-run, the opposite is true for the short-run. This finding may point towards the short-run speculative behavior of market participants, while at the same incorporating a negative volatility-driven risk premium in the return of financial companies' stocks. A similar result is found in the case of the foreign exchange rate, with a negative long-run and a positive short-run impact of a EUR appreciation against the USD. This result is not necessarily surprising and we decipher here the augmented negative long-term impact of increases in the price of oil that, coupled with an appreciating currency, increase the systemic risk at macroeconomic level. At the same time, the positive short-term exposure to a EUR appreciation may highlight more the speculative actions in the financial market. Rather interesting, we see that a higher weight of oil and oil products in the countries' imports represents good news for financial companies' stock prices over the long-term and bad news over the short-term; this might be connected to a market perception that an increase in the importance of oil imports is related to a growth process of the economy that eventually boosts stock prices, including the ones of financial companies. Else ways, the short-run negative coefficients may be more likely linked to the increase in risk that is perceived once imports of oil and related products surge.

For what concerns the specific short-term exposure of EU financial companies to oil price changes, the distribution of cross-section coefficients across companies does not reveal any specific pattern based on signs, origin of headquarters, or financial services industry—see Table 6. We find, though, that the number of cross-sections that show more positive than negative coefficients is higher than the cross-sections with the reverse situation (34 compared to 20); of them, 28 companies display statistically significant short-term coefficients for oil price in all panel specifications, of which 19 are positive (companies from all financial services industries and from Finland, Germany, Hungary, and Italy) and nine are negative (companies from the Diversified insurance, Investment services, Regional banks and Consumer financial services, and from Austria, Belgium, France, Germany, Netherlands, Poland, Spain, Sweden, and United Kingdom). These results indicate that EU financial companies' exposure to oil price risk has specificities across the financial services sector but also across countries; at the same time, many of these companies are conglomerates with diversified operations around the world in whose case the oil exposure might be well diversified as a component of global market risk and, as a result, no exposure is identified by the model.

Summarizing our findings, we consider the most important result of our research the fact that oil price displayed statistically significant coefficients when we excluded from the equations other variables, such as the local stock market index, HICP, or the EUR/USD exchange rate. This result indirectly highlights the pervasive exposure of economies from EU to risk factors through the financial sector channels, which raises alarming challenges from the perspective of macroeconomic and financial policies. In other words, our results reinforce the role of the EU financial sector as systemic risk transmitter.

In the case of panel regressions' standard errors (SE), smaller values are better because they indicate that the observations are closer to the fitted line. Contrarily, in the case of log likelihood (LL), the higher the value, the better the model. Taking these into account, we consider Model 1 (SE value is 0.159) as the best Model from our panel ARDL estimations. This model includes all variables and uses VIX to designate financial market volatility.

## 5. Conclusions

Our study investigates the relationship between financial companies' stock prices and oil price using a sample of 76 financial companies headquartered in EU and included in the Forbes 2000 Ranking of the World's Largest Public Companies. The macroeconomic set of explanatory variables includes Brent crude oil prices, local stock market indices, EUR/USD exchange rate, oil imports dependency, HICP, VIX, and VSTOXX. We employ panel data as the base econometric model but also an ARDL specification that handles better cross-section specificities and provides a long-run versus a short-run perspective on financial companies' exposure to oil price risk. The investigation of the most important financial institutions in the European Union in terms of their stock price performance in relation to changes in oil prices and the application of the panel ARDL methodology are, in our opinion, the most important contributions we make to the debate that already exists in the literature. Thus, we provide financial institutions' managers with a better grasp of the risk triggers that influence their performance and point towards directions to improve the design of their hedging policies. Moreover, governments and authorities are offered an enlarged view over the links between macroeconomic risks and financial sector performance, which allows for ameliorated measures of economic policy that protect against shocks.

We find that financial companies headquartered in the European Union are ubiquitously exposed to oil price risk, but this exposure is a long-run one and comes hand in hand with the exposure of financial companies to real economy risk factors. Concurrently, the short-run exposure to oil price changes is less strong compared to the long-run exposure and bears specificities across companies, financial services industries, and countries. Moreover, our results suggest that market investors, although displaying a speculation-driven behavior over the short-term, adjust their risk premiums and valuation of financial companies' stock prices and returns over the long-run in order to incorporate macroeconomic risk circumstances.

Overall, our research reinforces the previous findings on the fundamental role of financial companies for the EU financial stability and highlights their relevance as transmitters of economic shocks, even outside the traditional financial landscape. At the same time, oil price is able to generate turbulences at economy-wide level, as long as even industries that are not oil producers and/or consumers display sensitivities to its fluctuations. These findings open several avenues for future research. One possible direction refers to the inclusion in our estimations of the non-oil energy sources, given the increase consumption of renewable primary energy in the European Union. An extension of our study to the most important EU companies from other industries is another possible line of future research, coupled with the inclusion of similar companies from other countries, in order to examine the robustness of our findings at an international level. As well, more sophisticated methodologies may be employed with the aim of providing further reliability to our estimates and of consolidating the knowledge on the exposure of the various economic sectors to risk sources.

**Author Contributions:** The authors contributed equally to this work.

**Funding:** This research received no external funding.

# Appendix A

**Table A1.** Financial companies by industry and country of headquarters.

| Industry | Country of Financial Company's Headquarters | Name of the Company | Market Value (2018) | Sales (2018) |
|---|---|---|---|---|
| Consumer Financial Services | Sweden | Fastighets Balder | $5.8 B | $772 M |
| | United Kingdom | Melrose Industries | $12.4 B | $11.5 B |
| Diversified Insurance | Austria | Vienna Insurance Group | $3.6 B | $11.9 B |
| | Austria | Uniqa | $3.3 B | $6.7 B |
| | Belgium | Aegeas | $10.2 B | $12.3 B |
| | Finland | Sampo | $25.8 B | $9.5 B |
| | France | AXA Group | $63.6 B | $139.7 B |
| | France | CNP Assurances | $16.5 B | $47.6 B |
| | France | Scor | $8.1 B | $16.9 B |
| | Germany | Allianz | $102.3 B | $118.8 B |
| | Germany | Munich Re | $36.3 B | $62.9 B |
| | Germany | Nuernberger Beteiligungs | $922 M | $5.1 B |
| | Italy | Generali Group | $30.1 B | $92.1 B |
| | Italy | Unipol Gruppo | $3.7 B | $16.4 B |
| | Italy | Cattolica Assicurazioni | $1.6 B | $7.1 B |
| | Netherlands | Aegon | $10.7 B | $30.9 B |
| | Spain | Mapfre | $9.2 B | $25.2 B |
| | United Kingdom | Willis Towers Watson | $22.9 B | $8.5 B |
| Investment Services | Belgium | Sofina | $6.8 B | $145 M |
| | France | Wendel | $6.1 B | $9.9 B |
| | Germany | Deutsche Boerse | $24.2 B | $3.7 B |
| | Germany | Wuestenrot & Wuerttembergische | $2 B | $7.5 B |
| | Netherlands | Exor | $15.9 B | $169.1 B |
| | Sweden | Investor AB | $36.1 B | $4.7 B |
| | Sweden | Industrivarden | $10.1 B | $624 M |
| | United Kingdom | London Stock Exchange | $22.4 B | $2.8 B |
| | United Kingdom | Investec | $6.3 B | $6 B |
| | United Kingdom | 3i Group | $13.7 B | $590 M |
| | United Kingdom | St. James's Place | $7.7 B | $1.6 B |
| | United Kingdom | TP ICAP | $2.1 B | $2.4 B |
| Major Banks | Austria | Erste Group Bank | $17 B | $11.5 B |
| | Belgium | Dexia | $9 M | $11 B |
| | Denmark | Danske Bank | $16.6 B | $15.4 B |
| | France | BNP Paribas | $68.7 B | $101.6 B |
| | France | Societe Generale | $24.5 B | $49.5 B |
| | France | Natixis | $18.6 B | $17.5 B |
| | Germany | Deutsche Bank | $18.1 B | $42.3 B |
| | Netherlands | ING Group | $52.7 B | $39.4 B |
| | Spain | Santander | $84.1 B | $89.5 B |
| | Sweden | Nordea Bank | $33.9 B | $15 B |
| | Sweden | SEB AB | $21.2 B | $8 B |
| | Sweden | Svenska Handelsbanken | $21.8 B | $7.4 B |
| | Sweden | Swedbank | $18.6 B | $6.6 B |
| | United Kingdom | HSBC Holdings | $175.5 B | $64.3 B |
| | United Kingdom | Lloyds Banking Group | $60.9 B | $35.2 B |
| | United Kingdom | Royal Bank of Scotland | $41.3 B | $22.2 B |
| | United Kingdom | Barclays | $38 B | $28.2 B |
| | United Kingdom | Standard Chartered | $28.9 B | $24 B |
| Regional Banks | Austria | Raiffeisen Bank International | $8.9 B | $8.8 B |
| | Belgium | KBC Group | $31.3 B | $15 B |
| | Denmark | Jyske Bank | $3.4 B | $2.4 B |
| | France | Credit Agricole | $38.4 B | $52.2 B |
| | Germany | Commerzbank | $11.4 B | $15.2 B |
| | Greece | National Bank of Greece | $2 B | $2 B |
| | Greece | Piraeus Bank | $873 M | $2.9 B |
| | Greece | Alpha Bank | $2.6 B | $3.8 B |
| | Greece | Eurobank Ergasias | $1.8 B | $2.2 B |
| | Hungary | OTP Bank | $12 B | $4.4 B |
| | Ireland | AIB Group | $13 B | $3.8 B |
| | Italy | Bank of Ireland | $7.2 B | $5.9 B |
| | Italy | Intesa Sanpaolo | $45.9 B | $28.1 B |
| | Italy | Unicredit | $32 B | $31.1 B |
| | Italy | Mediobanca | $9.4 B | $3.5 B |
| | Italy | UBI Banca | $3.6 B | $5.6 B |
| | Italy | Banco BPM | $3.6 B | $6.1 B |
| | Italy | BPER Banca | $2.3 B | $2.9 B |
| | Italy | Credito Emiliano | $1.7 B | $4 B |
| | Italy | Banca Popolare di Sondrio | $1.9 B | $2.3 B |
| | Poland | PKO Bank Polski | $1.3 B | $1.3 B |
| | Poland | Bank Pekao | $12.9 B | $4.7 B |
| | Portugal | Banco Comercial Portugues | $8.1 B | $2.6 B |
| | Spain | BBVA-Banco Bilbao Vizcaya | $4.2 B | $3.2 B |
| | Spain | CaixaBank | $41.6 B | $28.3 B |
| | Spain | Banco de Sabadell | $19.9 B | $12.4 B |
| | Spain | Bankinter | $6.4 B | $7.8 B |
| Thrifts and Mortgage Finance | Germany | Aareal Bank | $7.4 B | $2.6 B |

Data source: [23].

## References


1. A Study on Oil Dependency in the EU. A Report for Transport and Environment. Cambridge Econometrics. 2016. Available online: https://www.camecon.com/wp-content/uploads/2016/11/Study-on-EU-oil-dependency-v1.4_Final.pdf (accessed on 4 September 2019).
2. Hamao, Y. An empirical examination of the arbitrage pricing theory: Using Japanese data. *Jpn. World Econ.* **1988**, *1*, 45–61. [CrossRef]
3. Kaneko, T.; Lee, B.S. Relative importance of economic factors in the U.S. and Japanese stock markets. *J. Jpn. Int. Econ.* **1995**, *9*, 290–307. [CrossRef]
4. Huang, R.D.; Masulis, R.W.; Stoll, H.R. Energy shocks and financial markets. *J. Futures Mark.* **1996**, *16*, 1–27. [CrossRef]
5. Jones, C.M.; Kaul, G. Oil and the stock markets. *J. Financ.* **1996**, *51*, 63–491. [CrossRef]
6. Cong, R.G.; Wei, Y.M.; Jiao, J.L.; Fan, Y. Relationships between oil price shocks and stock market: An empirical analysis from China. *Energy Policy* **2008**, *36*, 3544–3553. [CrossRef]
7. Jammazi, R.; Aloui, C. Wavelet decomposition and regime shifts: Assessing the effects of crude oil shocks on stock market returns. *Energy Policy* **2010**, *38*, 1415–1435. [CrossRef]
8. Sadorsky, P. Oil price shocks and stock market activity. *Energy Econ.* **1999**, *21*, 449–469. [CrossRef]
9. Ciner, C. Energy shocks and financial markets: Nonlinear linkages. *SNDE* **2001**, *5*, 1079. [CrossRef]
10. Cevik, E.I.; Atukeren, E.; Korkmaz, T. Oil Prices and Global Stock Markets: A Time-Varying Causality-In-Mean and Causality-in-Variance Analysis. *Energies* **2018**, *11*, 2848. [CrossRef]
11. Dominioni, G.; Romano, A.; Sotis, C. A Quantitative Study of the Interactions between Oil Price and Renewable Energy Sources Stock Prices. *Energies* **2019**, *12*, 1693. [CrossRef]
12. Reboredo, A.; Rivera-Castro, M.A. A wavelet decomposition approach to crude oil price and exchange rate dependence. *Econ. Model.* **2013**, *32*, 42–57. [CrossRef]
13. Lin, J.B.; Tsai, W. The Relations of Oil Price Change with Fear Gauges in Global Political and Economic Environment. *Energies* **2019**, *12*, 2982. [CrossRef]
14. Arouri, M.E.H.; Nguyen, D.K. Oil prices, stock markets and portfolio investment: Evidence from sector analysis in Europe over the last decade. *Energy Policy* **2010**, *38*, 4528–4539. [CrossRef]
15. Gogineni, S. Oil and the stock market: An industry level analysis. *Financ. Rev.* **2010**, *45*, 995–1010. [CrossRef]
16. Arouri, M.E.H. Does crude oil move stock markets in Europe? A sector investigation. *Econ. Model.* **2011**, *28*, 1716–1725. [CrossRef]
17. Degiannakis, S.; Filis, G.; Floros, C. Oil and stock returns: Evidence from European industrial sector indices in a time-varying environment. *J. Int. Financ. Mark. Inst. Money* **2013**, *26*, 175–191. [CrossRef]
18. Narayan, P.K.; Sharma, S.S. New evidence on oil price and firm returns. *J. Bank. Financ.* **2011**, *35*, 3253–3262. [CrossRef]
19. Scholtens, B.; Yurtsever, C. Oil price shocks and European industries. *Energy Econ.* **2012**, *34*, 1187–1195. [CrossRef]
20. Bagirov, M.; Mateus, C. Oil prices, stock markets and firm performance: Evidence from Europe. *Int. Rev. Econ. Financ.* **2019**, *61*, 270–288. [CrossRef]
21. Dhaoui, A.; Khraief, N. Empirical Linkage Between Oil Price and Stock Market Returns and Volatility: Evidence from International Developed Markets. *Economics Discussion Papers.* 2014. No. 2014-12, Kiel Institute for the World Economy. Available online: http://www.economics-ejournal.org/economics/discussionpapers/2014-12 (accessed on 4 September 2019).
22. Jinghzen, L.; Klinkowska, O. Impact of oil price changes on stock returns of UK oil and gas companies: A wavelet-based analysis. *SSRN Electron. J.* **2017**, *43*. [CrossRef]
23. Forbes. Available online: https://www.forbes.com/global2000/list/#tab:overall (accessed on 12 August 2019).
24. Celebi, K.; Honig, M. The Impact of Macroeconomic Factors on the German Stock Market: Evidence for the Crisis, Pre- and Post-Crisis Periods. *J. Financ. Stud.* **2019**, *7*, 18. [CrossRef]
25. Bae, J.; Kim, C.J.; Nelson, C.R. Why are stock returns and volatility negatively correlated? *J. Empir. Financ.* **2007**, *14*, 41–58. [CrossRef]
26. Blau, B.M.; Whitby, R.J. Range based volatility, expected stock returns and the low volatility anomaly. *PLoS ONE* **2017**, *12*, e0188517. [CrossRef] [PubMed]

27. Pesaran, M.H.; Shin, Y. Generalized Impulse Response Analysis in Linear Multivariate Models. *Econ. Lett.* **1998**, *58*, 17–29. [CrossRef]
28. Pesaran, M.H.; Shin, Y.; Smith, R.J. Bounds testing approaches to the analysis of level relationships. *J. Appl. Econ.* **2001**, *16*, 289–326. [CrossRef]
29. Pesaran, M.H.; Shin, Y.; Smith, R.P. Pooled Mean Group Estimation of Dynamic Heterogeneous Panels. *J. Am. Stat. Assoc.* **1999**, *94*, 621–634. [CrossRef]
30. Arellano, M.; Bond, S. Some Tests of Specification for Panel Data: Monte Carlo Evidence and an Application to Employment Equations. *Rev. Econ. Stud.* **1991**, *58*, 277–297. [CrossRef]
31. Baek, J.; Choi, Y.J. Does Foreign Direct Investment Harm the Environment in Developing Countries? Dynamic Panel Analysis of Latin American Countries. *Economies* **2017**, *5*, 39. [CrossRef]
32. He, G.; Bai, L.; Ren, H. Analyst coverage and future stock price crash risk. *J. Appl. Account. Res.* **2019**, *20*, 63–77. [CrossRef]
33. Larcker, D.F.; Rusticus, T.O. On the use of instrumental variables in accounting research. *J. Account. Econ.* **2010**, *49*, 186–205. [CrossRef]
34. Nkoro, E.; Uko, A.K. Autoregressive Distributed Lag (ARDL) cointegration technique: Application and interpretation. *J. Stat. Econom. Methods* **2016**, *5*, 63–91.
35. Koyck, L.M. *Distributed Lags and Investment Analysis*; North-Holland: Amsterdam, The Netherland, 1954.
36. Almon, S. The distributed lag between capital appropriations and net expenditures. *Econometrica* **1965**, *33*, 178–196. [CrossRef]
37. Frances, P.H.; van Oest, R. *On the Econometrics of the Koyck Model*; Report 2004–07; Econometric Institute, Erasmus University: Rotterdam, The Netherland, 2004.
38. Donggyou, L.; Jungho, B. Stock prices of renewable energy firms: Are there asymmetric responses to oil price changes? *Economies* **2018**, *6*, 59.
39. Armeanu, D.S.; Joldes, C.C.; Gherghina, S.C. On the linkage between the energy market and stock returns: Evidence from Romania. *Energies* **2019**, *12*, 1463. [CrossRef]
40. Al-Hajj, E.; Al-Mulali, U.; Solarin, S.A. The influence of oil price shocks on stock market returns: Fresh evidence from Malaysia. *Int. J. Energy Econ. Policy* **2017**, *7*, 235–244.
41. Khan, M.K.; Teng, J.-H.; Khan, M.I. Asymmetric impact of oil prices on stock returns in Shanghai stock exchange: Evidence from asymmetric ARDL model. *PLoS ONE* **2019**, *14*, e0218289. [CrossRef]
42. Shrestha, M.B.; Bhatta, G.R. Selecting appropriate methodological framework for time series data analysis. *J. Financ. Data Sci.* **2018**, *4*, 71–89. [CrossRef]
43. Engle, R.F.; Granger, C.W.J. Co-integration and error correction: Representation, estimation, and testing. *Econometrica* **1987**, *55*, 251–276. [CrossRef]
44. Chudik, A.; Pesaran, M.H. Common correlated effects estimation of heterogeneous dynamic panel data models with weakly exogenous regressors. *J. Econom.* **2015**, *188*, 393–420. [CrossRef]

*Article*

# Energy Performance Certificates—The Role of the Energy Price

**Jon Olaf Olaussen** [1,*], **Are Oust** [1], **Jan Tore Solstad** [1] **and Lena Kristiansen** [2]

[1]   NTNU Business School, Norwegian University of Science and Technology, 7491 Trondheim, Norway;
     are.oust@ntnu.no (A.O.); jan.t.solstad@ntnu.no (J.T.S.)
[2]   Sparebank 1 Regnskapshuset SMN, Postboks 4799 Torgaard, 7467 Trondheim, Norway;
     lena.kristiansen@smnregnskap.no
*    Correspondence: jon.o.olaussen@ntnu.no

Received: 22 August 2019; Accepted: 16 September 2019; Published: 17 September 2019

**Abstract:** Energy performance certificates (EPCs) were introduced to give property buyers better information about the energy efficiency of dwellings and provide incentives to make energy-efficient investments. Previous studies on the effect of EPCs on property value have yielded divergent results, with some studies finding that energy labels affect property values, but others finding that energy labels have little or no effect. The present paper takes the analysis one step further. Using data on energy prices in combination with transaction data from Oslo, we conclude that not only the energy label, but also the energy performance of dwellings in general, has little to no effect on transaction prices. This result is in line with the inferences of several survey studies, which indicate that when people buy a dwelling, they pay considerably less attention to its energy performance compared with other factors, such as the location, neighborhood, size, garden, and the number of bedrooms.

**Keywords:** energy performance certificates; PV energy cost; PV energy savings; house prices; environmental regulation

---

## 1. Introduction

In July 2010, Norway implemented the energy labeling system for houses and dwellings, and energy performance certification became fully mandatory. Since then, all houses and dwellings for sale are required to have an energy performance certificate (EPC). The motivation for using EPCs is to provide information to buyers and tenants about the energy performance of buildings. Reliable information on energy consumption is supposed to improve the functioning of real estate markets and create incentives to invest in energy efficiency. The information provided to potential buyers by the EPC is intended to stimulate energy efficiency investments because the consequent improved energy performance will potentially increase the sale prices and rents of buildings [1].

The EPC reflects the expected energy consumption of a building, which enables buyers to account for the expected current and future energy costs when assessing their willingness to pay for a residence. However, the expected energy costs are not only a function of energy consumption. Energy prices and the interest rate will also influence the energy costs, with the latter working through the discounting of future values into present values. Hence, energy performance will potentially influence the transaction prices of dwellings and houses, in combination with the influence of energy prices and the interest rate.

The empirical literature has drawn contrasting conclusions concerning the role of EPCs in energy conservation [2]. In the commercial segment, Eichholtz et al., found that US office buildings with a "green rating" sold for prices about 16% higher than did those without such ratings [3]. In a study applying hedonic regression on residential dwellings in the Netherlands, Brounen and Kok found a price premium for houses labeled as more energy efficient [4]. (A hedonic regression breaks down the house price into its constituent characteristics and obtains estimates of the contributory value of each characteristic.) Fuerst et al., used both hedonic and augmented repeat sales regressions, and found a significant EPC premium for dwellings sold in England [5]. In addition, a report to the European Commission concluded that EPCs have a significant effect on property prices and rents in selected European Union (EU) countries [1].

Other studies indicate that EPCs have a weak or negligible impact on transaction prices. Interestingly, Murphy investigated the case of the Netherlands, that is the same housing market as Brounen and Kok [4,6]. By applying an online questionnaire, she studied the role of the EPC in the transaction process of buildings. Contrary to Brounen and Kok, she concluded that few householders pay attention to the EPC and stated that the EPC would not have the planned impact, even if the system was fully implemented [4]. Similar surveys from the UK by Laine, and from Germany by Amecke came to the same conclusion, namely that EPCs only have a minor or negligible effect on price negotiations and investment decisions [7,8]. In a similar manner, Backhaus et al. performed in-depth interviews with homeowners in 10 European countries, as well as a large survey among homeowners in five European countries, and found that EPCs have a modest or negligible impact on homeowners' purchase decisions [9]. In a hedonic model for single family housing in Sweden, Wahlström found no price premium of EPCs, but rather a price premium for housing attributes that improve the energy efficiency [10].

Olaussen et al. carried out a statistical study resembling that of Brounen and Kok for the case of Norway [4,11]. However, they reached a similar conclusion to that of Murphy [6]. Performing a hedonic regression analysis based on housing transactions in Oslo, the capital of Norway, they concluded that there was no price premium caused by the energy label itself. Indeed, they suggested that the positive price premium of the EPCs found in the former studies was the result of the methodological design rather than evidence of the impact of EPCs.

Olaussen et al. took advantage of the fact that the EPC system was implemented in Norway by the government "overnight" on July 1 2010. This meant they had a quasi-natural experimental design with pre- and post- EPC data [11]. For each dwelling that was sold before the implementation of the EPCs in 2010, they identified the energy label that the same dwelling was given when resold in 2014. Interestingly, when using the energy labels of dwellings resold in 2014 as a variable in a hedonic regression for dwellings sold before the introduction of the EPC system, they found the same positive relationship between energy labels and the transaction prices. This means that the positive price effect of the energy label was present even before it was implemented, which strongly suggests that the studies that found positive price effects from the energy label captured something other than the effects of the label itself.

However, Olaussen et al. and many other earlier studies on the impact of EPCs on transaction prices did not account for changes in the energy prices in their analyses [11]. Indeed, looking at the development of the energy price in Oslo, a trend is observed whereby the energy price increases, reaching a peak in 2010, and then decreases. As 2010 was the year in which the EPC was implemented in Norway, it is possible that the lower energy price in the post-label period actually neutralizes the potential price premium of the EPC implementation. To gain a more comprehensive understanding of the impact of EPCs on transaction prices, we include a time series for the energy price in the analysis.

Based on the energy price time series, as well as extracting information about the expected energy consumption of buildings from their assigned energy label, we calculate the expected annual energy cost of buildings. Moreover, following Olaussen et al., we utilize the fact that energy labels were implemented overnight on 1 July 2010 [11]. Instead of focusing on labels, we now focus directly on the energy consumption. Therefore, to each dwelling sold before the implementation of the EPCs, we assign the same expected annual energy use that was calculated for that dwelling when it was resold after 2010. Thus, using a hedonic regression, we can assess whether the post-label impact of the expected annual energy cost on transaction prices is stronger than the pre-label impact. Our results indicate that the impact of the expected energy costs on transaction prices is more moderate after the implementation of the EPCs.

The paper is organized as follows. Section 2 outlines the energy labeling system and some descriptive statistics. Section 3 describes the method, and the results of the empirical analysis are presented in Section 4. A discussion and concluding remarks are provided in Section 5.

## 2. The Energy Labeling System for Dwellings and Houses

The energy performance of buildings directive (EPBD) is the EU's main legislative instrument to improve the energy performance of buildings [12]. Based on the EPBD, the EPC system was implemented gradually throughout different member states from 2006. The final deadline for implementing an EPC system in the member states was 2009. A recasting of the EPBD [13] in 2010 strengthened the role of EPCs by demanding that when buildings "are offered for sale or for rent, the energy performance indicator of the energy performance certificate of the building or the building unit, as applicable, is stated in the advertisements in commercial media" [13] (p. 24), rather than at the time of signing a purchase agreement or rental contract [1].

In most EU member states, the energy performance ratings are expressed on a letter scale, for instance, from A to G, where A is very efficient and G very inefficient, and this is also the case in Norway. As improved energy performance of buildings are supposed to increase sales prices and rents, the EPC is intended to generate incentives among owners to invest in improving energy efficiency [1]. Still, the implementation of EPCs has been slow in EU, and it has been argued that both the implementation and quality of certification schemes vary from country to country [1].

As mentioned in the introduction, the EPC system was fully implemented in Norway on 1 July 2010. The Ministry of Petroleum and Energy and the Ministry of Local Government and Regional Development were given overall responsibility for the introduction, with the Norwegian Water Resources and Energy Directorate (NVE) as the authority in charge of the certification and inspection schemes [14]. The EPC system was fully mandatory from the beginning, that is, from July 2010, and since then all transactions involving houses and dwellings in Norway must be accompanied by an EPC.

The EPC is a legal document that is required to be shown to the buyer. However, parts of the certificate, for example the Energy Label, can be used as a simplifying short version [14]. Among other details, the document contains data identifying the building, the agent issuing the certificate, the energy label (which represents the calculated delivered energy need) on a scale from A to G, the heating grade (which represents to what extent heating of space and water can be done with renewable energy sources) represented by color grades, advice on how to save energy, and some general recommendations to the buyer [14]. (Since most Norwegian homes use electric heating, the Norwegian EPC system focuses on energy consumption.)

The operational liability of the EPC system in Norway is with ENOVA. (ENOVA is a state-owned company owned by the Ministry of petroleum and energy until July 2018 and then by the Ministry of climate and environment. ENOVA is responsible for funding cost efficient changes in energy production and consumption.) For the owners of existing buildings, there is a self-assessment option in the certification scheme. Normally, these certificates are less detailed than those carried out by professionals, and the cost of the certification process for these buildings is typically at least NOK 1000 (NOK = Norwegian kroner). The hedonic regressions in the studies from the different countries we have referred to in the introduction show surprisingly similar values regardless of whether the EPC scheme was based on self-assessments or assessments made by experts. This includes both the assessment of the quality of the dwelling with respect to energy efficiency and the extra advertising costs associated with selling when energy label information is included. When it comes to new buildings, a qualified expert is required for certification and, hence, it is more expensive than for existing buildings. The quality assurance aspect of the Norwegian certification system is attended to by controls in the market, where wrong inputs may be considered a breach of contract. In such cases, a fine may be issued. The transaction process is supervised by the Norwegian Water Resources and Energy Directorate (NVE), which supervises whether EPCs are presented at sale, whether the EPCs reflect the actual energy efficiency standard of the building, and whether the experts meet the competence requirements [14].

## 3. Methods

We utilize two methods to analyze the effect of EPCs on the value of dwellings when taking the present value of the energy price into account. Both methods build on Gordon's dividend model in hedonic regression [15]. In the first method, the dividend model is utilized to calculate the expected value added of each energy label. This expected value added is compared with the estimate for the actual value added, which is estimated based on the hedonic regression. The second method is a hedonic regression model in which the energy price and the rate of discount are included by using the present value of the expected energy cost as an explanatory variable.

### 3.1. Calculating Expected Value Added

Gordon and Shapiro's growth model is written as follows [15]:

$$PV_0 = \frac{D_1}{r - g} \tag{1}$$

where $PV_0$ is the items value at time $t = 0$, $D_1$ is the expected dividend at time $t = 1$, and $r$ is the demand on return. When we use this model with respect to the dwelling and energy consumption, we can define $PV_0$ as the future energy cost and $D_t$ as the yearly energy cost of the dwelling. If the yearly energy cost is expected to grow with a yearly rate of $g$, we can rewrite the model as follows (The formula calculates the maximum present value (PV) of energy cost, and may overestimate the theoretical costs for dwellings that have a small remainder life expectancy. We expect most of these dwellings to fall in the G category.):

$$PV_0 = \frac{D_1}{r - g} = \frac{D_0(1 + g)}{r - g} \tag{2}$$

To find the yearly energy cost $D_0$, we calculate the maximum energy consumption per square meter in the different energy label categories. This energy consumption is calculated based on the demands from the Norwegian Water Resources and Energy Directorate. The formula for the different labels is presented in Table 1. The only requirement for the G category is that the energy consumption is higher than in the F category. We calculate this category by assuming that energy consumption is 25% higher than in the F category, which is based on the average difference in energy consumption between the F and G categories. Table 1 shows the maximum energy consumption associated with each energy label category.

**Table 1.** Formula for calculating the maximum energy consumption associated with each energy label category [16].

| Type of Dwelling | Delivered Energy per m² Heated (kWh/m²) | | | | | | |
|---|---|---|---|---|---|---|---|
| | **A** $\leq$ | **B** $\leq$ | **C** $\leq$ | **D** $\leq$ | **E** $\leq$ | **F** $\leq$ | **G** No Limit |
| Apartment | 85 +600/A | 95 +1000/A | 110 +1500/A | 135 +2200/A | 160 +3000/A | 200 +4000/A | >F |
| Other dwellings | 95 +800/A | 120 +1600/A | 145 +2500/A | 175 +4100/A | 205 +5800/A | 250 +8000/A | >F |

For instance, a 100-square meter apartment must not exceed the maximum limit of 105 kWh/m² (95 + 10,000/100) for energy consumption to earn a grade B on its EPC. A dwelling without a heat pump and solar energy that was built in accordance with the minimum requirements of the building regulations will normally achieve a grade C. Grade B may be earned, for example, by installing a heat pump to utilize solar energy or by improving the insulation of windows. Grade A is only achieved by dwellings wherein all measures of energy efficiency are adopted. Few dwellings achieve grades A or B. Tables 2 and 3 show the distribution of energy grades for small houses and apartments, respectively. Table 4 presents the distribution after 2010 and the calculated distribution before 2010 based on our data.

**Table 2.** The distribution (number of new certificates) of energy labels for small houses [17].

| Year | A | B | C | D | E | F | G |
|---|---|---|---|---|---|---|---|
| 2014 | 310 | 3274 | 6635 | 7689 | 7709 | 10,802 | 17,417 |
| 2013 | 318 | 2079 | 5415 | 10,587 | 10,148 | 14,535 | 13,228 |
| 2012 | 221 | 1323 | 5017 | 13,129 | 11,660 | 16,546 | 9504 |
| 2011 | 45 | 819 | 4030 | 13,515 | 11,991 | 14,063 | 6310 |
| 2010 | 20 | 340 | 3000 | 9701 | 8665 | 7394 | 2239 |

**Table 3.** The distribution (number of new certificates) of energy labels for apartments [17].

| Year | A | B | C | D | E | F | G |
|---|---|---|---|---|---|---|---|
| 2014 | 889 | 3495 | 4867 | 7954 | 7375 | 9171 | 11,180 |
| 2013 | 339 | 2524 | 4813 | 9466 | 7100 | 10,555 | 11,352 |
| 2012 | 122 | 1624 | 4424 | 11,356 | 6529 | 11,849 | 11,222 |
| 2011 | 97 | 999 | 3377 | 8412 | 5887 | 9596 | 6724 |
| 2010 | 0 | 169 | 1501 | 5144 | 4118 | 4848 | 2982 |

**Table 4.** EPCs and sale year.

| Year | A | B | C | D | E | F | G | Total |
|---|---|---|---|---|---|---|---|---|
| 2014 | 2 | 40 | 209 | 367 | 306 | 474 | 633 | 2031 |
| 2013 | 1 | 11 | 44 | 58 | 45 | 52 | 98 | 309 |
| 2012 | 0 | 12 | 37 | 42 | 28 | 44 | 55 | 218 |
| 2011 | 0 | 7 | 33 | 48 | 14 | 60 | 69 | 231 |
| 2010 (July–Dec) | 0 | 0 | 3 | 14 | 6 | 15 | 29 | 67 |
| 2010 (Jan–June) | 0 | 0 | 6 | 24 | 10 | 22 | 27 | 89 |
| 2009 | 0 | 0 | 25 | 48 | 33 | 44 | 52 | 202 |
| 2008 | 0 | 2 | 20 | 55 | 30 | 40 | 61 | 208 |
| 2007 | 0 | 0 | 16 | 58 | 28 | 41 | 65 | 208 |
| 2006 | 0 | 1 | 29 | 53 | 31 | 59 | 68 | 241 |
| 2005 | 0 | 2 | 15 | 42 | 30 | 55 | 64 | 208 |
| 2004 | 0 | 0 | 9 | 41 | 23 | 52 | 52 | 177 |
| 2003 | 0 | 0 | 6 | 30 | 34 | 45 | 66 | 181 |
| 2002 | 0 | 0 | 2 | 29 | 31 | 58 | 63 | 183 |
| 2001 | 0 | 0 | 1 | 8 | 15 | 35 | 29 | 88 |
| 2000 | 0 | 0 | 1 | 8 | 12 | 19 | 15 | 55 |
| Total | 3 | 75 | 456 | 925 | 676 | 1115 | 1446 | 4693 |

Note: The table shows the number of dwellings sold in a given year with a given EPC grade after July 2010 and the implicit EPC grade before July 2010.

Then, the energy consumption per square meter (kWh/m$^2$) is used to calculate the yearly energy cost for each dwelling as follows:

$$Yearly\ energy\ cost = \frac{kwh}{m^2} \cdot m^2 \cdot p_t^e \tag{3}$$

where $P_t^e$ is the energy cost at time t measured in NOK/kwh. By applying the present value for the grade A dwellings and subtracting the present value of a dwelling in the grade B category, we find the expected value added of a grade A labeled dwelling. This value added can be compared with the value added found in the market transaction data to test whether the expected energy label premium is achieved in the market.

### 3.2. Estimating the Actual Energy Label Premium

The real energy label price premium is estimated based on a hedonic regression model and real estate transaction data. The hedonic model is used to control for heterogeneity with respect to different characteristics, with dummy variables for the different energy labels included as follows:

$$lnP_n^t = \beta_o^t + \sum_{k=1}^{K} \beta_k z_{nk}^t + \varepsilon_n^t \tag{4}$$

Here, the logarithm of the dwelling price per square meter, P, is explained by a set of explanatory variables $z_{nk}^t$. The explanatory variables $z$ comprise age, location, dwelling type, energy label, and dwelling size, and $\varepsilon_n^t$ is the error term.

First, the explanatory variables are the energy labels from A to G, with F as the reference (baseline) energy label. F is chosen instead of G as the baseline because of the unique characteristics of the G category. We found that the G category includes all dwellings where sellers neglect to identify the energy label. For example, if the owner of a C label dwelling neglects to go through the labeling process, the dwelling will automatically get a G label. Second, the age variable measures the difference between the year of the sale and the construction year of the dwelling. As this difference probably is of less importance the older the dwelling is, we measure the age variable by 1/(sale year − construction year). This accounts for the fact that the age of a building is a relatively more important factor if we compare a brand-new dwelling with a one-year-old dwelling than if we compare a 20-year-old with a 21-year-old dwelling. Because of the way the variable is constructed, we expect it to be positively corelated with the house price. Third, we include dummy variables for location based on the different city districts in Oslo (St. Hanshaugen, Gamle Oslo, Grynerløkka og Sagene, Outer Oslo West, Outer Oslo East), where the district Frogner is used as the baseline (it would have been preferable to include smaller, and hence more urban, districts, but the number was chosen based on the number of observations. See Marmolejo-Duarte and Chen (2019) [18]). Fourth, we control for dwelling type, where we separate single-family houses, townhouses, and semidetached houses with dummies, and where apartment is the baseline category. Fifth, we also include dummy variables for different size of the dwellings. Small is a dummy for dwellings between 50–80 m$^2$, medium is dummy for dwellings from 81–120 m$^2$, and large is dummy for dwellings >120 m$^2$. Hence, the baseline size is <50 m$^2$. We use the log-linear (semilog) functional form in the regressions because it makes it easier to interpret the coefficients and because the semilog functional form is known to mitigate the problem of heteroscedasticity [19]. In total, we observe (T + 1) periods. Note that if we ignore the year dummies and the time subscript, we are left with a standard hedonic model. Based on Equation (4), we are hence able to estimate two models, the post-label hedonic model, Model 1, and the pre-label hedonic model, Model 2.

Based on the results from the model, we estimate the price per square meter for the reference dwelling, which is a dwelling <50 m$^2$ located in Frogner. We set the age of this dwelling at five years and calculate the square meter price for all energy label categories. The price difference between the different labels is the actual value added achieved in the market compared with the expected value added.

### 3.3. Hedonic Model With Energy Price and Rate of Discount

To examine how the energy label, energy price, and rate of discount affect the price of the dwelling, a hedonic model is constructed where these three factors are represented through the expected energy cost. Hence, we apply the same model as described above, but replace the energy label dummies with the expected value of the energy costs given by:

$$LN(PV\ energy\ cost) = LN\left(\frac{kWh \cdot P_t^e}{r_t - g_t}\right) \tag{5}$$

Here, the logarithm of the present value per square meter, PV, is given by taking the logarithm of energy consumption per square meter times the expected energy cost per square meter, divided by the discount rate minus the growth rate. Figure 1 illustrates the development in energy price and discount rate over the time period 2000–2014.

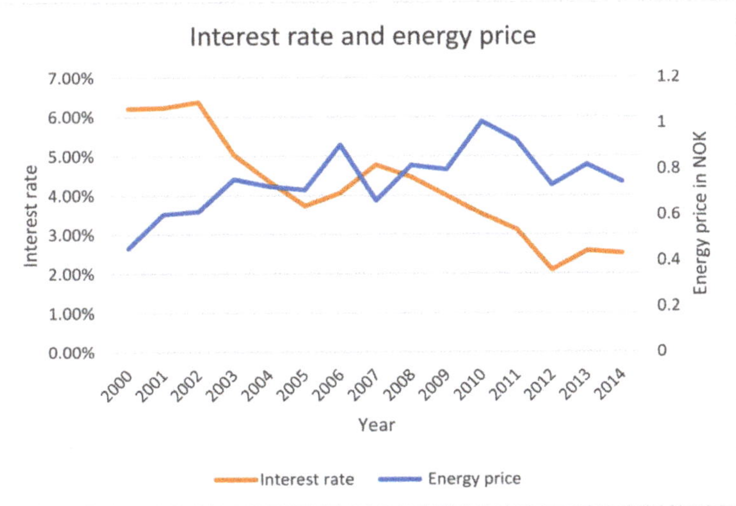


**Figure 1.** Energy price and interest rate year 2000–2014. The interest rate is given by the Norwegian 10-year government bond and the energy price is from the energy price area of Oslo [20–22].

## 4. Results

The yearly energy costs are used to calculate the present value of the different energy labels. In Figure 2, this is illustrated with an example from a 40 m$^2$ apartment. Note that there is a distinct difference between present values in 2009 and 2014. This difference is due to the discount rate being lower in 2014 than in 2009. This effect dominates, even if the energy price is slightly higher in 2009 than in 2014 (see Figure 1).

The expected price premiums of the energy labels in 2014 are presented in Table 5 and those for 2009 in Table 6. The price premiums are given per square meter. For instance, if we take an apartment of 40 m$^2$ with the energy label C, the expected price premium is NOK 682 per square meter, compared with a similar apartment with the energy label D. The same comparison in 2009 (Table 6) yields a price difference of NOK 502.

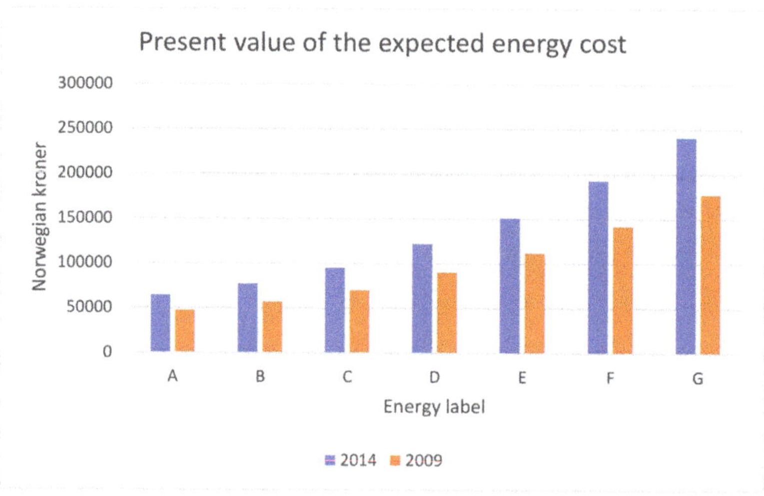



**Figure 2.** Present value of the expected energy cost for different EPCs. The figure shows the present value of the expected energy cost for a 40 m$^2$ apartment in 2009 and 2014 for different EPC grades in Norwegian kroner (NOK). (NOK 1 = €0.11 (per 31.12.2014)).

**Table 5.** Expected price premium per m$^2$ from different EPCs in 2014 in NOK.

| | A | B | C | D | E | F | G |
|---|---|---|---|---|---|---|---|
| **B** | 321 | | | | | | |
| **C** | 763 | 442 | | | | | |
| **D** | 1445 | 1124 | 682 | | | | |
| **E** | 2168 | 1846 | 1405 | 723 | | | |
| **F** | 3211 | 2890 | 2449 | 1766 | 1044 | | |
| **G** | 4416 | 4094 | 3653 | 2970 | 2248 | 1204 | |

**Table 6.** Expected price premium per m$^2$ from different EPCs in 2009 in NOK.

| | A | B | C | D | E | F | G |
|---|---|---|---|---|---|---|---|
| **B** | 236 | | | | | | |
| **C** | 561 | 325 | | | | | |
| **D** | 1064 | 827 | 502 | | | | |
| **E** | 1596 | 1359 | 1034 | 532 | | | |
| **F** | 2364 | 2128 | 1802 | 1300 | 768 | | |
| **G** | 3250 | 3014 | 2689 | 2187 | 1655 | 886 | |

The difference between the expected and actual value added is interesting. First, for 2014, we find that the actual value added (Table 7) is higher than the expected price premium (Table 5). This implies that dwellings with better energy labels receive a higher premium than can be explained by the energy costs; that is, a value added beyond the cost savings expected from a more energy-efficient dwelling. The pattern is confirmed in the 2009 tables (Tables 7 and 8). This means that, even before the energy labels were available to buyers, there was a price premium beyond what could be explained by the energy cost. However, these results are dependent on the rate of discount and sensitivity analysis shows that if the rate of discount in 2014 was set at 4%, the difference between the actual and expected price premium is much lower.

**Table 7.** Estimated (actual) price premium per m$^2$ from different EPCs in 2014 in NOK.

| | A | B | C | D | E | F | G |
|---|---|---|---|---|---|---|---|
| B | | | | | | | |
| C | | 2715 | | | | | |
| D | | 5667 | 2952 (4%) | | | | |
| E | | 11,486 | 8771 (12%) | 5819 (8%) | | | |
| F | | 12,229 | 9514 (13%) | 6562 (9%) | 743 | | |
| G | | 8203 | 5488 | 2536 | −3283 | −4026 | |

**Table 8.** Estimated (actual) price premium per m$^2$ from different EPCs in 2009 in NOK.

| | A | B | C | D | E | F | G |
|---|---|---|---|---|---|---|---|
| B | | | | | | | |
| C | | 8456 | | | | | |
| D | | 9182 | 726 (1%) | | | | |
| E | | 12,714 | 4258 (9%) | 3531 (7%) | | | |
| F | | 14,132 | 5676 (12%) | 4950 (11%) | 1418 | | |
| G | | 13,048 | 4593 | 3866 | 335 | −1084 | |

Table 9 presents the results from the hedonic models for the period after the introduction of EPCs (post-label), Model 1, and before the introduction of labels (pre-label), Model 2. All coefficients have the expected sign and are significant at the 1% level, except for the two location dummies for the districts of St. Hanshaugen and Outer Oslo West in Model 2, which are significant at the 5% level. The most interesting result in this analysis is the present value of the energy cost per square meter, which is positive and significant at the 1% level in both 2014 and 2009. The difference between the coefficients is rather small and not significantly different at the 1% level. Note that the 1% confidence intervals for the 2014 coefficient (0.061–0.119) and the 2009 coefficient (0.029–0.133) overlap significantly. Note also that the overall results do not change if we substitute the present value of the energy cost per square meter with the expected energy cost per square meter, or if we look at different dwelling types separately (these results are not reported in the paper).

**Table 9.** Energy costs and dwelling prices. Hedonic models, dependent variable: natural logarithm of transaction prices per m$^2$.

| | Post-Label<br>Model 1: 2014<br>Coef. (Std. Err.) | Pre-Label<br>Model 2: Before July 2010<br>Coef. (Std. Err.) |
|---|---|---|
| Ln PV energy cost | −0.090 *** (0.011) | −0.081 *** (0.020) |
| Age | 0.132 *** (0.021) | 0.080 *** (0.020) |
| St. Hanshaugen | −0.115 *** (0.026) | −0.072 ** (0.034) |
| Gamle Oslo | −0.283 *** (0.027) | −0.207 *** (0.033) |
| Grynerløkka og Sagene | −0.249 *** (0.024) | −0.140 *** (0.028) |
| Outer Oslo West | −0.198 *** (0.023) | −0.078 ** (0.027) |
| Outer Oslo East | −0.502 *** (0.023) | −0.368 *** (0.027) |
| Single-family houses | 0.105 *** (0.019) | 0.090 *** (0.027) |
| Townhouses | 0.050 *** (0.017) | 0.090 *** (0.022) |
| Semidetached houses | 0.087 *** (0.018) | 0.070 *** (0.025) |
| Small | −0.125 *** (0.011) | −0.078 *** (0.014) |
| Medium | −0.120 *** (0.015) | −0.107 *** (0.019) |
| Large | −0.220 *** (0.021) | −0.183 *** (0.027) |
| Constant | 12.352 *** (0.143) | 12.097 *** (0.026) |
| Adj R-squared | 0.47 | 0.40 |
| Number of observations | 2789 | 1608 |

Note: ***, **, signal significance at the 1% and 5%levels, respectively. See Section 3.2 above for variable definitions.

House prices are in fixed 2014 prices, and every dwelling price is multiplied by the house price index value for 2014 divided by the house price index value of the year of the transaction. Model 1 hence consists of buildings sold in 2011–2014, in 2014 prices.

*4.1. Robustness Check*

The nature of the potential causal relationship between energy labels and sales prices is crucial for our analysis. As a robustness check to test this relationship, we utilize the natural experiment that took place when the energy labels became mandatory in July 2010. The data allow us to compare the transaction prices of dwellings sold before and after the introduction of the EPC system in July 2010. If energy labels affect the sale prices, then two houses sold in, for example, 2008, for approximately the same price, should have approximately the same price as each other when resold after July 2010 if they were given the same energy label. On the other hand, if one of them received a higher energy label than the other, it should, ceteris paribus, have a higher resale price.

### 4.1.1. The Weighted Repeat Sales Method

The robustness check is performed with the weighted repeat sales method. The following model is applied [23,24]:

$$\ln(p_n^t/p_n^s) = \sum_{t=0}^{T} \gamma_t D_n^t + \mu_n^t,$$ (6)

where $P_n^t$ is the price at the time of the resale, $p_n^s$ is the price of the previous sale, $D_n^t$ is a dummy variable with the value 1 in the period in which the resale occurs, –1 in the period in which the previous sale occurs, and 0 otherwise. $\mu_n^t$ is the error term. To account for the possibility that the residual variance increases with increasing time intervals between sales, we apply the weighted repeat sales (WRS) method developed in [23].

The data does not contain enough observations before the introduction of the energy performance certificates in 2010 to create indices for energy label A and energy label B. To remove the house price trend, we divide the indices with a repeated sales index constructed based on all the dwellings in the dataset. We use a simple Dickey–Fuller test, to test whether variables are stationary (Table 10). All the variables have one unit root and are thus differentiated to make them stationary.

**Table 10.** Dickey–Fuller tests for unit root of all variables.

| Variables | Levels | | First Differences | |
|---|---|---|---|---|
| | Test Statistic | Critical Value | Test Statistic | Critical Value |
| PV energy cost | −2.269 | −3.000 | −3.042 | −3.000 |
| C | −5.207 | −3.000 | −6.935 | −3.000 |
| D | −4.922 | −3.000 | −8.628 | −3.000 |
| E | −2.302 | −3.000 | −5.713 | −3.000 |
| F | −3.642 | −3.000 | −5.763 | −3.000 |
| G | −2.194 | −3.000 | −4.313 | −3.000 |

Note: The 5% interpolated Dickey–Fuller critical values are used. No lags are included in the test. Ln means that natural logarithms have been used. C = index with dwellings with energy label C; D = index with dwellings with energy label D; E = index with dwellings with energy label E; F = index with dwellings with energy label F; and G = index with dwellings with energy label G.

We use a Durbin Watson test and a Portmanteau test for white noise which shows indication of autocorrelation AR (1). To reduce the problem of autocorrelation, we apply a Prais–Winsten regression [25].

Our regression is:

$$Y' = \beta_0(1 - \rho) + \sum \beta_j x'_{jt} + \sum \delta_j s_{jt} + \varepsilon_{jt}$$ (7)

where $\beta_j$ is the coefficient for the *j*th explanatory variable *x*, $\delta_j$ is the coefficient for the *j*th dummy variable *s*, and $\varepsilon_{jt}$ is the error term. The symbol ' indicates the transformation of the variables. The explanatory variable is the present value of the energy cost in the different categories. In addition, we use a dummy for the time when the energy labeling was made mandatory, from July to December 2010.

### 4.1.2. Repeat Sales Results

We start to explore the effect of introducing energy labels by constructing price indices for the different labels and let them all have a value of 100 in the year 2000 (Figure 3). The figure shows some price variations, but do not indicate a price effect from the energy performance certificates in July 2010. If energy labeling has the price effect found in the hedonic data, we should expect a kink with an increasing slope after July 2010 for the most energy efficient energy labels. However, it is difficult to ascertain any shift taking place in July 2010.

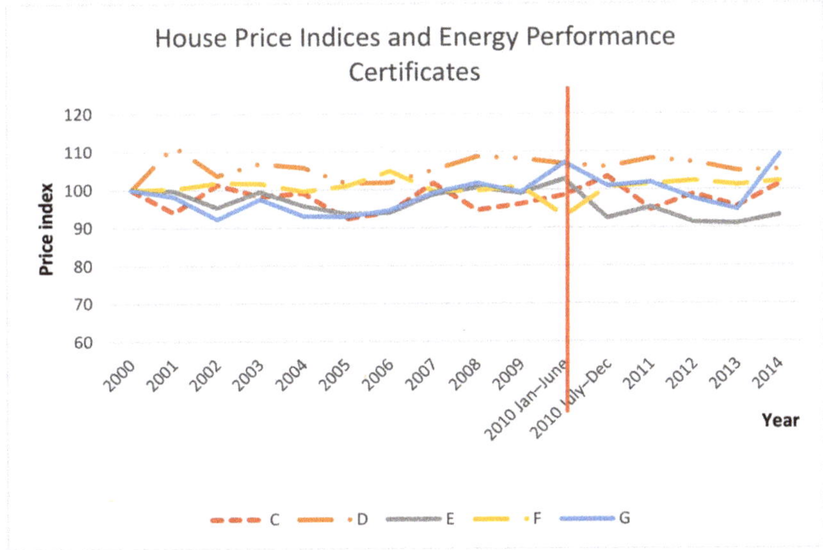


**Figure 3.** Dwelling price indices in different energy label categories. All of the indices start at 100 in year 2000 (Note: Fixed house price indices between 2000 and 2014, with trend removed. All indices start at 100 in 2000. As energy labeling was made mandatory on 1 July 2010, the year 2010 has been given two data points in the indices, one for January–June and one for July–December. The vertical line indicates when the energy labeling became mandatory. C = index for dwellings with energy label C; D = index for dwellings with energy label D; E = index for dwellings with energy label E; F = index for dwellings with energy label F; and G = index for dwellings with energy label G).

In Table 11 we test for the effect of introducing energy labels controlling for the present value of the energy cost. The dependent variable is the house price in the different energy label categories, and where we regress on the main index as well a dummy variable for the second part of 2010, when the energy label was made mandatory. The adjusted R-squares are all negative, while the Durbin Watson statistics, transformed after using the Prais–Winsten regression, range from 1.59 to 2.40, which means that we keep the null hypothesis of zero autocorrelation. (With $n = 15$ and $k = 2$, the retained H0 critical values range from 1.25 to 2.75.) If energy labeling has the price effect found in the post-label hedonic data, we should expect significant dummy coefficients in Table 11. However, none of the dummies are significant, nor the present value of energy cost. Hence, despite the strong label effect demonstrated in the hedonic post-label model (Model 1), just as in the pre-label hedonic regression (Model 2), we find no evidence to support the price premium effect. We also find no price effect from the present value of energy cost.

**Table 11.** House price under different energy labels.

|  | Ln C | Ln D | Ln E | Ln F | Ln G |
|---|---|---|---|---|---|
| Dummy 2010 July–Dec | −0.001 | −0.004 | −0.015 | 0.012 | −0.001 |
| PV energy cost | 0.000 | 0.000 | 0.000 | 0.000 | 0.000 |
| Adj. $R^2$ | −0.154 | −0.079 | −0.051 | −0.042 | −0.138 |
| DW transf. | 2.134 | 1.587 | 1.797 | 2.399 | 1.754 |

Note: We compare how well the dummy for the period when energy labeling was made mandatory (July–December 2010) together with the PV of energy cost is able to explain the house prices indexes for different energy labels. Ln C = logarithmic house price index with dwellings with energy label C; Ln D = logarithmic house price index with dwellings with energy label D; Ln E = logarithmic house price index with dwellings with energy label E; Ln F = logarithmic house price index with dwellings with energy label F; and Ln G = logarithmic house price index with dwellings with energy label G. DW transf. refers to the Durbin–Watson statistic, transformed after using the Prais–Winsten regression.

## 5. Discussion and Concluding Remarks

The energy performance certificate system was introduced in Europe to provide buyers with better information about the energy performance of dwellings. In part, the aim of this policy was to provide better valuations of dwellings when they are sold and to give buyers incentives to purchase energy-efficient dwellings. Earlier studies in this area have yielded contradictory conclusions. Brounen and Kok found that there was a significant price premium associated with energy labels in the real estate market in the Netherlands [4], whereas other studies, such as Murphy, found little or no effect of energy labels in the same market [6]. The present paper follows up the study by Olaussen et al. of the Norwegian real estate market [11]. Replicating the hedonic model by Brounen and Kok for Norwegian data, Olaussen et al. found the same results as Brounen and Kok [4,11]. However, when running a fixed effect model with data before and after the introduction of energy labels in 2010, they found that something other than the energy label must explain the apparent price premium. One potential explanation for this is that the energy efficiency of the dwelling was known to the buyers even before the labeling system was issued. To test for this, we use the energy price over time to see if the cost of energy may be the underlying explanation. By controlling for the present value of the expected energy consumption, we find no evidence of energy costs being important for the energy label premium.

By applying data for energy prices and the rate of discount, and the associated demands for the different energy label categories, we calculate the expected price premium that dwellings with better energy labels should achieve compared with similar dwellings with lower energy labels. Then, these price premiums are compared with the actual price premiums estimated in the hedonic models. The analyses show that the actual price premiums are much higher than the expected price premiums based on the energy cost differences. Moreover, we find this difference both before and after the energy label system was introduced. In addition, we find no significant differences in the actual price premium before and after the introduction of the energy labels in 2010. The same results are provided by the robustness check, in which we apply the repeated sales method; that is, we find that the present value of energy costs has no effect on the price of dwellings.

These results support previous studies that showed that the energy label does not affect the price of dwellings at the time of sale [11,26,27]. This is in line with the inferences of several survey studies, which indicate that when people buy a dwelling, they pay considerably less attention to its energy performance compared with other factors, such as the availability of garden and outdoor space, the location, the neighborhood, and the size of the property. Hence, there are reasons to believe that, when energy labels have been associated with price premiums in other studies, this results from factors other than the energy labels themselves. One explanation for our result may be that the buyers are well informed about the energy efficiency of the dwellings even without the energy labels and, hence, were already well informed before the energy label system was introduced. Another explanation may be that we have omitted explanatory variables in our models. Potential omitted variables may be the standard of the dwelling, for example, how recently it was renovated, or different amenities associated with the building. This explanation is in line with [26–28]. These omitted variables were visible to buyers before the energy label system was introduced, and it is quite likely that, e.g., the dwelling standard is closely correlated with the energy efficiency. Hence, it may be that the price premium associated with the energy label is explained by the standard of the dwelling. However, data regarding when dwellings have been renovated are not easily accessible. A detailed, in-depth study of potential omitted variables correlated with EPCs may be a fruitful path for future research.

**Author Contributions:** All authors have contributed equally, L.K. prepared an initial first draft, which was completed, corrected, reviewed and revised by J.O.O., A.O., and J.T.S.

**Funding:** This research received no external funding.

**Conflicts of Interest:** The authors declare no conflict of interests.

## References

1. Bio Intelligence Service; Lyons, R.; Institute for European Environmental Policy. Energy Performance Certificates in Buildings and Their Impact on Transaction Prices and Rents in Selected EU Countries; Final Report Prepared for the European Commission (DG Energy); European Commission: 2013. Available online: https://ec.europa.eu/energy/sites/ener/files/documents/20130619-energy_performance_certificates_in_buildings.pdf (accessed on 6 April 2018).
2. Ramos, A.; Labandeira, X.; Löschel, A. Pro-environmental Housholds and Energy Efficiency in Spain. *Environ. Resour. Econ.* **2016**, *63*, 367–393. [CrossRef]
3. Eichholtz, P.; Kok, N.; Quigley, J.M. Doing well by doing good? Green office buildings. *Am. Econ. Rev.* **2010**, *100*, 2494–2511. [CrossRef]
4. Brounen, D.; Kok, N. On the economics of energy labels in the housing market. *J. Environ. Econ. Manag.* **2011**, *62*, 166–179. [CrossRef]
5. Fuerst, F.; McAllister, P.; Nanada, A.; Wyatt, P. Does energy efficiency matter to home-buyers? An investigation of EPC ratings and transaction prices in England. *Energy Econ.* **2015**, *48*, 145–156. [CrossRef]
6. Murphy, L. The influence of the Energy Performance Certificate: The Dutch case. *Energy Policy* **2014**, *67*, 664–672. [CrossRef]
7. Laine, L. Room for Improvement: The Impact of EPCs on Consumer Decision-Making. 2011. Available online: http://www.consumerfutures.org.uk (accessed on 12 April 2018).
8. Amecke, H. The impact of energy performance certificates: A survey of German home owners. *Energy Policy* **2012**, *46*, 4–14. [CrossRef]
9. Backhaus, J.; Tigchelaar, C.; de Best-Waldhober, M. Key Findings and Policy Recommendations to Improve Effectiveness of Energy Performance Certificates and the Energy Performance of Buildings Directive. 2011. Available online: https://www.ecn.nl/publications/BEE/0 (accessed on 5 April 2015).
10. Wahlström, M. Doing good but not that well? A dilemma for energy conserving homeowners. *Energy Econ.* **2015**, *60*, 197–205. [CrossRef]
11. Olaussen, J.O.; Oust, A.; Solstad, J.T. Energy Performance Certificates—Informing the informed or the indifferent? *Energy Policy* **2017**, *111*, 246–254. [CrossRef]
12. Directive 2002/91/Ec of The European Parliament and of The Council of 16 December 2002 on the Energy Performance of Buildings. Available online: https://eur-lex.europa.eu/LexUriServ/LexUriServ.do?uri=OJ:L:2003:001:0065:0071:EN:PDF (accessed on 5 October 2018).
13. Directive 2010/31/Eu of The European Parliament and of the Council of 19 May 2010 on the Energy Performance of Buildings (Recast). Available online: https://eur-lex.europa.eu/LexUriServ/LexUriServ.do?uri=OJ:L:2010:153:0013:0035:EN:PDF (accessed on 5 October 2018).
14. Isachsen, O.; Rode, W.; Grini, G. Implementation of the EPBD in Norway. Status in November 2010. Country Reports on EPBD Implementation, Concerted Action EPBD. 2011. Available online: https://www.buildup.eu/en/practices/publications/implementation-epbd-norway-status-november-2010 (accessed on 2 April 2017).

15. Gordon, M.J.; Shapiro, E. Capital equipment analysis: The required rate of profit. *Manag. Sci.* **1956**, *3*, 102–110. [CrossRef]

16. Energimerking.no. Karakterskalaen. Available online: https://www.energimerking.no/no/energimerking-bygg/om-energimerkesystemet-og-regelverket/karakterskalaen/ (accessed on 29 February 2016).

17. Energimerking.no. Energimerkestatistikk. Available online: https://www.energimerking.no/no/energimerking-bygg/energimerkestatistikk/ (accessed on 29 February 2016).

18. Marmolejo-Duarte, C.; Chen, A. The evolution of energy efficiency impact on housing prices. An analysis for Metropolitan Barcelona. *Revista de la Construcción* **2019**, *18*, 156–166. [CrossRef]

19. Malpezzi, S. Hedonic pricing models: A selective and applied review. In *Housing Economics and Public Policy*; O'Sullivan, T., Gibb, K., Eds.; Blackwell Science: Oxford, UK, 2003; pp. 67–89.

20. Nordpoolgroup.com. Day-Ahead Prices. Available online: https://www.nordpoolgroup.com/Market-data1/Dayahead/Area-Prices/ALL1/Yearly/?view=table/ (accessed on 29 February 2016).

21. NVE. no. Utvikling av Gjennomsnittlig Nettleie for Husholdninger. Available online: https://www.nve.no/reguleringsmyndigheten-for-energi-rme-marked-og-monopol/nettjenester/nettleie/nettleiestatistikk/utvikling-av-gjennomsnittlig-nettleie-for-husholdninger-1993-dd/ (accessed on 29 February 2016).

22. Norges-bank.no. Rentestatistikk. Available online: https://www.norges-bank.no/tema/Statistikk/Rentestatistikk/ (accessed on 29 February 2016).

23. Case, K.E.; Shiller, R.J. Prices of single family homes since 1970: New indexes for four cities. *N. Engl. Econ. Rev.* **1987**, 45–56.

24. Case, K.E.; Shiller, R.J. The efficiency of the market for single-family homes. *Am. Econ. Rev.* **1989**, *79*, 125–137.

25. Prais, S.J.; Winsten, C.B. *Trend Estimators and Serial Correlation*; Discussion Paper No. 383.; Cowles Commission Chicago: Chicago, IL, USA, 1954.

26. Fregonara, E.; Rolando, D.; Semeraro, P. Energy Performance Certificates in the Turin real estate market. *J. Eur. Real Estate Res.* **2017**, *10*, 149–169. [CrossRef]

27. Marmolejo-Duarte, C.; Chen, A. The Uneven Price Impact of Energy Efficiency Ratings on Housing Segments and Implications for Public Policy and Private Markets. *Sustainability* **2019**, *11*, 372. [CrossRef]

28. Marmolejo-Duarte, C. The incidence of the energy rating on residential values: An analysis for the multifamily market in Barcelona. *Informes de la Construcción* **2016**, *68*, 543.

*Article*

# Analysis of Regional Differences and Influencing Factors on China's Carbon Emission Efficiency in 2005–2015

**Liangen Zeng [1], Haiyan Lu [1,2,*], Yenping Liu [1], Yang Zhou [3] and Haoyu Hu [1]**

[1]   College of Urban and Environmental Sciences, Peking University, Beijing 100871, China
[2]   Freie Universität Berlin, Department of History and Cultural Studies, Institute of Sinology, Fabeckstraße 23-25, 14195 Berlin, Germany
[3]   Shenzhen Environmental Science and New Energy Technology Engineering Laboratory, Tsinghua-Berkeley Shenzhen Institute, Shenzhen 518055, China
*   Correspondence: haiyanlu@zedat.fu-berlin.de; Tel.: +49-176-3713-9505

Received: 4 June 2019; Accepted: 23 July 2019; Published: 9 August 2019

**Abstract:** With the challenge to reach targets of carbon emission reduction at the regional level, it is necessary to analyze the regional differences and influencing factors on China's carbon emission efficiency. Based on statistics from 2005 to 2015, carbon emission efficiency and the differences in 30 provinces of China were rated by the Modified Undesirable Epsilon-based measure (EBM) Data Envelopment Analysis (DEA) Model. Additionally, we further analyzed the influencing factors of carbon emission efficiency's differences in the Tobit model. We found that the overall carbon emission efficiency was relatively low in China. The level of carbon emission efficiency is the highest in the East region, followed by the Central and West regions. As for the influencing factors, industrial structure, external development, and science and technology level had a significant positive relationship with carbon emission efficiency, whereas government intervention and energy intensity demonstrated a negative correlation with carbon emission efficiency. The contributions of this paper include two aspects. First, we used the Modified Undesirable EBM DEA Model, which is more accurate than traditional methods. Secondly, based on the data's unit root testing and cointegration, the paper verified the influencing factors of carbon emission efficiency by the Tobit model, which avoids the spurious regression. Based on the results, we also provide several policy implications for policymakers to improve carbon emission efficiency in different regions.

**Keywords:** carbon emission efficiency; regional differences; influencing factors; the Modified undesirable EBM DEA model; Tobit model

---

## 1. Introduction

Global warming has attracted the attention of politicians and scholars as it has severely affected the survival and development of human beings. According to the assessment results of the UN Intergovernmental Panel on Climate Change (IPCC), greenhouse gases, specifically $CO_2$, are the main cause of global warming [1,2]. Hence, carbon emission reduction has become a consensus by the international community. As the largest carbon emitter, China has more pressure to reduce its carbon emission. According to the 2018 BP Statistical Review of World Energy, the average growth rate per annum of $CO_2$ emission is 3.2% in mainland China from 2006 to 2016. Its $CO_2$ emissions reach 9232.6 million tons in 2017, occupying 27.6 percent of the global emission amount [3]. Based on the Global Carbon Budget 2016 released by the Global Carbon Project (GCP), the carbon dioxide emissions per unit economic outputs of China is 0.65 kg $CO_2$ per-year dollars, which is 1.8 times that of the United States and 2.8 times that of the European Union [4]. Therefore, the carbon dioxide emissions per unit GDP in China is also higher than that of developed countries.

As the carbon emission reduction has become an urgent task, China pledged to reduce carbon dioxide emissions per unit of GDP at the Copenhagen Climate Change Conference in December 2009 [5,6]. The national government also set up a target of reducing carbon emissions per unit of GDP by 20% in its thirteenth Five-Year Social and Economic Plan [7]. However, China has relied on an energy-intensive, heavy, industry-based developmental pattern for decades [8,9]. The path dependence of its economic structure makes it challenging to achieve this goal, and one suitable solution is to improve carbon emission efficiency.

However, China is a vast country with many regions, owning multifaceted features in its economic structure and resource endowment. The national government needs to formulate policies about carbon emission reduction at the regional level. The evaluation of the regional differences and influencing factors on the Chinese carbon emission efficiency has attracted the attention of scholars recently. Wang and Zhou et al. in 2013 found that $CO_2$ emission performance on the provincial was the higher in southeastern coastal areas but lower in central and western inland regions. They also stated that their carbon emission performance has increased by different rates after 2001 [10]. Their evidence is also consistent with the recent finding on the regional low-carbon economic development by Chu and Geng et al. in 2019. The recent study also showed that the eastern region still had a higher degree of low-carbon economic development than the central and western regions [11]. Zhong et al. in 2012 also discovered that China's total-factor carbon emission performance contains significant regional characteristics, and low carbon emission performance in the central and west regions [12].

Until now, most of the studies have divided Chinese 30 provinces into three areas (East, Central, and West regions) based on geographical position rather than on economic characteristics, which cannot accurately reflect the state of regional carbon emission efficiency in China. In 2015, the Development Research Center of the State Council (DEC) classified Chinese provinces into eight economic regions (Northern coast, Eastern coast, Southern coast, Northeast, Middle Yellow River, Middle Yangtze River, Southwest, and Northwest) based on their geographical position, economic development level, and resource endowment [13]. This paper has chosen to investigate the disparity of carbon emission efficiency in the above eight economic regions proposed by the DRC.

As for the method to measure the carbon emission efficiency, the Charnes–Cooper–Rhodes (CCR); Banker, Chames, and Cooper Mode (BCC); or slacks-based measure (SBM) model have been widely adopted by scholars [14–16]. However, these methods have their advantages as well as insufficiencies [16–23]. Therefore, we used the Modified Undesirable Epsilon-based measure (EBM) DEA Model to calculate the carbon emission efficiency to overcome their shortcomings [16,19,23,24]. Based on the analysis of the regional difference in the carbon emission efficiency, we verified the influencing factors by the Tobit model [25]. This research aims to analyze the regional differences and influencing factors on China's carbon emission efficiency from 2005 to 2015, based on the Modified Undesirable EBM DEA Model and Tobit model.

The basic structure of the rest of the paper is as follows. Section 2 will review the definition, measurement, and influencing factors of carbon emission efficiency. Section 3 will introduce the research method. Section 4 will present indicator selection and its data source. Section 5 will analyze the carbon emission efficiency in eight Chinese regions by the used the Modified Undesirable EBM DEA Model. Section 6 will adopt a Tobit model to verify the influencing factors of carbon emission efficiency. Finally, Section 7 will conclude and state the research limitations.

## 2. Literature Review

In recent years, carbon emission efficiency has become an important research issue in academic circles, and the main topics cover the definition, measurement, and influencing factors of carbon emission efficiency.

At present, there is no unified agreement on the definition of carbon emission efficiency by scholars. The definition of carbon emission efficiency stems from the method of calculation, such as the single-factor indicator method or total-factor indicator method [6]. The dominant types of single-factor indictor method include carbon dioxide emissions per unit of GDP, carbon dioxide emissions per unit of energy consumption, and carbon dioxide emissions per capita. Specifically, Kaya and Yokobori first proposed the concept of carbon production efficiency in 1993, which is the ratio of total GDP to the total amount of carbon emission [26]. Meilnik and Goldembergbet proposed the concept of the carbonization index in 2009, which is the ratio that the total amount of carbon emission to the total consumption of energy [27]. Sun (2005) proposed the carbon intensity, which calculates the ratio of $CO_2$ emissions to GDP [28]. Based on the concept of frontier production, the total-factor indicator method for carbon emission consists of both input and output indicators. Meanwhile, some scholars regarded carbon emission as an undesirable output indicator to evaluate carbon emission efficiency [29–33].

The main methods of evaluating carbon emission efficiency include parametric and nonparametric ones. As for the parametric methods, Aigner and Lovell proposed the Stochastic Frontier Approach (SFA) by using a stochastic frontier production function to evaluate the technical efficiency in 1977 [34]. Later, SFA has become one of the most commonly used parametric methods. The disadvantage of this parameter method is the requirement for a particular functional form assumption for the frontier. Thus, possible incorrect functional forms can cause inaccurate results [10,35–39].

As for the nonparametric method, the DEA is a common nonparametric method for evaluating the relative efficiency of several Decision-Making Units (DMUs). Unlike SFA, DEA is a deterministic method that also has many advantages, which has the capability of handling multiple inputs and outputs [14]. In other words, the DEA keeps the input and output of DMUs unchanged by adopting the effective sample [40]. On the other hand, DEA does not require any specification of the functional form of the frontier, therefore more scholars use the DEA approach and its various modified modes to study carbon emission efficiency. These models they adopted can be classified into three types, radial model, nonradial model, and Directional Distance Function Model (DDFM)s [41].

(A) Radial model, namely CCR or BCC, which are the basic models of the DEA approach proposed in 1978 and 1984, respectively. After widely application in the literature, these methods have also been adopted in the Chinese context recently. Wei et al. in 2010 used the CCR model for measuring changes in total carbon emission efficiency of provinces in China from 1986 to 2008 [42]. Zhong et al., in 2012, applied BCC model to measure static carbon emission performance of China's 29 provinces from 1995 to 2009 [12]. However, CCR or BCC model requires that all the inputs change in the same proportion, which ignores nonradial slacks and is contrary to reality [17,19,23,24].

(B) Nonradial model (SBM). To solve the weakness of CCR or BCC model, Tone proposed a nonradial Slacks-Based Measure that was able to calculate efficiency with slacks of input and output variables. SBM has been widely applied to evaluate carbon emissions efficiency and abatement potential recently. Chu and Geng et al. in 2019 applied the SBM to measure the carbon emission efficiency in 30 provinces of China from 2005 to 2017 [11]. Some other scholars also have conducted relevant research using SBM DEA model [21,43,44].

(C) Directional Distance Function Model (DDFM). The DDFM approach was put forward by Chung et al. (1997), which allows proportional expansion of desirable outputs and shrinking of undesirable outputs and inputs [41]. Some scholars attempted to apply these models to calculate the carbon emissions efficiency [45,46].

In addition to the definition and measurement of the carbon emission efficiency, influencing factors also have attracted academic attention. The common factors adopted include industrial structure, energy structure, openness, and one of primary research methods is the measurement method. The widely adopted one is the Economic Measurement Method, which employed both the spatial econometric model and the nonspatial econometric models. The spatial econometric model includes Spatial Lag Model (SLM) [47], Spatial Error Model (SEM) [47], and Spatial Durbin Model (SDM) [48], which are suitable to the case of the DUMs with obvious spatial correlations. For instance, Chuai et al. in 2012 analyzed the SLM between carbon emissions from energy consumption and their influencing factors in Chinese regions from 1997 to 2009. They stated that GDP and population are the two leading factors, which can strengthen the spatial autocorrelation of carbon emissions [49]. Ma and

Chen et al. in 2015 established the SEM to perform an empirical study on the influence factors of carbon emissions efficiency by using panel data from 30 provinces from 1998 to 2011. They argued that the economic scale, industry structure, and energy consumption structure hurt carbon emission efficiency, while opening-up, enterprise ownership structure, and government intervention play a positive role in efficiency [50]. Cheng et al. in 2014 applied the SDM to examine the dominating factors of China's carbon intensity from energy consumption from 1997 to 2010 [51]. The leading approach of the nonspatial econometric model is the Tobit model [25], also called the "limited dependent variable model", or the "check model". It considers the trend of continuous variable variation with limited dependent variables [25,26,52]. Wang at al. in 2019, used the Tobit model to analyze the influencing factors of carbon emission performance and the technology gap ratio of carbon emission in 30 provinces in China. They stated that influencing factors have various impacts on the carbon emission efficiency in the Chinese regions [53]. Some other scholars also investigated the carbon emission efficiency by using Tobit studies recently [54–56].

The above studies have provided a useful reference for further study on China's carbon emission efficiency. However, these studies also have the following limitations. From the perspective of the calculation method, Radial models, such as the CCR or BCC model, fail to consider the effect of nonradial slacks on the technical efficiency and cannot realize the factor decomposition in evaluating the efficiency, which can lead to biased estimation results. For the nonradial SBM-DEA model, the slacks are not necessarily proportional to the inputs or outputs, and the DUMs can lose the proportionality in the original inputs or outputs [17,19,23,24].

Based on the above problems, the improvements in this paper are as follows. We used the nonoriented the Modified Undesirable EBM DEA Model to investigate carbon emission efficiency. The Modified Undesirable EBM DEA Model is based on the EBM model. Tone introduced an EBM approach in 2010 to combine the radial model and nonradial model, which is more in line with reality [23]. As the EBM model cannot solve the problem related to undesirable outputs, Li and Chiu et al. (2019) extended the EBM model into the Modified Undesirable EBM DEA Model to deal with them [17,19,24].

## 3. Research Method

### 3.1. The Definition of Carbon Emission Efficiency

In a broad sense, carbon emission is an abbreviated concept of greenhouse gas emissions, which includes $CO_2$ (carbon dioxide), $CH_4$ (methane), $N_2O$ (nitrous oxide), HFCs (hydrofluorocarbons), PFCs (perfluorocarbons), and $SF_6$ (sulfur hexafluoride) [57]. Generally, carbon emission is regarded as $CO_2$ emission because carbon dioxide in the greenhouse effect is the principal greenhouse gas. As $CO_2$ is also the main greenhouse gas in China, this paper chose the latter definition.

As the single-factor indictor method does not consider the coupling between the various production factors, it ignores the influence of various factors such as labor force and energy consumption [6]. To improve the evaluation result accuracy, we used the total-factor indicator method. Carbon emission efficiency is defined as a production system for creating more goods output and less $CO_2$ emissions while consuming fewer resources, such as labor, energy, and capital.

### 3.2. The Modified Undesirable EBM DEA Model

Because Tone and Tsutsui's EBM did not consider any undesirable factors [23], Li and Chiu et al. combined the EBM DEA and an undesirable factor into the Modified Undesirable EBM DEA Model [17,19,24]. The paper applies it for the evaluation of the carbon emission efficiency of 30 provinces across mainland Chinese from 2005 to 2015.

Suppose $n$ $DMU_k$ ($k = 1,2, \ldots, n$) and $m$ type inputs $X_j$ ($x_{1j}, x_{2j}, \ldots, x_{mj}$) can produce $s$ type outputs $Y_j$ ($y_{1j}, y_{2j}, \ldots, y_{mj}$). Li and Chiu et al.'s nonoriented, Modified Undesirable EBM DEA Model evaluates the technical efficiency $\gamma^*$ of DMU ($X_o$, $Y_o$) by solving the following linear program [17,19,21,23,24].

$$\gamma^* = \min_{0\varphi,\lambda,s^-,s^g,s^b} \frac{\theta - \varepsilon x \sum_{i=1}^{m} \frac{\omega_i^- s_i^-}{x_i 0}}{\varphi + \varepsilon_y \sum_{i=1}^{s1} \frac{\omega_i^{+s1} s_i^g}{y_i 0} + \varepsilon_y \sum_{i=1}^{s2} \frac{\omega_i^{-s2} s_i^b}{y_i 0}}$$

$$s.t \begin{cases} X\lambda - \theta X_0 + s^- = 0 \\ Y^g \lambda - \varphi Y_0 - s^g = 0 \\ Z^b \lambda - \varphi Y_0 - s^b = 0 \\ \lambda 1 + \cdots + \lambda n = 1 \\ \lambda \geq 0, s^- \geq 0, s^g \geq 0, s^b \geq 0, \varphi \geq 1, \theta \leq 1 \end{cases} \tag{1}$$

where $x_{i0}$ and $y_{i0}$ are the $i$th input and the $i$th output when calculating the $o$th DMU, respectively; $s_i^-$ stands for the slack variable of input; $s^g$ and $s^b$ are the slacks of desired output and undesired output, respectively; and $w_i^-$ is the weight of input $i$, which satisfies $\sum \omega_i^{-1} = 1 \left( \forall i \omega_i^- \geq 0 \right)$. $w_i^{+s1}$ and $w_i^{-s2}$ indicate the weights of the desired output $i$ and the undesired output $i$, respectively, which satisfy $\sum \omega_i^{+s1} + \sum \omega_i^{-s2} = 1 \left( \forall i \omega_i^+ \geq 0 \right)$. $\varepsilon_x$ represents the combination of radial $\theta$ and nonradial slack, and $\varepsilon_y$ denotes the combination of radial $\varphi$ and nonradial slack. $\gamma^*$, which is the optimal solution in the EBM model and stands for the technical efficiency value of the DMU. With the value range between 0 and 1, the DMU is in the efficient state (if $\gamma^* = 1$) or the nonefficient state (if $\gamma^* < 1$). An inefficient DMU can reach the production frontier by reducing inputs and undesirable outputs or expanding desirable outputs.

### 3.3. Tobit Model

As defined above, the value of efficiency from the Modified Undesirable EBM DEA Model falls between the interval 0 and 1, which makes $Y$ a limited dependent variable. If the Ordinary Least Squares (OLS) model is used to calculate the parameter, estimating results will be biased and consistent [6,25,40]. We utilized the Tobit regression model (1958) to analyze the influencing factors of carbon emission efficiency, which can estimate the parameters by using maximum likelihood estimation [25]. The structural equation of Tobit model was given as

$$Y^* = \beta X_i + ui$$
$$Y_i = \begin{cases} Y_i^* & \text{if} & Y_i^* > 0 \\ 0 & \text{if} & Y_i^* \leq 0 \end{cases} \tag{2}$$

In Equation (2), $i$ stands for the $i$th DMU. $Y^*$ is the latent variable and $Y_i$ stands for a limited dependent variable. $Y_i$ is the latent variable, $X_i$ is the explanatory variable, $\beta$ represents the correlation coefficient, and $u$ is the random error with the distribution of $N(0, \sigma^2)$. We calculated the regression coefficients by using maximum likelihood estimation in the Stata12.0 software.

## 4. Data Source and Indicator Selection

This paper investigated 30 provinces, municipalities, or autonomous regions (except Tibet) in China from 2005 to 2015. We selected the annual data of capital stock, labor force, and energy consumption as three inputs according to production processes and the prior research results [6,11,21,43,44]. We treated gross domestic product (GDP) as a desirable output and $CO_2$ emission as an undesirable output. The inputs and outputs are explained as follows.

1.  Capital stock. The paper estimated the capital stock by using the perpetual inventory method, defined as follows [58].

$$K_{i,t} = I_{i,t} + (1 - \delta_{i,t}) K_{i,t} - 1 \tag{3}$$

where $K_{i,t}$ and $I_{i,t}$ stand for the capital stock and the gross fixed capital formation of the $i$th province in the $t$th year, respectively. $\delta$ is the depreciation rate of capital stock, and is set to 9.6% in accordance with previous studies [59]. The provincial data of capital stock were converted into the 2005 constant price. The capital stock data of provinces in 2005 was expressed as follows

$$\text{Captial stock in 2005} = \frac{\text{the gross fixed captial fomation in 2005}}{10\%} \tag{4}$$

All data of the gross fixed capital formation as well as the price index of fixed-asset investment in Chinese provinces were from the China Statistical Yearbook (2006–2016).

2. Labor force. The paper adopted the total amount of employees in three industries as the labor force variable. The data on the provincial level were collected from the statistical yearbook (2006–2016).

3. Energy consumption. This paper chose the energy consumption of each province as the input index. The data were collected from China's Energy Statistical Yearbook (2006–2016).

4. GDP. To diminish the impact of inflation, we convert the provincial GDP into the 2005 constant price. The data came from the China Statistical Yearbook (2006–2016).

5. $CO_2$ emissions. This paper estimated the $CO_2$ emissions generated by the burning of fossil energy and the emissions from the process of cement production, which is consistent with the previous studies [60,61].

This paper calculated the carbon emissions from seven types of fossil energy, such as coal, coke, gasoline, kerosene diesel, fuel oil, and natural gas referring to the National Greenhouse Gas Emissions Inventory introduced by IPCC in 2006. The types of the fuels are classified according to the prior research results [6,8,11,20]. The formula for calculating $CO_2$ emissions from fossil fuels is [62]

$$C_E = \sum_{i=7}^{7} \left( E_i \times ALCV_i \times CCF_i \times COF_i \times \frac{44}{12} \right) \tag{5}$$

where $C_E$ represents the total $CO_2$ emissions, ALCV stands for the average low calorific value, CCF denotes the carbon content factor, and COF is carbon oxidation factor. The number (44/12) represents the ratio of the molecular weight of $CO_2$ (44) to the molecular weight of carbon. The subscript $i$ stands for the energy source. The data of energy consumption were collected from the China Energy Statistical Yearbook (2006–2016). The data of the average low-order calorific, carbon content factor, and carbon oxidation factor were from the China Energy Statistical Yearbook and National Greenhouse Gas Emission Inventory Guide (2006) [62] and Guidelines for Provincial Greenhouse Gas Inventories in China (2011) [63], as shown in Table 1.

**Table 1.** The carbon emission factors of various types of fossil fuels.

| Fuel Type | Coal | Coke | Gasoline | Kerosene | Diesel | Fuel Oil | Natural Gas |
|---|---|---|---|---|---|---|---|
| ALCV (kj/kg) | 20,908 | 28,435 | 43,070 | 43,070 | 42,652 | 41,816 | 38,931 |
| CCF (kg/Tj) | 95,333 | 107,000 | 70,000 | 71,500 | 74,100 | 77,400 | 56,100 |
| COF(%) | 92.30 | 92.80 | 98.60 | 98.00 | 98.20 | 98.50 | 99.00 |

The formula for calculating carbon emissions from the process of cement production is

$$C_C = Q \times EF_C \tag{6}$$

In Formula (6), $C_C$ represents the total amount of $CO_2$ emissions during the process of cement production, $Q$ stands for the total amount of cement production, and $EF_C$ represents the carbon emissions emission coefficient of cement production, the value of $EF_C$ is 527 kgCO$_2$/t [64]. The all indicators were listed in Table 2.

**Table 2.** Carbon emission efficiency measurement index system.

| Indicator Type | Primary Indicators | Secondary Indicators |
|---|---|---|
| Input indicator | Capital | Capital stock (unit: 100 million yuan) |
| | Labor | Total number of employees in three industries (unit: 10,000) |
| | Energy | Total energy consumption (unit: 10,000 tons of standard coal) |
| Output indicator | Desired outcomes | GDP (unit: 100 million yuan) |
| | Undesired outcomes | CO$_2$ (emissions unit: $10^4$ tons) |

## 5. Analysis of Regional Differences in Carbon Efficiency

### 5.1. Overall Characteristics of Chinese Carbon Emission Efficiency

The carbon emission efficiencies of 30 provinces of China during the period of 2005 to 2015 were calculated by Equations (1)–(3). The results are listed in Table 3.

**Table 3.** Carbon emission efficiency of 30 provinces in China from 2005 to 2015.

| Regions | 2005 | 2006 | 2007 | 2008 | 2009 | 2010 |
|---|---|---|---|---|---|---|
| Beijing | 1 | 1 | 1 | 1 | 1 | 1 |
| Tianjing | 0.8580 | 0.8276 | 0.8165 | 0.8083 | 0.7897 | 0.8050 |
| Hebei | 0.6260 | 0.5545 | 0.5517 | 0.5337 | 0.5415 | 0.5603 |
| Shanxi | 0.5719 | 0.4794 | 0.4806 | 0.4601 | 0.4598 | 0.4779 |
| Inner Mongoria | 0.4858 | 0.4605 | 0.4672 | 0.4620 | 0.4814 | 0.4984 |
| Liaoning | 0.6123 | 0.5778 | 0.5754 | 0.5572 | 0.5723 | 0.6046 |
| Jilin | 0.6085 | 0.5731 | 0.5716 | 0.5452 | 0.5433 | 0.5535 |
| Heilongjiang | 1 | 1 | 0.6864 | 0.6563 | 0.6715 | 0.6919 |
| Shanghai | 1 | 1 | 1 | 1 | 1 | 1 |
| Jiangsu | 0.8052 | 0.7765 | 0.7710 | 0.7304 | 0.7319 | 0.7546 |
| Zhejiang | 0.8400 | 0.8029 | 0.7874 | 0.7423 | 0.7319 | 0.7622 |
| Anhui | 0.6770 | 0.6056 | 0.5962 | 0.5601 | 0.5706 | 0.6024 |
| Fujian | 0.8219 | 0.8044 | 0.7919 | 0.7459 | 0.7278 | 0.7585 |
| Jiangxi | 0.6771 | 0.6282 | 0.6195 | 0.5953 | 0.6026 | 0.6220 |
| Shandong | 0.6444 | 0.6208 | 0.6235 | 0.6001 | 0.6118 | 0.6368 |
| Henan | 0.6553 | 0.5852 | 0.5763 | 0.5504 | 0.5511 | 0.5567 |
| Hubei | 0.6484 | 0.5766 | 0.5726 | 0.5622 | 0.5852 | 0.6069 |
| Hunan | 0.7000 | 0.6077 | 0.6006 | 0.5779 | 0.6019 | 0.6287 |
| Guangdong | 1 | 1 | 1 | 1 | 1 | 1 |
| Guangxi | 0.6613 | 0.6081 | 0.5975 | 0.5725 | 0.5704 | 0.5569 |
| Hainan | 0.8204 | 0.7686 | 0.7502 | 0.6841 | 0.6761 | 0.6974 |
| Chongqing | 0.5854 | 0.5670 | 0.5732 | 0.5586 | 0.5835 | 0.6178 |
| Sichuan | 0.6362 | 0.5655 | 0.5582 | 0.5243 | 0.5494 | 0.5835 |
| Guizhou | 0.4972 | 0.4089 | 0.4082 | 0.3868 | 0.4142 | 0.4466 |
| Yunnan | 0.5873 | 0.5052 | 0.5011 | 0.4786 | 0.4940 | 0.5027 |
| Shaanxi | 0.5993 | 0.5480 | 0.5409 | 0.5290 | 0.5323 | 0.5355 |
| Gansu | 0.5664 | 0.4806 | 0.4731 | 0.4477 | 0.4767 | 0.4989 |
| Qinghai | 0.4256 | 0.3845 | 0.3809 | 0.3790 | 0.3951 | 0.4160 |
| Ningxia | 0.3693 | 0.3311 | 0.3308 | 0.3260 | 0.3386 | 0.3469 |
| Xinjiang | 0.5726 | 0.5192 | 0.5147 | 0.4922 | 0.5005 | 0.5093 |

**Table 3.** *Cont.*

| Regions | 2005 | 2006 | 2007 | 2008 | 2009 | 2010 |
|---|---|---|---|---|---|---|
| East | 0.8207 | 0.7939 | 0.7880 | 0.7638 | 0.7621 | 0.7799 |
| Central | 0.6923 | 0.5944 | 0.5880 | 0.5634 | 0.5732 | 0.5925 |
| West | 0.5442 | 0.4890 | 0.4860 | 0.4688 | 0.4851 | 0.5011 |
| China | 0.6851 | 0.6289 | 0.6239 | 0.6022 | 0.6102 | 0.6277 |

| Regions | 2011 | 2012 | 2013 | 2014 | 2015 | mean |
|---|---|---|---|---|---|---|
| Beijing | 1 | 1 | 1 | 1 | 1 | 1 |
| Tianjing | 0.8087 | 0.8000 | 1 | 1 | 1 | 0.8649 |
| Hebei | 0.5603 | 0.5382 | 0.5798 | 0.5586 | 0.5424 | 0.5588 |
| Shanxi | 0.4802 | 0.4593 | 0.4966 | 0.4662 | 0.4388 | 0.4792 |
| Inner Mongoria | 0.4883 | 0.4595 | 0.5063 | 0.4819 | 0.4820 | 0.4794 |
| Liaoning | 0.6010 | 0.5680 | 0.6081 | 0.5759 | 0.5701 | 0.5839 |
| Jilin | 0.5435 | 0.5301 | 0.5561 | 0.5343 | 0.5255 | 0.5532 |
| Heilongjiang | 0.6986 | 0.6601 | 0.7093 | 0.6697 | 0.6368 | 0.7073 |
| Shanghai | 1 | 1 | 1 | 1 | 1 | 1 |
| Jiangsu | 0.7330 | 0.7153 | 0.7750 | 0.7736 | 0.7674 | 0.7576 |
| Zhejiang | 0.7442 | 0.7295 | 0.7812 | 0.7761 | 0.7650 | 0.7693 |
| Anhui | 0.6139 | 0.5903 | 0.6330 | 0.6256 | 0.6018 | 0.6070 |
| Fujian | 0.7281 | 0.7029 | 0.7603 | 0.7390 | 0.7198 | 0.7546 |
| Jiangxi | 0.6264 | 0.6109 | 0.6444 | 0.6458 | 0.6302 | 0.6275 |
| Shandong | 0.6397 | 0.6138 | 0.6782 | 0.6527 | 0.6350 | 0.6324 |
| Henan | 0.5528 | 0.5414 | 0.5619 | 0.5407 | 0.5243 | 0.5633 |
| Hubei | 0.6153 | 0.5970 | 0.6525 | 0.6353 | 0.6178 | 0.6063 |
| Hunan | 0.6386 | 0.6200 | 0.6654 | 0.6523 | 0.6337 | 0.6297 |
| Guangdong | 1 | 1 | 1 | 1 | 1 | 1 |
| Guangxi | 0.5397 | 0.5117 | 0.5327 | 0.5195 | 0.5077 | 0.5616 |
| Hainan | 0.6644 | 0.6078 | 0.6229 | 0.5868 | 0.5600 | 0.6762 |
| Chongqing | 0.6405 | 0.6290 | 0.7162 | 0.6967 | 0.6913 | 0.6236 |
| Sichuan | 0.6162 | 0.6067 | 0.6528 | 0.6438 | 0.6328 | 0.5972 |
| Guizhou | 0.4655 | 0.4451 | 0.4845 | 0.4717 | 0.4447 | 0.4430 |
| Yunnan | 0.5013 | 0.4750 | 0.5008 | 0.4775 | 0.4559 | 0.4981 |
| Shaanxi | 0.5367 | 0.5156 | 0.5468 | 0.5237 | 0.5192 | 0.5388 |
| Gansu | 0.5112 | 0.4952 | 0.5381 | 0.5266 | 0.5033 | 0.5016 |
| Qinghai | 0.4163 | 0.3889 | 0.3923 | 0.3582 | 0.3352 | 0.3884 |
| Ningxia | 0.3430 | 0.3284 | 0.3478 | 0.3203 | 0.2973 | 0.3345 |
| Xinjiang | 0.5035 | 0.4655 | 0.4852 | 0.4478 | 0.4218 | 0.4938 |
| East | 0.7709 | 0.7523 | 0.8005 | 0.7875 | 0.7782 | 0.7816 |
| Central | 0.5962 | 0.5761 | 0.6149 | 0.5962 | 0.5761 | 0.5967 |
| West | 0.5057 | 0.4837 | 0.5185 | 0.4971 | 0.4810 | 0.4964 |
| China | 0.6270 | 0.6068 | 0.6476 | 0.6300 | 0.6153 | 0.6277 |

According to Table 3 and Figure 1, the carbon emission efficiency of 30 provinces was generally low from 2005 to 2015, with an average value of 0.6277. Only 11 provinces (Beijing, Shanghai, Guangdong, Tianjin, Zhejiang, Jiangsu, Fujian, Heilongjiang, Hainan, Shandong, and Hunan) performed better than the national average efficiency value. Yunnan, Xinjiang, Inner Mongolia, Shanxi, Guizhou, Qinghai, and Ningxia owned very low levels of carbon emission efficiency, which all were less than 0.5, indicating that most provinces still face the challenge of energy saving and emission reduction.

**Figure 1.** The schematic diagram of carbon emission efficiency intervals in 30 provinces of China.

## 5.2. Regional Differences of Carbon Emission Efficiency: East, Central, and West

Figure 2 shows that (1) the eastern regions had the highest level of carbon emission efficiency—0.7816—followed by the central region (0.5967) and the western region (0.4964). (2) As for the eastern region, the carbon emission efficiency values for Beijing, Shanghai, and Guangdong in the eastern zone were 1, which showed that these provinces are the front runners in carbon emission efficiency and can benchmark for others. The carbon emission efficiency of the eastern provinces' was higher than the national average, except Hebei and Liaoning. (3) For the central region, Heilongjiang had the highest level of carbon emission efficiency (0.7072), which was little below the frontier. Shanxi had the minimum level of carbon emission efficiency (0.4792), which performed slightly worse than Hebei (the one with the lowest level in the eastern region). As for the western region, Chongqing had the highest level of carbon emission efficiency in the region, but its performance was worse than that of Heilongjiang. Ningxia had the lowest average efficiency in the whole country.

## 5.3. Regional Differences in Carbon Emission Efficiency: A Comparison of Eight Comprehensive Economic Regions

In this subsequent analysis, Chinese provinces are divided into eight regions (Figure 3) in line with the division of the State Council to analyze the regional differences in carbon efficiency further. The carbon emission efficiency of eight regions is shown in Table 4.

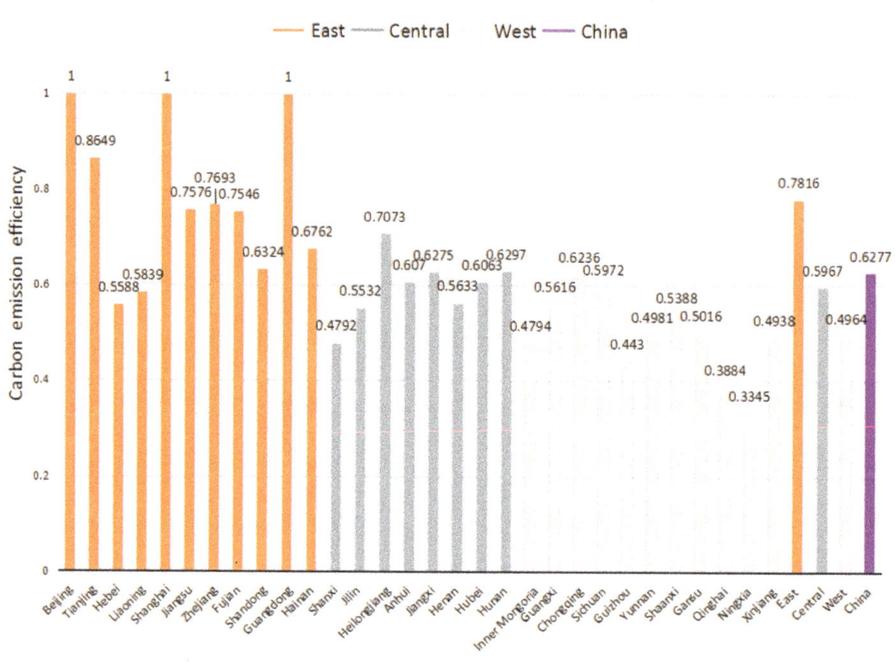

Figure 2. Carbon emission efficiency in China's three major economic zones.

Figure 3. The schematic diagram of the eight economic zones of China.

**Table 4.** Carbon emission efficiency in the Chinese eight regions.

| Regions | Provinces | Value | Regions | Provinces | Value |
|---|---|---|---|---|---|
| Northern coast | Beijing | 1 | Middle Yellow River | Shanxi | 0.4792 |
| | Tianjin | 0.8649 | | Inner Mongolia | 0.4794 |
| | Hebei | 0.5588 | | Henan | 0.5633 |
| | Shandong | 0.6324 | | Shaanxi | 0.5388 |
| | mean | 0.7640 | | Mean | 0.5152 |
| Eastern coast | Shanghai | 1 | Middle Yangtze River | Anhui | 0.6070 |
| | Jiangsu | 0.7576 | | Jiangxi | 0.6275 |
| | Zhejiang | 0.7693 | | Hubei | 0.6063 |
| | mean | 0.8423 | | Hunan | 0.6297 |
| | | | | Mean | 0.6176 |
| Southern coast | Fujian | 0.7546 | Southwest | Guangxi | 0.5616 |
| | Guangdong | 1 | | Chongqing | 0.6236 |
| | Hainan | 0.6762 | | Sichuan | 0.5972 |
| | mean | 0.8103 | | Guizhou | 0.4430 |
| | | | | Yunnan | 0.4981 |
| | | | | Mean | 0.5447 |
| Northeast | Liaoning | 0.5839 | Northwest | Ningxia | 0.5016 |
| | Jilin | 0.5532 | | Gansu | 0.3884 |
| | Heilongjiang | 0.7073 | | Qinghai | 0.3345 |
| | mean | 0.6148 | | Xinjiang | 0.4938 |
| | | | | Mean | 0.4296 |

Table 4 shows that (1) from 2005 to 2015, Northern, Eastern, and Southern coastal regions had a higher level of carbon emission efficiency than other regions, with an average efficiency of 0.7640, 0.8102, and 0.8423, respectively. Northeast and Middle Yangtze River had an average carbon emission efficiency of 0.6148 and 0.6176, respectively. Middle Yellow River, Southwest, and Northwest kept the carbon emission efficiency at a low level, and these regions' average efficiencies were 0.5152, 0.5447, and 0.4296, respectively. (2) In a comparison of the provinces with the highest efficiency level in every region, the leading ones in Northern, Eastern, and Southern coastal regions were Beijing, Shanghai, and Guangdong, all of which are on the production frontier. Heilongjiang had the highest value of carbon emission efficiency in Northeast, with an average value of 0.7073, which was a little more than the national average while far below those of Beijing, Shanghai, and Guangdong. The provinces with higher carbon emission efficiency in the Middle Yangtze River and Southwest were Hunan and Chongqing. Those provinces' average values were all smaller than 0.7 and far below the average level of frontier provinces. The most efficient provinces of Middle Yellow River and Northwest were Henan and Ningxia, and they held the value of more than 0.5, far from the frontier level. (3) The gap of carbon emission efficiency in the interior of the Northern coast and Southern coast was more substantial than other regions. The efficiency of Hebei and Hainan were lower than the average level of Northern and Southern coasts. The difference in carbon emission efficiency in the interior of the Middle Yangtze River was the smallest of the eight regions.

Several factors have contributed to these regional differences. First, Northern, Eastern, and Southern coastal regions are in the eastern part of China, which are the earliest areas exposed to open-up policy, and are more conducive to the introduction of foreign advanced technology, intelligence, equipment, experience, and capital. On the one hand, many universities and research institutes in the coastal regions play a crucial role in innovation activities. On the other hand, coastal regions own well-established transportation infrastructure, solid industrial foundation, and abundant human capital, which have attracted the introduction of other production factors from abroad. Therefore, a high level of production technology in the coastal regions results in high carbon emissions efficiency.

Second, many provinces in the Northeast and Middle Yangtze River belong to the Central area in China. The openness, education, and science and technology level of these provinces is lower than those of the coastal regions. As the old industrial bases in China, the Northeast region's economic pillar is heavy industry, which is responsible for much of $CO_2$, $SO_2$, and COD emissions. Driven by the "Rise of Central China" Strategy (The Rise of Central China Plan is a policy to accelerate the development of its central regions in 2004, which covers six provinces: Shanxi, Henan, Anhui, Hubei, Hunan, and Jiangxi.), Middle Yangtze River has set up four national industrial transfer demonstration zones, which introduced many manufacturing industries from the coastal areas. These industries are characterized by high pollution and energy consumption, which leads to high carbon emissions inevitably.

Third, many provinces in the Middle Yellow River, Southwest, and Northwest belong to Western China. Their unfavorable geographic locations limit the introduction of foreign advanced technology, intelligence, equipment, experience, and capital. With the deepening of the reform and opening-up policy, the brain drain in these regions can be found by more talents moving to the eastern and central parts of China for better opportunities, which increase the difficulty improve the production technology and carbon emission efficiency.

## 6. Influence Factors of Carbon Emission Efficiency Based on the Tobit Regression

### 6.1. Determinants of Carbon Emission Efficiency

Based on the regional characteristics of carbon emission efficiency, we used the Tobit regression model to analyze their influencing factors. The influencing factors were selected by referring to previous studies, including government regulations, industrial structure, foreign trade level, foreign capital utilization level, energy intensity, and science and technology [6,8,11,53,65–89]. Furthermore, we compared the results with our prejudgments of each influencing factor. The data were collected from the China Statistical Yearbook (2006–2016).

#### 6.1.1. Government Intervention in the Economy

We chose the government intervention in the economy as an independent variable, consistent with previous studies [6,11]. In China, the government's macro-control policies have a significant impact on economic performance. For instance, the macro-control policies influence the resources allocation and industrial transfer among regions, which also impact the carbon emission reduction possesses indirectly. Additionally, if the macro-control policies promote the development of energy-saving industries, they will contribute to carbon emission efficiency. However, there is no agreed indicator for the government intervention policies, and we have adopted the proportion of the local financial expenditure to GDP as the proxy for the level of government intervention.

#### 6.1.2. Industrial Structure

More recent attention has also focused on the relationship between industrial structure and carbon emission efficiency [66–71]. The higher proportion of tertiary industry in industrial structure can result in low energy consumption and carbon emissions. Therefore, this paper postulated that industrial structure is one factor to explain regional differences in carbon emission efficiency. This paper used the proportion of the local value-added of the tertiary industry to GDP as the proxy for the industrial structure.

### 6.1.3. Foreign Trade Level

In the process of globalization, China's foreign trade volume has escalated after the opening-up policy in 1978. A growing body of literature has investigated the relationship between carbon emission efficiency and foreign trade level [11,72,73]. The opinions about the relationship between carbon emission efficiency and foreign trade level are contradictory in literature. Some scholars claimed foreign trade is a driving force to promote the carbon emission efficiency, because Chinese enterprises are under the pressure of international competition to improve their product and service quality. Their improvement efforts facilitate energy saving and resources consumption. Other scholars insisted that the industries in many Midwest provinces still export products with high resources consumption, which increase corresponding carbon emission. Therefore, the effect of foreign trade level worth further investigating, and we adopted the proportion of the local import and export trade to GDP as its proxy.

### 6.1.4. The Foreign Capital Utilization Level

The absorption of foreign capitals is an essential factor in the economic development and technological progression of developing countries. Currently, a considerable amount of research has been published on this topic [74–78]. On the one hand, the foreign capital may accompany a transfer of pollution-intensive industries from developed countries to developing countries less stringent environmental regulations, which gives rise to the problem of carbon leakage (the Pollution Haven Hypothesis) [79]. On the other hand, FDI is a "complex" of capital, technology, organization, and marketing networks [80]. FDI may also introduce advanced production technologies and management methods from aboard, which improve carbon emission efficiency indirectly. To investigate the impact of foreign capital utilization level, we adopted the proportion of local foreign direct investment to GDP to evaluate the degree of foreign capital utilization.

### 6.1.5. Energy Intensity

Some scholars also emphasized the impact of energy intensity on carbon emission efficiency [81–85]. Energy intensity is calculated as units of energy per unit of GDP. Low energy intensity indicates a lower cost of converting energy into GDP and accompanied by higher carbon emission efficiency. As energy intensity has a stronger impact on $CO_2$ emissions, this paper has chosen energy intensity as one of the influencing factors for carbon emission efficiency.

### 6.1.6. Science and Technology Level

The recently published studies also described the relationship between science and technology level and carbon emission efficiency [11,86–88]. Adopting advanced technology, equipment, manufacturing models, or procedures can improve energy usage efficiency with a low cost, and thus enhance carbon emission efficiency. One precondition to improving science and technology level is to increase research and development (R&D) input. Liu and Xia et al. in 2018 stated that increasing technological expenditure can promote carbon emission efficiency growth [89]. Wang and Zhao et al. in 2019 found that R&D investments have a vital role in $CO_2$ emission reduction [53]. Therefore, this paper chose the local R&D expenditure to GDP as a proxy for the science and technology level.

Table 5 will show the concrete definition of variable index.

**Table 5.** Influencing factors.

| Explanatory Variable | Variables' Definition and Unit | Prejudgment | Key References |
|---|---|---|---|
| Government intervention in the economy (GIE) | The proportion of the local financial expenditure to GDP (%) | Unknown | [6,11] |
| Industrial structure (IS) | The proportion of the local added value of the tertiary industry to GDP (%) | Positive | [66–71] |
| Foreign trade level (FTL) | The proportion of the local import and export trade to GDP (%) | Unknown | [11,72,73] |
| Foreign capital utilization level (FCUL) | The proportion of the local foreign direct investment to GDP (%) | Unknown | [74–80] |
| Energy intensity (EI) | 10,000 yuan GDP standard coal consumption (yuan/tons of standard coal) | Negative | [81–85] |
| Science and technology level (STL) | The proportion of the local R&D expenditure to GDP (%) | Positive | [11,53,86–89] |

### 6.2. Unit Root Test and Cointegration Test

Before proceeding to any econometric analysis, we tested the existence of unit roots in all variables first. In a non-stationary time series, the outstanding regression relationship between a variable and another random variable may result in spurious regression [90]. This paper adopted the LLC [91], IPS [92], Fisher-ADF [93], and PP-ADF [94] panel unit root test to determine the stability of the variables. In Table 6, results showed that four variables are not smooth level through the test, but all became stationary at the 1% significance level and rejected the null hypothesis of "existing unit root" at the significance level within 1% after taking first differences. This implies that all variables are stationary at the first difference, and there can be a long-term equilibrium relationship among all the variables.

**Table 6.** Panel unit root test results.

|  | LLC | IPS | Fisher-ADF | PP-ADF |
|---|---|---|---|---|
| CEE | −13.9912 *** | −8.15403 *** | 155.223 *** | 202.149 *** |
| GIE | −1.15188 | 4.99367 | 20.0210 | 36.1092 |
| IS | 6.28823 | 7.88391 | 21.5548 | 19.3982 |
| FTL | −2.48395 *** | 1.40910 | 41.2505 | 44.3308 |
| FCUL | −7.04262 *** | −2.20561 ** | 88.6931 *** | 150.166 *** |
| EI | −2.24391 *** | 4.94331 | 26.1251 | 37.3777 |
| STL | −2.86846 *** | 1.52624 | 48.4451 | 77.5646 * |
| $\Delta$CEE | −19.0895 *** | −11.2225 *** | 221.242 *** | 371.516 *** |
| $\Delta$GIE | −11.0323 *** | −6.73075 *** | 155.873 *** | 190.270 *** |
| $\Delta$IS | −5.75004 *** | −1.35639 * | 78.7522 * | 101.827 *** |
| $\Delta$FTL | −12.2580 *** | −5.97896 *** | 150.454 *** | 208.309 *** |
| $\Delta$FCUL | −11.9671 *** | −5.06193 *** | 133.507 *** | 156.930 *** |
| $\Delta$EI | −15.1595 *** | −9.84822 *** | 205.206 *** | 242.850 *** |
| $\Delta$STL | −14.9755 *** | −8.24486 *** | 187.069 *** | 244.140 *** |

Note: ***, **, * representing variables significant at 1%, 5%, and 10%, respectively.

If all variables remain non-stationary until the first-order difference, this analysis proceeds with the cointegration test. The paper used the Pedroni panel cointegration to determine whether the panel data had a cointegration relationship [95]. The test results are shown in Table 7.

**Table 7.** Panel cointegration test results.

|  | Panel-V-Stat | Panel-Rho-Stat | Panel-PP-Stat | Panel-ADF-Stat |
|---|---|---|---|---|
| within-dimension | −1694.92 | 4.901894 | −15.11433 *** | −2.657887 *** |
| between-dimension |  | 7.692484 | −22.66525 *** | −2.098071 *** |

Note: *** representing variables significant at 1%, respectively.

Table 7 presented the panel cointegration test results for all variables. These results showed that four of the six statistics rejected the null hypothesis of no cointegration at the 1% significance level. Hence, the paper concluded that there was a constant long-run equilibrium relationship among carbon emission efficiency, government intervention in the economy, industrial structure, foreign trade level, foreign capital utilization level, energy intensity, and science and technology level in China from 2005 to 2015.

*6.3. Explaining Carbon Emission Efficiency: Tobit Regression Results*

Based on the above results, this paper evaluated the impact of above influencing factors on the carbon emission efficiency in China. The Tobit regression model assumed that

$$CEE_{i,t} = \beta_0 + \beta_1 GIEit + \beta_2 ISit + \beta_3 FTLit + \beta_4 FCULit + \beta_5 EIit + \beta_6 STLit + uit \tag{7}$$

where $CEE_{i,t}$ represents the carbon emission efficiency value of the ith province in the tth year, $\beta_0, \beta_1, \beta_2, \dots, \beta_6$ stands for the unknown coefficients, and $u_{i,t}$ is a random disturbance term. The parameters were estimated by Stata12.0 software. The results are given in Table 8.

**Table 8.** Tobit regression results.

| Variable | Coefficient | Std. Err. | Z-Statistic | P > \|z\| | [95% Conf. Interval] | |
|---|---|---|---|---|---|---|
| DGI | −0.461083 *** | 0.0585006 | −7.88 | 0.000 | −0.5757416 | −0.3464233 |
| IS | 0.0076522 | 0.0836948 | 0.09 | 0.927 | −0.1563865 | 0.171691 |
| FTL | 0.0881033 *** | 0.0230223 | 3.83 | 0.000 | 0.0429804 | 0.1332263 |
| FCUL | 0.8143128 *** | 0.236718 | 3.44 | 0.000 | 0.350354 | 6.832712 |
| EI | −0.043391 *** | 0.0110785 | −3.92 | 0.003 | −0.0651047 | −0.0216776 |
| STL | 4.95859 *** | 0.9562014 | 5.19 | 0.003 | 3.084472 | 6.629005 |

Note: *** representing variables significant at 1%, respectively.

Government intervention was significant at the 1% level, with a regression coefficient of 0.461083, which indicates that government intervention influences carbon emission efficiency negatively. This result is consistent with a study in 2011 [96] and a recent one in 2019 [11]. From 2005 to 2015, the proportion of financial expenditure to GDP gradually increased from 18.11% to 25.63%, as the Chinese government took the proactive fiscal policy measures to intervene in the market economy. The results showed that the government's excessive intervention in the economy was not conducive to the improvement of carbon emission efficiency.

The industrial structure is positively related to carbon emission efficiency in a non-significant way, implying that industrial structure only improves carbon emission efficiency to a certain extent. There was a significant positive correlation between foreign trade and carbon emission efficiency with a coefficient rate of 0.8143128. This finding is consistent with the study by Zhu and Du et al. in 2013 [97], as well as Wang and Ma's in 2018 [55]. This result supported that the development of foreign trade can improve China's carbon emission efficiency.

Foreign capital utilization level plays a positive role in improving carbon emission efficiency, and the result is significant at the 1% level, which implies the Pollution Haven Hypothesis fails in the Chinese context. The results echo the findings by Perkins and Neumayer in 2009 [98], as well as Wang and Ye in 2019 [99], suggesting the Chinese provinces need to take measures to attract more FDI.

The estimated coefficient of the energy intensity is significantly negative (at the 1% level), which is in accord with our assumption that the decline in energy intensity can improve carbon emission efficiency, also consistent with the policy of energy intensity reduction in its Thirteenth Five-Year plan in 2016. The coefficient score of science and technology was significantly positive, indicating that the science and technology development was conducive to the improvement of carbon emission efficiency.

## 7. Conclusions and Discussions

This paper presented the analysis of regional differences and influencing factors on China's carbon emission efficiency from 2005 to 2015 by adopting the Modified Undesirable EBM DEA Model and Tobit model. The results showed that (1) most Chinese provinces generally have low carbon emission efficiencies levels. The average value of 30 provinces' carbon emission efficiency was 0.6277. Only 11 provinces had a higher level than the national average level. Therefore, energy saving and emission reduction in China are still challenging for governments. (2) Significant differences in carbon emission efficiency exist across the three regions in China. The Eastern region ranked the top, with the average efficiency was 0.7816, followed by the Central region (0.5967) and the Western region (0.4964). (3) Northern, Eastern, and Southern coastal regions had a higher level of carbon emission efficiency, followed by Northeast and Middle Yangtze River and Middle Yellow River. The Southwest and Northwest had the lowest level of carbon emission efficiency. (4) In the results from Tobit regression, foreign trade level, foreign capital utilization level, and science and technology level all had significant positive effects on carbon emission efficiency. Government intervention and energy intensity negatively affected the carbon emission efficiency in a significant way. The relationship between industrial structure and carbon emission efficiency had a non-significant positive result. The theoretical contribution of this manuscript is to adopting the Modified Undesirable EBM DEA Model, which is more accurate than traditional methods, such as CCR, BCC or SBM model. Moreover, based on the data's unit root testing and cointegration, the paper verified the influencing factors of carbon emission efficiency by the Tobit model, which avoids the spurious regression.

Based on the above analysis, we raised several policy implications to improve carbon emission efficiency. When making full use of foreign trade and FDI, governments can reduce the import and export of high-energy and emission products and support the foreign trade of high-tech and environmentally friendly products. The governments can also optimize the energy structure and increase investment in science and technology. The government can adjust industrial structure by promoting the tertiary sector and reduce the development of manufacturing with high pollution and emission. Additionally, governments can promote low carbon industries by providing subsidies or special findings.

As the regions have different carbon emission efficiencies, the local governments should make use of their strengths. The provincial and municipal governments of the eastern provinces should fully tap the potential of energy saving and emission reduction based on their original economic development level. As for the local governments of the central and western provinces, they can make use of their industrial legacies and combine their advantages with the import of foreign capital. In this foreign capital attraction process, the local governments should restrict pollution-intensive industries with environmental regulations, and attract high-tech industries through FDI. The provincial and municipal governments should strengthen the monitoring and supervision of pollutant emissions with improved evaluation system and punishment mechanism for carbon emission efficiency.

However, the limitations of the research remain in the data collection. We calculated the provincial capital stock and $CO_2$ emissions due to lacking data in the provincial statistical yearbook. The accuracy of the research result can be improved if the central and local Statistical Bureau to disclose statistics on the provincial capital stock and $CO_2$ emissions. Future research can analyze various industrial sectors' carbon emission efficiency in the next study period, which further reflects the regional differences in carbon emission efficiency in various industrial sectors.

**Author Contributions:** Conceptualization, L.Z.; Methodology, L.Z., H.H. and Y.L.; Software, L.Z.; Validation, H.L.; Formal Analysis, L.Z. and H.L.; Investigation, L.Z. and Y.Z.; Resources, H.H.; Data Curation, L.Z. and Y.L.; Writing – Original Draft Preparation, L.Z. and H.L.; Writing – Review & Editing, L.Z., H.L. and Y.Z.; Visualization, L.Z. and Y.L.; Project Administration, L.Z. and Y.L.

**Acknowledgments:** The authors are grateful for the supervision and help from Prof. Pengjun Zhao. This research was funded by the NSFC, project No. 41571147. The publication of this article was funded by Freie Universität Berlin.

**Conflicts of Interest:** The authors declare no conflict of interest.

## References


1. IPCC. *Climate Change 2007: Synthesis Report*; Cambridge University Press: Cambridge, UK, 2007.
2. Edenhofer, O.; Pichs-Madruga, R.; Sokona, Y.; Farahani, E.; Kadner, S.; Kadner, K.; Seyboth, A.; Adler, I.; Baum, S.; Myhre, G.; et al. *Climate Change 2014: Mitigation of Climate Change*; Working Group III Contribution to the IPCC Fifth Assessment Report; Cambridge University Press: Cambridge, UK, 2015.
3. BP. *Statistical Review of World Energy*; BP: London, UK, 2018.
4. The Global Carbon Project (GCP). Global Carbon Budget 2017. *Earth Syst. Sci. Data* **2018**, *10*, 405–448. Available online: https://www.earth-syst-sci-data.net/10/405/2018/ (accessed on 25 July 2019). [CrossRef]
5. Xie, L.; Chen, C.L.; Yu, Y.H. Dynamic Assessment of Environmental Efficiency in Chinese Industry: A Multiple DEA Model with a Gini Criterion Approach. *Sustainability* **2019**, *11*, 2294. [CrossRef]
6. Wang, S.; Wang, H.; Zhang, L.; Dang, J. Provincial Carbon Emissions Efficiency and Its Influencing Factors in China. *Sustainability* **2019**, *11*, 2355. [CrossRef]
7. *The 13th Five-Year Plan for Economic and Social Development of the People's Republic of China (2016–2020)*; The Central Compilation and Translation Bureau: Beijing, China, 2016.
8. Meng, Z.S.; Wang, H.A.; Wang, B.N. Empirical Analysis of Carbon Emission Accounting and Influencing Factors of Energy Consumption in China. *Int. J. Environ. Res. Public Health* **2018**, *15*, 2467. [CrossRef] [PubMed]
9. Feng, F.; Peng, L.L. Is There Any Difference in the Effect of Different R and D Sources on Carbon Intensity in China? *Sustainability* **2019**, *11*, 1701. [CrossRef]
10. Wang, Q.W.; Zhou, P.; Shen, N.; Wang, S.S. Measuring Carbon Dioxide Emission Performance in Chinese Provinces: A Parametric Approach. *Renew. Sustain. Energy Rev. Dev.* **2013**, *21*, 324–330. [CrossRef]
11. Chu, X.X.; Geng, H.; Guo, W. How Does Energy Misallocation Affect Carbon Emission Efficiency in China? An Empirical Study Based on the Spatial Econometric Model. *Sustainability* **2019**, *11*, 2115. [CrossRef]
12. Zhong, Y.Y.; Zhong, W.Z. China's Regional Total Factor Carbon Emission Performance and Influencing Factors Analysis. *J. Bus. Econ.* **2012**, *1*, 85–96. [CrossRef]
13. Wang, M. *Key Issues in China's Development*; China Development Press: Beijing, China, 2005.
14. Charnes, A.; Cooper, W.W.; Rhodes, E. Measuring the Efficiency of Decision Making Units. *Eur. J. Oper. Res.* **1978**, *2*, 429–444. [CrossRef]
15. Banker, R.D.; Charnes, R.F.; Cooper, W.W. Some Models for Estimating Technical and Scale Inefficiencies in Data Envelopment Analysis. *Manag. Sci.* **1984**, *30*, 1078–1092. [CrossRef]
16. Tone, K. A Slacks-based Measure of Efficiency in Data Envelopment Analysis. *Eur. J. Oper. Res.* **2001**, *130*, 498–509. [CrossRef]
17. Li, Y.; Chiu, Y.H.; Lin, T.Y. The Impact of Economic Growth and Air Pollution on Public Health in 31 Chinese Cities. *Int. J. Environ. Res. Public Health* **2019**, *16*, 393. [CrossRef]
18. Qu, Y.; Yu, Y.; Appolloni, A.; Li, M.; Liu, Y. Measuring Green Growth Efficiency for Chinese Manufacturing Industries. *Sustainability* **2017**, *9*, 637. [CrossRef]
19. Li, Y.; Chiu, Y.H.; Lin, T.Y. Energy and Environmental Efficiency in Different Chinese Regions. *Sustainability* **2019**, *11*, 1216. [CrossRef]
20. Tian, Z.; Ren, F.R.; Xiao, Q.W.; Chiu, Y.H.; Lin, T.Y. Cross-Regional Comparative Study on Carbon Emission Efficiency of China's Yangtze River Economic Belt Based on the Meta-Frontier. *Int. J. Environ. Res. Public Health* **2019**, *16*, 619. [CrossRef]
21. Yang, L.; Wang, K.L.; Geng, J.C. China's Regional Ecological Energy Efficiency and Energy Saving and Pollution Abatement Potentials: An Empirical Analysis Using Epsilon-Based Measure Model. *J. Clean. Prod.* **2018**, *194*, 300–308. [CrossRef]

22. Cui, Q.; Li, Y. CNG2020 Strategy and Airline Efficiency: A Network Epsilon-Based Measure with Managerial Disposability. *Int. J. Sustain. Transp.* **2018**, *12*, 313–323. [CrossRef]
23. Tone, K.; Tsutsui, M. An Epsilon-Based Measure of Efficiency in DEA-A Third Pole of Technical Efficiency. European. *J. Oper. Res.* **2010**, *207*, 1554–1563. [CrossRef]
24. Li, Y.; Chiu, Y.H.; Lu, L.C. New Energy Development and Pollution Emissions in China. *Int. J. Environ. Res. Public Health* **2019**, *16*, 1764. [CrossRef]
25. Tobin, J. Estimation of relationships for limited dependent variables. *Econometrica* **1958**, *26*, 24–36. Available online: https://search.proquest.com/docview/214674441?OpenUrlRefId=info:xri/sid:baidu&accountid=13151 (accessed on 25 July 2019). [CrossRef]
26. Kaya, Y.; Yokobori, K. *Global Environment, Energy, and Economic Development*; United Nations University: Tokyo, Japan, 1993.
27. Mielnik, O.; Goldemberg, J. Communication the Evolution of the Carbonization Index in Developing Countries. *Energy Policy* **1999**, *27*, 307–308. [CrossRef]
28. Sun, J.W. The Decrease of $CO_2$ Emission Intensity is Decarbonization at National and Global Levels. *Energy Policy* **2005**, *33*, 975–978. [CrossRef]
29. Li, Z.L. Analysis on Regional Carbon Emission Efficiency Measurement and Influence Factors in China. Master's Thesis, China University of Mining & Technology, Beijing, China, 2014.
30. Zofit, J.L.; Prieto, A.M. Environmental Efficiency and Regulatory Standards: The Case of $CO_2$ Emissions from OECD Industries. *Resour. Energy Econ.* **2001**, *23*, 63–83. [CrossRef]
31. Wang, K.; Wei, Y.M.; Zhang, X. A Comparative Analysis of China's Regional Energy and Emission. Performance: Which is the Better Way to Deal with Undesirable Outputs? *Energy Policy* **2012**, *46*, 574–584. [CrossRef]
32. Zhou, G.; Chung, W.; Zhang, Y. Measuring Energy Efficiency Performance of China's Transport Sector: A Data Envelopment Analysis Approach. *Expert Syst. Appl.* **2014**, *41*, 709–722. [CrossRef]
33. Yao, X.; Zhou, H.; Zhang, A.; Li, A.J. Regional Energy Efficiency, Carbon Emission Performance and Technology Gaps in China: A Meta-frontier Non-Radial Directional Distance Function Analysis. *Energy Policy* **2015**, *84*, 142–154. [CrossRef]
34. Aigner, D.; Lovell, C.A.K.; Schmidt, P. Formulation and Estimation of Stochastic Frontier Production Function Models. *J. Econ.* **1977**, *6*, 21–37. [CrossRef]
35. Lee, S.; Kim, C. Estimation of Association between Healthcare System Efficiency and Policy Factors for Public Health. *Appl. Sci.* **2018**, *8*, 2674. [CrossRef]
36. Greene, W. Reconsidering heterogeneity in panel data estimators of the stochastic frontier model. *J. Econ.* **2005**, *126*, 269–303. [CrossRef]
37. Cullinane, K.; Song, D.W.; Gray, R. A stochastic frontier model of the efficiency of major container terminals in Asia: Assessing the influence of administrative and ownership structures. *Transp. Res. Part A* **2002**, *36*, 743–762. [CrossRef]
38. Zheng, J.; Zhang, H.Q.; Xing, Z.C. Re-Examining Regional Total-Factor Water Efficiency and Its Determinants in China: A Parametric Distance Function Approach. *Water* **2018**, *10*, 1286. [CrossRef]
39. Cullinane, K.; Wang, T.F.; Song, D.W.; Ji, P. The technical efficiency of container ports: Comparing data envelopment analysis and stochastic frontier analysis. *Transp. Res. Part A Policy Pract.* **2006**, *40*, 354–374. [CrossRef]
40. Wang, S.Q.; Zhou, L.; Wang, H.; Li, X.C. Water Use Efficiency and Its Influencing Factors in China: Based on the Data Envelopment Analysis (DEA)-Tobit Model. *Water* **2018**, *10*, 832. [CrossRef]
41. Chung, Y.H.; Fare, R.; Grosskopf, S. Productivity and Undesirable Outputs: A Directional Distance Function Approach. *J. Environ. Manag.* **1997**, *51*, 229–240. [CrossRef]
42. Wei, M.; Cao, M.F.; Jiang, J.R. Determinants of Long-run Carbon Emission Performance. *J. Quant. Tech. Econ.* **2010**, *9*, 43–52+81. Available online: http://en.cnki.com.cn/Article_en/CJFDTOTAL-SLJY201009006.htm (accessed on 25 July 2019).
43. Wang, Y.; Duan, F.M.; Ma, X.J.; He, L.C. Carbon Emissions Efficiency in China: Key Facts from Regional and Industrial Sector. *J. Clean. Prod.* **2019**, *206*, 850–869. [CrossRef]
44. Li, K.; Lin, B. Metafroniter Energy Efficiency with $CO_2$ Emissions and its Convergence Analysis for China. *Energy Econ.* **2015**, *48*, 230–241. [CrossRef]

45. Huo, T.; Ren, H.; Cai, W.; Feng, W.; Tang, M.; Zhou, N. The Total-Factor Energy Productivity Growth of China's Construction Industry: Evidence from the Regional Level. *Nat. Hazards* **2018**, 1–24. [CrossRef]

46. Wei, Y.G.; Li, Y.; Wu, M.Y.; Li, Y.B. The Decomposition of Total-Factor $CO_2$ Emission Efficiency of 97 Contracting Countries in Paris Agreement. *Energy Econ.* **2019**, *78*, 365–378. [CrossRef]

47. Anselin, L. *Spatial Econometrics: Methods and Models*; Kluwer Academic Publishers: Dordrecht, The Netherlands, 1988.

48. Lesage, J.; Pace, R.K. *Introduction to Spatial Econometrics*; CRC Press: New York, NY, USA, 2009.

49. Chuai, X.W.; Huang, X.J.; Wang, W.J.; Wen, J.Q.; Chen, Q.; Peng, J.W. Spatial Econometric Analysis of Carbon Emissions from Energy Consumption in China. *J. Geogr. Sci.* **2012**, *22*, 630–642. Available online: https://link.springer.com/article/10.1007/s11442-012-0952-z (accessed on 25 July 2019). [CrossRef]

50. Ma, D.L.; Chen, Z.C.; Wang, L. Spatial Econometrics Research on Inter-Provincial Carbon Emissions Efficiency in China. *Renew. Sustain. Energy Rev.* **2015**, *25*, 67–77. Available online: http://en.cnki.com.cn/Article_en/CJFDTotal-ZGRZ201501010.htm (accessed on 25 July 2019).

51. Cheng, Y.Q.; Wang, Z.Y.; Ye, X.Y.; Wei, D. Spatiotemporal Dynamics of Carbon Intensity from Energy Consumption in China. *J. Geogr. Sci.* **2014**, *24*, 631–650. Available online: http://en.cnki.com.cn/Article_en/CJFDTotal-ZGDE201404004.htm (accessed on 25 July 2019). [CrossRef]

52. Yi, T.; Tong, L.; Qiu, M.H.; Liu, J.P. Analysis of Driving Factors of Photovoltaic Power Generation Efficiency: A Case Study in China. *Energies* **2019**, *12*, 355. [CrossRef]

53. Wang, Y.; Zhao, T.; Wang, J.A.; Guo, F.; Kan, X.; Yuan, R. Spatial Analysis on Carbon Emission Abatement Capacity at Provincial Level in China from 1997 to 2014: An Empirical Study Based on SDM Model. *Atmos. Pollut. Res.* **2019**, *10*, 97–104. [CrossRef]

54. Shi, S.A.; Xia, L.; Meng, M. Energy Efficiency and Its Driving Factors in China's Three Economic Regions. *Sustainability* **2017**, *9*, 2059. [CrossRef]

55. Wang, S.J.; Ma, Y.Y. Influencing Factors and Regional Discrepancies of the Efficiency of Carbon Dioxide Emissions in Jiangsu, China. *Ecol. Indic.* **2018**, *90*, 460–468. [CrossRef]

56. Gai, M.; Cao, G.Y.; Tian, C.S.; Ke, L.N. Decoupling Analysis of Energy Carbon Emissions and Regional Economic Growth in the Liaoning Coastal Economic Belt. *Resour. Sci.* **2014**, *36*, 1267–1277. Available online: http://www.cnki.com.cn/Article/CJFDTOTAL-ZRZY201406020.htm (accessed on 25 July 2019).

57. Takakai, F.; Kobayashi, M.; Sato, T.; Yasuda, K.; Kaneta, Y. Effects of Forage Rice Cultivation on Carbon and Greenhouse Gas Balances in a Rice Paddy Field. *Atmosphere* **2018**, *9*, 504. [CrossRef]

58. Goldsmith, R.M. A Perpetual Inventory of National Wealth. *Stud. Income Wealth* **1951**, *14*, 5–73. Available online: http://www.nber.org/books/unkn51-2 (accessed on 25 July 2019).

59. Zhang, J.; Wu, G.Y.; Zhang, J.P. The Estimation of China's Provincial Capital Stock: 1952–2000. *Econ. Res. J.* **2004**, *10*, 35–44. Available online: http://en.cnki.com.cn/Article_en/CJFDTOTAL-JJYJ200410004.htm (accessed on 25 July 2019).

60. Zhao, Z.Y.; Yang, C.F. The Decomposition Analysis on the Driving Factors of China's Carbon Emission. *Chin. Soft Sci.* **2012**, *6*, 175–183. Available online: http://en.cnki.com.cn/Article_en/CJFDTotal-SEJJ201106005.htm (accessed on 25 July 2019).

61. Hu, Y.; Liu, J.F.; Hu, W.; Zhang, Z. Regional Variance, Trend Evolution and Factors of China's Carbon Emission Intensity Based on 30 Provinces (cities and districts)' 1997–2012 Panel Data. *Resour. Dev. Mark.* **2016**, *18*, 7–13. Available online: http://www.en.cnki.com.cn/Article_en/CJFDTotal-ZIYU201605002.htm (accessed on 25 July 2019).

62. IPCC. *Guidelines for National Greenhouse Gas Inventories*; Cambridge University Press (IPCC Secretariat): Cambridge, UK, 2006.

63. National Development and Reform Commission. Guidelines for Provincial Greenhouse Gas Inventories in China. Available online: http://www.tanjiaoyi.com/article-27200-1.html (accessed on 31 May 2019).

64. Liu, J.P.; Ge, H.X.; Wang, Y.R.; Chen, J.Z.; Wang, H. Dynamic Evaluation of Carbon Emission Efficiency in China's Provinces Based on Cross Efficiency DEA Model. *Resour. Dev. Mark.* **2017**, *33*, 1041–1057. Available online: http://www.en.cnki.com.cn/Article_en/CJFDTotal-ZTKB201709004.htm (accessed on July 2019).

65. Tang, D.C.; Zhang, Y.; Bethel, B.J. An Analysis of Disparities and Driving Factors of Carbon Emissions in the Yangtze River Economic Belt. *Sustainability* **2019**, *11*, 2362. [CrossRef]

66. Li, Y.M.; Zhang, L.; Cheng, X.L. A Decomposition Model and Reduction Approaches for Carbon Dioxide Emissions in China. *Resour. Sci.* **2010**, *32*, 218–222. Available online: http://en.cnki.com.cn/article_en/cjfdtotal-zrzy201002006.htm (accessed on 25 July 2019).

67. Zhang, Y.J.; Liu, Z.; Zhang, H.; Tan, T.-D. The impact of Economic Growth, Industrial Structure and Urbanization on Carbon Emission Intensity in China. *Nat. Hazards* **2014**, *73*, 579–595. Available online: https://link.springer.com/article/10.1007/s11069-014-1091-x (accessed on 25 July 2019). [CrossRef]

68. Cheng, Z.H.; Shi, X.A. Can Industrial Structural Adjustment Improve the Total-Factor Carbon Emission Performance in China? *Int. J. Environ. Res. Public Health* **2018**, *15*, 2291. [CrossRef]

69. Lin, B.Q.; Chen, T.; Li, Y.Q.; Song, H.H.; Ma, Z.X. Research on the Effects of Urbanization on Carbon Emissions Efficiency of Urban Agglomerations in China. *J. Clean. Prod.* **2018**, *197*, 1374–1781. [CrossRef]

70. Wang, S.; Liu, X.; Zhou, C.; Hu, J.; Ou, J. Examining the impacts of socioeconomic factors, urban form, and transportation networks on $CO_2$ emissions in China's megacities. *Appl. Energy* **2017**, *185*, 189–200. [CrossRef]

71. Mi, Z.; Wei, Y.M.; Wang, B.; Meng, J.; Liu, Z.; Shan, Y.; Liu, J.; Guan, D. Socioeconomic impact assessment of China's $CO_2$ emissions peak prior to 2030. *J. Clean. Prod.* **2017**, *142*, 2227–2236. [CrossRef]

72. Shui, B.; Harriss, R.C. The Role of $CO_2$ Embodiment in US–China Trade. *Energy Policy* **2006**, *34*, 4063–4068. [CrossRef]

73. Weber, C.L.; Peters, G.P.; Guan, D.; Hubacek, K. The Contribution of Chinese Exports to Climate Change. *Energy Policy* **2008**, *9*, 3572–3577. [CrossRef]

74. Lee, J.W. The Contribution of Foreign Direct Investment to Clean Energy Use, Carbon Emissions and Economic Growth. *Energy Policy* **2013**, *55*, 483–489. [CrossRef]

75. Pazienza, P. The impact of FDI in the OECD manufacturing sector on $CO_2$ emission: Evidence and policy issues. *Environ. Impact Assess. Rev.* **2019**, *77*, 66–68. [CrossRef]

76. Fan, B.; Zhang, Y.; Li, X.Z.; Miao, X. Trade Openness and Carbon Leakage: Empirical Evidence from China's Industrial Sector. *Energies* **2019**, *12*, 1101. [CrossRef]

77. Zhu, H.; Duan, L.; Guo, Y.; Yu, K. The effects of FDI, economic growth and energy consumption on carbon emissions in ASEAN-5: Evidence from panel quantile regression. *Econ. Model.* **2016**, *58*, 237–248. [CrossRef]

78. Tamazian, A.; Chousa, J.P.; Vadlamannati, K.C. Does higher economic and financial development lead to environmental degradation: Evidence from BRIC countries. *Energy Policy* **2009**, *37*, 246–253. [CrossRef]

79. Arik, L.; Taylor, M.S. Unmasking the Pollution Haven Effect. *Int. Econ. Rev.* **2008**, *49*, 223–254. [CrossRef]

80. Zhou, Y.; Fu, J.T.; Kong, Y.; Wu, R. How Foreign Direct Investment Influences Carbon Emissions, Based on the Empirical Analysis of Chinese Urban Data. *Sustainability* **2018**, *10*, 2163. [CrossRef]

81. Liu, N.; Ma, Z.J.; Kang, J.D. Changes in Carbon Intensity in China's Industrial Sector: Decomposition and Attribution Analysis. *Energy Policy* **2015**, *87*, 28–38. [CrossRef]

82. Fan, F.Y.; Lei, Y.L. Decomposition Analysis of Energy-Related Carbon Emissions from the Transportation Sector in Beijing. *Transp. Res. Part D Transp. Environ.* **2016**, *42*, 135–145. [CrossRef]

83. Shuai, C.Y.; Chen, X.; Wu, Y. Identifying the key Impact Factors of Carbon Emission in China. Results from a Largely Expanded Pool of Potential Impact Factors. *J. Clean. Prod.* **2018**, *175*, 612–623. [CrossRef]

84. Meng, B.; Xue, J.; Feng, K.; Guan, D.; Fu, X. China's inter-regional spillover of carbon emissions and domestic supply chains. *Energy Policy* **2013**, *61*, 1305–1321. [CrossRef]

85. Fan, T.; Luo, R.; Fan, Y.; Zhang, L.; Chang, X. Study on Influence Factors for carbon Dioxide Emissions in China's Chemical Industry with LMDI Method. *China Soft Sci.* **2013**, *3*, 166–174. Available online: http://en.cnki.com.cn/Article_en/CJFDTotal-ZGRK201303016.htm (accessed on 25 July 2019).

86. Zhang, P.Y.; He, J.J.; Hong, X.; Zhang, W.; Qin, C.Z.; Pang, B.; Li, Y.Y.; Liu, Y. Regional-Level Carbon Emissions Modelling and Scenario Analysis: A STIRPAT Case Study in Henan Province, China. *Sustainability* **2017**, *9*, 2342. [CrossRef]

87. Zhu, C.Z.; Gao, D.W. A Research on the Factors Influencing Carbon Emission of Transportation Industry in "the Belt and Road Initiative" Countries Based on Panel Data. *Energies* **2019**, *12*, 2405. [CrossRef]

88. Liu, W.D.; Xu, X.M.; Niu, D.X. Research on the Impact of Technology Import and Technology Innovation on the Peak of Carbon Emissions in China. *Technol. Econ. Manag. Res.* **2016**, *9*, 3–9. Available online: http://www.en.cnki.com.cn/Article_en/CJFDTOTAL-JXJG201609001.htm (accessed on 25 July 2019).

89. Liu, S.; Xia, X.H.; Chen, X.Y. Assessing Urban Carbon Emission Efficiency in China: Based on the Global Data Envelopment Analysis. *Energy Procedia* **2018**, *152*, 762–767. [CrossRef]

90. Gu, A.; Zhang, Y.; Pan, B.L. Relationship between Industrial Water Use and Economic Growth in China: Insights from an Environmental Kuznets Curve. *Water* **2017**, *9*, 556. [CrossRef]
91. Levin, A.; Lin, C.; Chu, C. Unit Root Tests in Panel Data: Asymptotic and Finite Sample Properties. *J. Econom.* **2002**, *108*, 1–24. Available online: https://www.sciencedirect.com/science/article/pii/S0304407601000987 (accessed on 25 July 2019). [CrossRef]
92. Im, K.S.; Pesaran, M.H.; Shin, Y.C. Testing for Unit Roots in Heterogeneous Panels. *J. Econ.* **2003**, *115*, 53–74. [CrossRef]
93. Dickey, D.A.; Fuller, W.A. Distribution of the Estimators for Autoregressive Time Series with A Unit Root. *J. Am. Stat. Assoc.* **1979**, *366*, 427–431. [CrossRef]
94. Phillips, P.C.B.; Perron, P. Testing for A Unit Root in Time Series Regression. *Biometrika* **1988**, *75*, 335–346. [CrossRef]
95. Pedroni, P. Critical Values for Cointegration Tests in Het-erogeneous Panels with Multiple Regressors. *Oxf. Bull. Econ. Stat.* **1999**, *61*, 653–670. Available online: https://www.scirp.org/reference/ReferencesPapers.aspx?ReferenceID=2340557 (accessed on 25 July 2019). [CrossRef]
96. Du, K.; Zou, C.Y. Regional Disparity, Affecting Factors and Convergence Analysis of Carbon Dioxide Emission Efficiency in China: On Stochastic Frontier Model and Panel Unit Root. *Zhejiang Soc. Sci.* **2011**, *11*, 32–43+156. Available online: http://en.cnki.com.cn/Article_en/CJFDTotal-ZJSH201111007.htm (accessed on 25 July 2019).
97. Zhu, D.J.; Du, K.R. Foreign Trade, Economic Growth and the Efficiency of Carbon Emission in China. *J. Shanxi Financ. Econ. Univ.* **2013**, *35*, 1–11. Available online: http://en.cnki.com.cn/Article_en/CJFDTOTAL-SXCJ201305000.htm (accessed on 25 July 2019).
98. Perkins, R.; Neumayer, E. Transnational Linkages and the Spillover of Environment-Efficiency into Developing Countries. *Glob. Environ. Chang.* **2009**, *19*, 375–383. [CrossRef]
99. Wang, Z.X.; Ye, D.J.; Zheng, H.H.; Lv, C.Y. The Influence of Market Reform on the $CO_2$ Emission Efficiency of China. *J. Clean. Prod.* **2019**, *225*, 236–247. [CrossRef]

*Article*

# The Symptoms of Illness: Does Israel Suffer from "Dutch Disease"?

**Ruslana Rachel Palatnik** [1,2,3,*], **Tchai Tavor** [1] **and Liran Voldman** [1]

[1]  Department of Economics and Management, The Max Stern Yezreel Valley College,
    Emek Yezreel 19300, Israel
[2]  SEED—The Sustainable Economic and Environmental Development Research Center, The Max Stern Yezreel
    Valley College, Emek Yezreel 19300, Israel
[3]  NRERC—Natural Resource and Environmental Research Center, University of Haifa, Haifa 31905, Israel
*  Correspondence: rachelpa@yvc.ac.il

Received: 8 May 2019; Accepted: 12 July 2019; Published: 18 July 2019

**Abstract:** The natural gas revolution in Israel started about two decades ago. Its numerous social impacts include moving to cleaner energy, improving energy security and the balance of trade, tightening international relations, and increasing tax revenue. However, "Dutch disease" phenomena—where the accelerated export of natural gas leads to the strengthening of the local currency, the subsequent weakening of other exporting industries, and rising unemployment—might suck Israel into the economic slowdown. This study examines whether the strengthening of the New Israeli Shekel (ILS) in recent years is a symptom of "Dutch disease". It is expected that the large-scale export of natural gas will start in 2021 with the development of the major offshore field "Leviathan". Notably, ILS has been appreciating for several years already. We employed the event study approach to analyze the fluctuations of the daily ILS/USD real exchange rate in the years 2009–2017, combined with the media announcements related to the gas discoveries published during this period. The results revealed that gas-related news does affect the exchange rate and appreciate ILS. GARCH analysis confirms the results.

**Keywords:** Dutch disease; natural gas; event study; real exchange rate; announcements; currency appreciation; export; expectations

---

## 1. Introduction

The past recent years have seen tremendous turmoil in regional and global energy markets, with volatile oil prices, geopolitical tensions over oil and natural gas (NG) supply, and tightened environmental regulation.

Up until recently, Israel was considered a resource-deprived country, especially with regard to fossil fuels. While traditionally relying on coal and oil imports, the last two decades have seen Israel diversifying its energy sources with the usage of NG. Israel only had its first commercially recoverable discovery of fossil fuel in 1999, with natural gas discoveries at the Noa and Mari-B fields in the Mediterranean. These fields are collectively known as Yam Tetis [1]. A major source of Israel's NG originated in Egypt, covering some 40% of Israeli demand. On the advent of civil unrest in Egypt in 2011, the el-Arish-Ashkelon pipeline, which delivered NG to Israel, was repeatedly sabotaged, effectively bringing imports to a halt [2]. This supply disruption inflicted heavy economic and environmental burden due to the need to switch to costly oil-based fuels in its electricity generation. However, by this time, significant NG discoveries (TAMAR field) gave Israel a foreseeable continuation of supply channels. In addition, the commissioning of liquefied natural gas (LNG) receiving capability in early 2013 (capable of supplying 1–3 billion cubic meters (BCM) a year) was able to negate the supply shortage from the depleted Yam Tetis and Egyptian gas to some extent. More significantly, a development rush of the Tamar field in 2013 allowed meeting the majority of current Israeli NG demand, and was set to supply between 50–80% of Israel's future consumption needs. The sharp increase in gas reserves enables Israel to pursue a new and unexpected path to energy independence [3].

In June 2010, the Leviathan structure located in deep water, 30 km west of Tamar, was found to contain the same gas-bearing Tamar Sands. The analysis indicated recoverable reserves of 500 BCM of gas in the Leviathan field. The giant Leviathan field was the largest discovery worldwide during the first decade of the 21st century. Exploration activity in the northern part of the Israeli Exclusive Economic Zone (EEZ) continued from 2011 to 2013. Additional amounts of NG were discovered in the Karish, Tanin, Dolphin, Tamar SW, and Aphrodita-Ishai fields (The Ministry of Energy).

Consequently, long a resource-poor country, Israel is now evaluated as having more natural gas than it needs for the next 30 years. As Israel's primary energy import bill before the NG discoveries was about \$10 billion—more than 5% of gross domestic product (GDP)—NG is expected to sharply improve the country's trade balance.

Debates about the nation's rights over its natural resources, as well the right of ownership by foreign entities, surround the economic dilemmas that the national revenue policy faces. The Sheshinski committee was given a charter to recommend changes in fiscal policy for Israel's natural-resource sector. The key committee recommendations [4] that were accepted by the government (Government decision 2762, 2011) included retaining the enacted 12.5% government royalties over oil and gas production, and the establishment of an excess profits tax of up to 50%. Hence, the maximum government take (GT)—i.e., the government share in the natural resources sales revenues—rose from about 30% to 62.5%.

The NG fields of Tamar, Leviathan, Karish, and Tanin were discovered and controlled by the United States (US)-based Noble Energy and Israel's Delek Group, effectively creating a cartel over the vast majority of the country's gas reserves. The issue of the gas cartel and the price of NG have become highly contentious [5]. The "Gas Framework" approved by the Knesset in 2016 called for Delek to divest itself of Tamar and for Noble to reduce its stake from 36% to 25% in six years. Following the "Gas Framework", Noble Energy and Delek sold the control of two smaller fields: Karish and Tanin. However, the two companies retain their holdings in the much bigger Leviathan field, while the government assured that it would not impose price supervision or annul contracts for NG already signed. Most importantly, the government committed to observing these terms for as long as 15 years.

The owners of Tamar field conditioned the investment in its development on a long-term contract with the Israeli Electric Corporation (IEC), which was a government-owned regulated monopoly at that time. This much-disputed contract served as an anchor for the entrepreneurs, assuring the demand for NG. With the discovery of Leviathan Field, it was clear that the largest fraction of domestic demand was already locked for Tamar field. The owners of Leviathan argued that the field would be developed in the nearby future only if exporting the NG would be allowed. Based on the Tzemach Committee's recommendations [6], the government set out the quantity of NG that would remain for domestic consumption in order to achieve energy security over time.

During 2018, an inter-ministerial team re-examined the gas export policy. The recommendations of this team included setting the amount of NG secured for the domestic market at about 500 BCM [7] while the remaining amount, estimated at about 340 BCM, would be available for export, and were adopted by government decision 4442. NG exports already underway or agreed with Jordan and Egypt total over 110 BCM (3.9 TCF).

The way the economy exploits the windfall of natural resources might adversely affect the pertained sectors, as well as the economy as a whole. One of several well-known studies on the subject of natural resources economics showed evidence to the relationship between countries' dependency on natural resources exports and their growth rates [8]. The study examined a sample of 97 developing economies and compared their natural resources exports to gross domestic product (GDP) ratio with growth per capita over the span of 20 years (1970–1989). The results showed that economies with a high ratio of natural resource exports to GDP tended to have low growth rate, even when controlling for other determinants of economic growth such as initial GDP, inequality, trade policy, government efficiency, and investment rates, which when combined form a phenomenon called the "Resource Curse". Such findings raise a debate regarding whether or not resource-abundant countries should be encouraged to exploit their resource bases.

The Dutch Disease is arguably the hallmark of the natural "Resource Curse" phenomenon. The Dutch disease is a scenario that can occur in small countries with an important resource extraction sector. The large-scale expansion of this sector generates large export revenues that are exchanged in domestic currency [9]. This demand appreciates the domestic currency, causing domestic goods to become expensive compared to foreign goods. Consequently, the country's international competitiveness suffers, hampering its exports of other goods and services [10]. Nevertheless, it is difficult to separate Dutch disease effects from the domestic and international macroeconomic conditions prevailing at the time of the shock. This is all the more so in the case of price-led energy booms, which might be accompanied by cross-economies recessions.

The question arises of whether this is actually a problem. [11] noted that some economists have claimed that the "disease" is merely an adaptation process that the economy goes through in light of its newfound wealth, all the more so when the source of increased inflows is permanent. At the same time, she noted that other economists argue that the damage caused by the transition of capital and labor between sectors is by itself a risk to economies' growth potential, requiring adequate policy measures to deal with such implications. When the booming sector is NG, oil, or minerals, the declining tradable sectors, according to the theory, would include manufacturing and agriculture [12]. In principle, such changes in the structure of production should be welfare improving, reflecting changes in demand associated with an improvement in national income [12]. However, they may be a matter of concern for policy makers if the declining sectors are thought to have some special characteristics that would stimulate growth and welfare in the long-term, such as increasing returns to scale, learning by doing, or positive technological externalities. This might be the source of concern for the Israeli economy, in which export comprises about 30% of the GDP, and is mainly composed of technology and labor-intensive goods and services (high-tech, medicine).

Recent empirical evidence and theoretical work provide strong support to a negative link between resource abundance and long-term growth [13,14]. The methodology for empirically assessing "Dutch Disease" usually involves the econometric analysis of time-series or panel data [15] for researching the correlation between resource abundance and the share of tradable sectors in the overall economy [16]. As Israel has only recently started the low-scale export of NG, this approach cannot be applied yet. However, expectations for major NG exports have been accumulating since 2010 with the discovery of the major offshore field "Leviathan".

The framework of this research aimed at investigating the much debated yet little explored effects of the newly introduced NG resources over the Israeli economy. Specifically, we estimated the implications of expected NG exports, as expressed through the media announcements, on the foreign exchange rate.

The aim of this paper was to analyze whether the Israeli economy showed symptoms of "Dutch Disease": an appreciation of local currency caused by massive natural resource exports. The export of NG started on a small scale only in 2017, but the expectations for large export potential have been escalating since 2010. Therefore, we employed the Event Study methodology instead of the commonly used econometric analysis of actual data, as it is not available yet. We investigated NG industry-related announcements and the fluctuations in the real exchange rate. Although this paper focused on the NG industry in Israel, both the methodology and the empirical results are of general interest. Technology advances allow discovering and developing natural reservoirs that previously were unknown or considered unprofitable. Moreover, shale oil and gas technologies revert the energy markets. Countries that were considered resource-deprived energy importers become net exporters. The implications of these transformations in terms of trade and economic growth are still ambiguous.

The rest of the paper is organized as follows. Section 2 presents the data and estimation strategy. Section 3 illustrated main empirical results. Section 4 generates the discussion.

## 2. Materials and Methods

### 2.1. Data

The present research considers 639 announcements published during the years 2008–2017 with regard to NG discoveries in Israel. The announcements were obtained from leading Israeli Internet news sites and from the publications of corporations and entities related to the energy sector and the NG industry in particular. Included among them are Ynet, Globes, Calcalist, Walla News, Bizportal, and Energy News. Each announcement in the database was classified according to a subtopic that characterizes it. The subtopic categories are Drilling, Exploration, Import, Export, Development, Infrastructure, Electricity, Companies, and Regulation. The announcements were also categorized according to the date and time of publication in order to prevent duplication from different sources of publication and maintain compatibility with the Event Study Methodology. The results of that methodology rely upon the examination of time periods.

For each announcement, 316 items of data regarding daily and intraday returns for the ILS to USD exchange rate were obtained from the Reuters system for the estimation window and the event window time periods. Our focus on the ILS to USD exchange rate is because all the NG contracts in Israel are signed in terms of USD per MMBTU (Measure of the energy content in a fuel). In addition, USD has the highest weight in Nominal Effective Exchange Rate in Israel [17]

Duplicate publications from different sources were removed from the initial collection of announcements, as were repeated announcements that had previously been published. After the additional screening, 296 "pure" articles remained. These were defined as announcements that were not additionally published with regard to the content of the initial publication during the 16 days preceding and the 16 days following.

The public awareness of the importance of NG in Israel began early in 2009 with the discovery of the Tamar gas field. The number of announcements published in 2010 was the largest in comparison to other years (53 pure announcements in 2010 versus on average 30 a year in 2009, 2011–2017). Most of the announcements that year concerned the Leviathan field, which was discovered at the time, and the Tamar field, which was discovered in the preceding year.

The database is divided into the four following analysis groups. (1) The first group is a general group including all 296 pure announcements in the sample. (2) Second, there is a group in which the sample is divided according to the subject area of the article. This second group included Panel A (68 Export announcements), Panel B (78 Exploration and Drilling announcements), Panel C (47 Companies and Development announcements), Panel D (72 Electricity and Infrastructure announcements), and Panel E (38 Import and Regulation announcements). In the second group, the research investigates whether announcements in specific areas have greater influence than those in other areas. (3) The third group consists of single announcements including 243 articles, as compared to multiple announcements including 43 articles. In this group, the research analyzed the influence of the announcements according to their frequency of publication. (4) The last group was divided according to the time period of publication. The sample is separated into two periods: old announcements (146 articles) and new announcements (150 articles). In this group, the research examines whether the announcements from recent years have a stronger impact on the real exchange rate than old announcements that were published prior to 1 January 2013. As mentioned above, the year 2013 is a focal point in the energy sector of Israel with the connection of Tamar field and the beginning of the local NG supply. The purpose of the division according to analysis groups is to characterize the influence of the publications on the exchange rate separately for each group, and to investigate the significance of the impact in each group.

Figure 1 presents the fluctuation in the daily ILS to USD representative exchange rate and some of the major events concerning the NG industry in Israel. The exchange rate policy of the Bank of Israel is based on the free movement of ILS's exchange rate against other currencies. With that, the bank maintains the option for "dirty float", i.e., to intervene in foreign currency trading in situations of extraordinary movements in the exchange rate that are not in line with fundamental economic conditions, or when the foreign exchange market is not functioning appropriately [18]

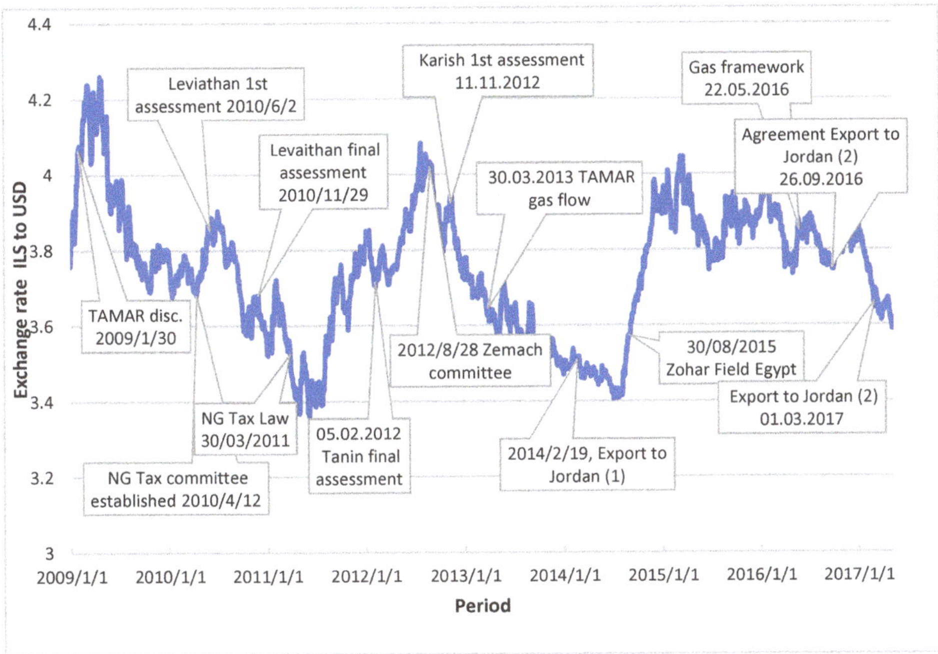


**Figure 1.** Timeline of natural gas (NG)-related events and the daily New Israeli Shekel (ILS) to United States dollar (USD) real exchange rate.

Evidently, between January 2009 and May 2017, the daily ILS against USD exchange rate reached as high as 4.30 ILS/USD on 23 April 2009 and as low as 3.35 ILS/USD on 7 June 2011. Notably, the former governor of the Bank of Israel claimed that the exchange rate decline in 2013–2014 was partly due to expectations of future trade balance increase due to NG exports: expectations that, in her opinion, overappreciated the currency [19].

*2.2. Methodology*

This study employed the Event Study Methodology followed by Generalized Auto-Regressive Conditional Heteroscedasticity (GARCH) analysis. The event study is commonly used in the literature regarding the stock market in order to examine the impact of newly published information on the stock rates. The present research adapts it to the area of macroeconomics and investigates whether and how the exchange rate is affected by the publication of announcements concerning NG in Israel.

For each event, the expected return was estimated by calculating the average exchange rate return over the estimation period. The expected return was used as the benchmark return in the normal situation to compare with the actual exchange rate return during the event window. The benchmark return represents the return that was not related to the event of interest. Next, we calculated the abnormal exchange rate return, which represents the difference between the actual return and the expected return. Afterwards, we calculated the average abnormal exchange rate return and aggregated the result. Then, the mean adjusted return (MAR) methodology was applied to analyze whether the announcement causes a statistically significant abnormal return.

In the present research, the estimation window begins 300 days before the publication of the article, and ends 17 days prior to its publication. The event window is defined from 16 days before publication of the article until 16 days subsequent to its publication. In the event study methodology, there is no universal rule on the lengths of the event windows. Over the years, many articles used the event study methodology and changed the size of the event window according to the research needs [20–25].

In addition, in the present method, the "event day" is defined as day zero: the day on which the announcement is published (Figure 2).

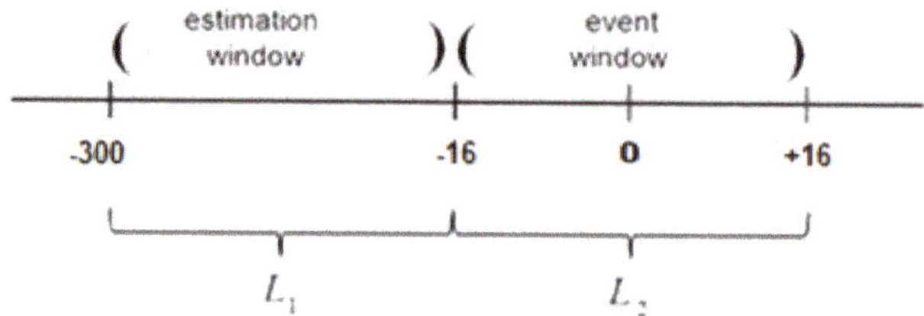


**Figure 2.** Estimation window (L1) for the period of time preceding the event (L2).

The development of the method used today began in the classic articles of [20,21]. They examined the publications of reports that focused on the influence of share splits, after taking into consideration the influence of a concurrently increased dividend. In order to study the influence, they compared the actual stock returns surrounding the date of the notice of the split to the expected return without the event. The role of the event study method in the present paper is to examine whether the rate of change in the ILS to USD exchange rate surrounding the publication of the announcement is identical to the normally expected return without the publication. The study follows the mean adjusted return (MAR) method based upon statistical expectations during the estimation period [22,26,27], so that the average exchange rate that will be obtained will also continue in the event window, and any change from the calculated rate will be called the abnormal real exchange rate.

The underlying hypothesis is that the expected change in the real exchange rate is equivalent to the actual real exchange rate. The present research tests the hypothesis using the t-test. The methodology in the present research is based on [28] and adapted to the calculation of the abnormal return of the real exchange rate. The detailed mathematical notation is presented in Appendix A.

Empirical work indicates that exchange-rate volatility behaves according to a GARCH )Generalized Auto-Regressive Conditional Heteroscedasticity) model that was developed by [29].According to the model, fluctuations in the exchange rate during a given period depend on fluctuations in the exchange rate in the preceding periods [30–33].

Recent studies confirmed that the GARCH(1,1) model is the most appropriate measure of exchange-rate volatility [34,35]. Additionally, research by [36] revealed that the exchange-rate series exhibits empirical regularities such as clustering volatility, non-stationarity, non-normality, and serial correlation, which justify the application of the GARCH methodology. Another recent study by [37] that used GARCH(1,1) found that exchange rate volatility affects both international trade and foreign direct investment (FDI) significantly but negatively in countries engaged in OBOR (One Belt One Road is a global development strategy adopted by the Chinese government involving infrastructure development and investments in 152 countries and international organizations in Asia, Europe, Africa, the Middle East, and the Americas). [38] showed that the GARCH(1,1) model was more effective than other complicated GARCH models when they took 330 ARCH-type specifications into consideration. Therefore, the GARCH(1,1) was utilized for the volatility measurement of exchange rate in the present study.

Then, the abnormal return for the real exchange rate $Ae_{i,t}$ of an announcement i on day t is defined as the difference between the actual return and the normal one.

$$Ae_{i,t} = e_{i,t} - E(e_{i,t}|I_{i,t})$$ (1)

where:

$Ae_{it}$—The abnormal return for real exchange rate of an announcement $i$ on day $t$.

$e_{it}$—The actually return for real exchange rate of an announcement $i$ on day $t$.

$E(e_{it}|I_t)$—The expected normal return for exchange rate of an announcement $i$, given information $I$ known at time $t$.

The expected normal return for exchange rate and its volatility $\sigma^2_{i,t}$ are presented in equations (2) and (3) as follows:

$$E(e_{i,t}|I_{i,t}) = \alpha_0 + \alpha_1 e_{i,t-1} + \varepsilon_{i,t} \tag{2}$$

$$\sigma^2_{i,t} = \beta_0 + \beta_1 \varepsilon^2_{i,t-1} + \gamma_1 \sigma^2_{i,t-1}, \ AAe_t \sim N\left(0, \sigma^2_{i,t}\right) \tag{3}$$

where $\beta_0$ is the constant term, $\varepsilon_{i,t}$ is the error term, $\beta_1$ is the coefficient for the lagged squared error at lag 1, and $\gamma_1$ is the coefficient for the lagged conditional variance at lag 1.

## 3. Results

The present research investigated whether an investor can utilize this information and yield an abnormal return during the period following the publication of the announcement. In this section, we present the results of event study analyses for the entire sample and for different groups of announcements, as well as the GARCH(1,1) estimation, to examine how the exchange rate was influenced by subjective evaluations of investors.

### 3.1. The Effect of Gas Discoveries on the Real Exchange Rate for the Entire Sample

First, the research used a MAR to examine the influence of NG discovery announcements on the real exchange rate in general, for the entire sample of 296 pure announcements. Table 1 and Figure 3 describe the cumulative average abnormal exchange rate return CAAe−16,+16 during the 33 days surrounding the time of the announcement, beginning from day (−16) prior to publication, and until day (16) following publication.

Table 1 describes the effect of gas discoveries on the abnormal real exchange rate for the entire sample. In the table, the cumulative average abnormal exchange rate return (CAAe), median cumulative abnormal exchange rate return (CAe), percentage of positive abnormal real exchange rate return, t-statistics, and number of observations are reported for the six event windows.

**Table 1.** The effect of gas discoveries on the real exchange rate for the entire sample of 296 pure announcements. CAAe: cumulative average abnormal exchange rate return, CAe: median cumulative abnormal exchange rate return.

| | (−16,−4) | (−3,−1) | (0) | (+1,+3) | (+4,+6) | (+7,+16) |
|---|---|---|---|---|---|---|
| CAAe | −0.04% | −0.09% | −0.06% * | −0.11% * | −0.08% | −0.01% |
| Median CAe | 0.00% | −0.04% | −0.09% | −0.24% | −0.16% | 0.18% |
| Percent Positive | 49.66% | 44.59% | 43.58% | 44.93% | 40.88% | 49.32% |
| t-statistics | −0.33 | −1.53 | −1.87 | −1.87 | −1.42 | −0.11 |
| N | 296 | 296 | 296 | 296 | 296 | 296 |

* 90% significance level.

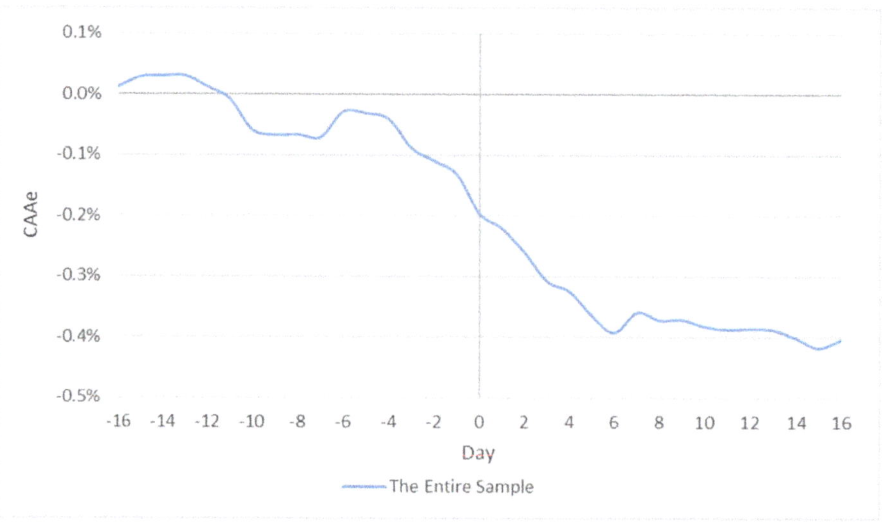

**Figure 3.** The cumulative average abnormal exchange rate return (CAAe) behavior during the 33-day event window surrounding the event day for the entire sample of 296 announcements.

Table 1 and Figure 3 show that during the period prior to publication of the announcement, the CAAe−16,−1 did not significantly decrease, which indicates that in general during this period, there was no leakage of information regarding the content of the announcement. Alternatively, the influence of publication of the announcement occurred a long time before the actual publication of the exchange rate. During this period, the Median CAe−16,−1 was almost unchanged, and 47% of the announcements had a positive CAe. On the day that the announcement was published, there was a significant decrease of 0.06% AAe0 in the exchange rate. This indicates that the announcements generally strengthened the ILS. The Median CAe0 decreased on this day by 0.09%, and 43.38% of the announcements had a positive CAe. Thus, it may be concluded that generally, the announcements provided new information to the investors on the day of publication and contributed to the appreciation of the shekel.

During the first three days (+1,+3) following the announcement, the abnormal appreciation increased and the CAAe+1,+3 decreased by an additional 0.11%. During this period, investors that were exposed to the first announcement were able to use a profit strategy by shorting the real exchange rate on the day of publication and closing the position at the end of the trading day on the third day following publication of the announcement. During this period, the Median CAe+1,+3 decreased by 0.24%, and 44.93% of the announcements had a positive CAe. In the subsequent period (+4,+16), the publication of announcements did not show a statistically significant impact on the exchange rate.

### 3.2. Analysis by Type of Announcement

This section examines the impact of NG-related announcements on the real exchange rate according to five groups of announcements: Panel A (Export), Panel B (Exploration and Drilling), Panel C (Companies and Development), Panel D (Electricity and Infrastructure), and Panel E (Import and Regulation), as presented in Section 2.1.

Table 2 and Figure 4 describe the CAAe−16,+16 by group during the 33 days surrounding the announcement, from day (−16) prior to publication and until day (16) following publication.

**Table 2.** The effect of gas discoveries on the exchange rate by announcement type.

| | (−16,−4) | (−3,−1) | (0) | (+1,+3) | (+4,+6) | (+7,+16) |
|---|---|---|---|---|---|---|
| *Panel A: Export* | | | | | | |
| CAAe | 0.25% | 0.05% | −0.12% ** | 0.02% | 0.03% | 0.04% |
| Median CAe | 0.31% | 0.00% | −0.10% | 0.00% | 0.09% | 0.34% |
| Percent Positive | 61.76% | 50.00% | 45.59% | 50.00% | 51.47% | 57.35% |
| t-statistics | 1.28 | 0.50 | −2.13 | 0.18 | 0.31 | 0.21 |
| N | 68 | 68 | 68 | 68 | 68 | 68 |
| *Panel B: Exploration and Drilling* | | | | | | |
| CAAe | −0.02% | −0.09% | 0.02% | 0.02% | −0.21% * | 0.13% |
| Median CAe | 0.01% | −0.06% | −0.03% | 0.02% | −0.27% | 0.06% |
| Percent Positive | 50.00% | 45.71% | 50.00% | 52.86% | 32.86% | 50.00% |
| t-statistics | −0.07 | −0.71 | 0.29 | 0.17 | −1.70 | 0.57 |
| N | 70 | 70 | 70 | 70 | 70 | 70 |
| *Panel C: Companies and Development* | | | | | | |
| CAAe | −0.29% | −0.18% | −0.15% ** | −0.17% | −0.30% ** | −0.46% ** |
| Median CAe | −0.33% | −0.17% | −0.16% | −0.08% | −0.30% | −0.55% |
| Percent Positive | 36.17% | 31.91% | 31.91% | 42.55% | 31.91% | 31.91% |
| t-statistics | −1.11 | −1.40 | −2.12 | −1.32 | −2.36 | −2.02 |
| N | 47 | 47 | 47 | 47 | 47 | 47 |
| *Panel D: Electricity and Infrastructure* | | | | | | |
| CAAe | −0.24% | −0.22% ** | −0.07% | −0.27% *** | −0.11% | 0.12% |
| Median CAe | −0.34% | −0.12% | −0.10% | −0.25% | −0.15% | 0.36% |
| Percent Positive | 47.22% | 43.06% | 44.44% | 34.72% | 41.67% | 52.78% |
| t-statistics | −1.01 | −2.00 | −1.08 | −2.43 | −1.01 | 0.57 |
| N | 72 | 72 | 72 | 72 | 72 | 72 |
| *Panel E: Import and Regulation* | | | | | | |
| CAAe | 0.04% | −0.01% | −0.03% | −0.23% | 0.27% | −0.03% |
| Median CAe | −0.14% | −0.03% | −0.12% | −0.16% | −0.03% | 0.19% |
| Percent Positive | 46.15% | 48.72% | 38.46% | 41.03% | 46.15% | 48.72% |
| t-statistics | 0.12 | −0.06 | −0.34 | −1.41 | 1.61 | −0.11 |
| N | 39 | 39 | 39 | 39 | 39 | 39 |

*** 99% significance level; ** 95% significance level; * 90% significance level.

The following table describes the effect of gas discoveries on the abnormal exchange rate by five announcement types: Panel A (Export), Panel B (Exploration and Drilling), Panel C (Companies and Development), Panel D (Electricity and Infrastructure), and Panel E (Import and Regulation).

In each panel, the cumulative average abnormal exchange rate (CAAe), median cumulative abnormal real exchange rate (CAe), percentage of positive abnormal real exchange rate, t-statistics, and number of observations are reported for the six event windows.

Figure 4 and Table 2 show that there was no information leakage in any of the groups except for the Electricity and Infrastructure group. In that group, the information leakage began three days before the publication of the announcement and continued until one day before publication of the announcement, which was accompanied by a decrease in CAAe−3,−1 by 0.22% (t-statistics = −2). This provides an indication that investors, who had inside information concerning the decisions in the area of electricity and infrastructure, were able to short the exchange rate three days before the information was exposed to the rest of the investors in the market. They potentially could close the position at the end of the trading day prior to publication of the announcement.

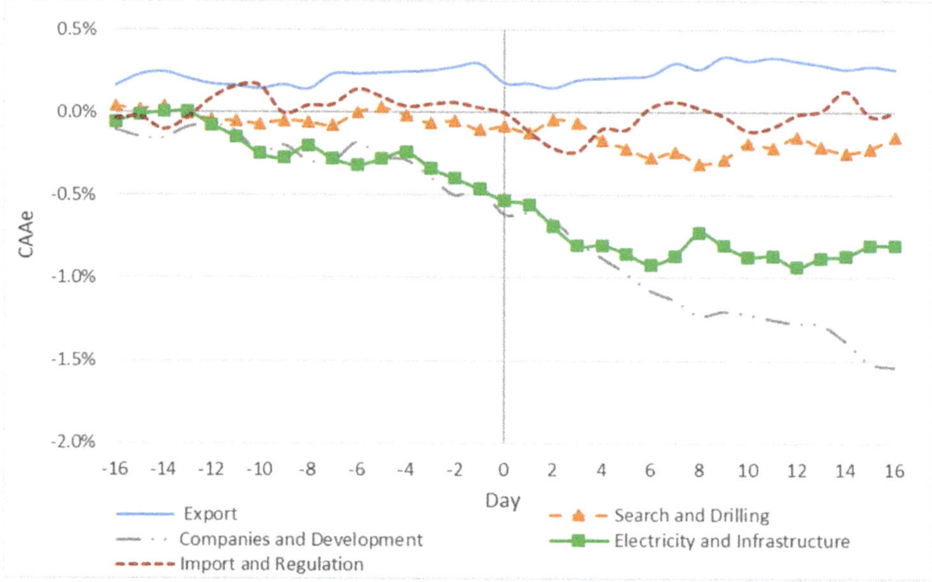

**Figure 4.** The CAAe behavior during the 33-day event window surrounding the event day for the entire sample of 296 announcements by five announcement types.

Also of interest in this period is that the Percent Positive shows the tendencies of the groups in Figure 4. In the Exploration and Drilling group and the Import and Regulation group, approximately 47% of the announcements yielded an increase in the real exchange rate. Therefore, the two graphs remained stable and without change in the real exchange rate. In the Companies and Development and Electricity and Infrastructure groups, approximately 39% of the announcements yielded devaluation; thus, there was a negative trend in the exchange rate in those two groups.

In the Export group and the Companies and Development group, there was a statistically significant decrease in AAe0 by 0.12% (t-statistics = −2.13) and 0.15% (t-statistics = −2.12) in the exchange rate. More importantly, the statistically significant abnormal appreciation of ILS on the day of announcements related to the prospects of exports and development of the reserves provides direct evidence for investors' strategy in view of the expected NG export and future currency appreciation.

In the Electricity and Infrastructure group during the first three days following the announcement (+1,+3), there was a statistically significant decline in the exchange rate, and the CAAe+1,+3 decreased by 0.27% (t-statistics = −2.43). During this period, investors who were exposed to the first announcement were able to use a profit strategy by shorting the exchange rate on the day of the announcement and closing the position at the end of the trading day on the third day following the publication of the announcement. During this period, the Median CAe+1,+3 decreased by 0.25%, and 34.72% of the announcements had a positive CAe.

In the Exploration and Drilling group, the statistically significant impact on the exchange rate occurred only during the first few days (+4,+6) after the announcement. Here, the exchange rate decreased, and the CAAe+4,+6 decreased by 0.21% (t-statistics = −1.70). During this period, the Median CAe+4,+6 decreased by 0.27%, and 32.86% of the announcements had a positive CAe.

In the Companies and Development group, the greatest influence was during the period following the announcement. During the 16 days following the announcement (+1, +16), the real exchange rate decreased and the CAAe+1,+16 decreased by 0.93% (t-statistics = −3.19). During this period, investors who were exposed to the first announcement were able to use a profit strategy by shorting the real exchange rate on the day of the announcement and closing the position at the end of the trading day on the 16th day following the publication. In this period, the Median CAe+1,+16 decreased by 0.93%, and 30.12% of the announcements had a positive CAe.

Figure 5 summarizes the results for the five groups surrounding the event day. The present research examined whether there was information leakage to investors who were insiders and associated with the decision makers during the period preceding the event. Figure 5 shows the information leakage only in the Electricity and Infrastructure group. The research further considers whether the announcement brought new information to investors on the event day. In the Export group and the Companies and Infrastructure group, the new information had a statistically significant impact and appreciated the ILS. During the period following the publication, investors sought a profit strategy. At that time, three groups had a statistically significant influence on the exchange rate. These were the Electricity and Infrastructure group, Exploration and Drilling group, and the group that had the most influence on profit strategy: Companies and Development.

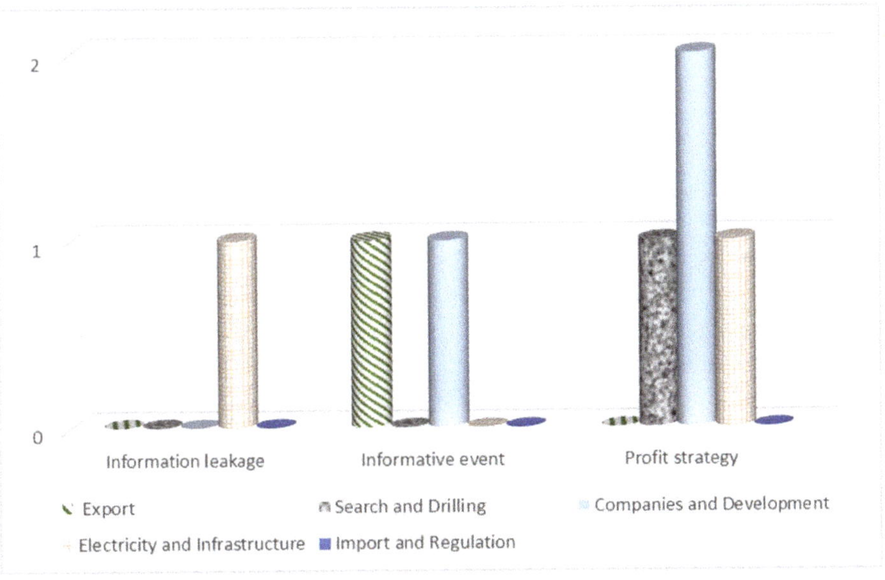


**Figure 5.** Summary of results according to five groups of announcements.

### 3.3. Old vs. New Announcements

This section examines whether announcements published in recent years influenced the exchange rate differently than announcements published earlier. For the purpose of this analysis, the sample was divided into two groups. The first group, referred to as Old, includes 146 old announcements that were published between 15 December 2008 and 31 December 2012. The second group, referred to as New, includes 150 announcements that were published between 1 January 2013 and 6 May 2017. In 2013, the Tamar gas field was connected and began delivering NG to Israel. Therefore, this group analyses whether the impact of the announcements became stronger after the supply of domestic NG began.

Figure 6 and Table 3 describe the CAAe−16,+16 during the 33 days surrounding the announcement, beginning from day (−16) prior to and until day (16) following the publication.

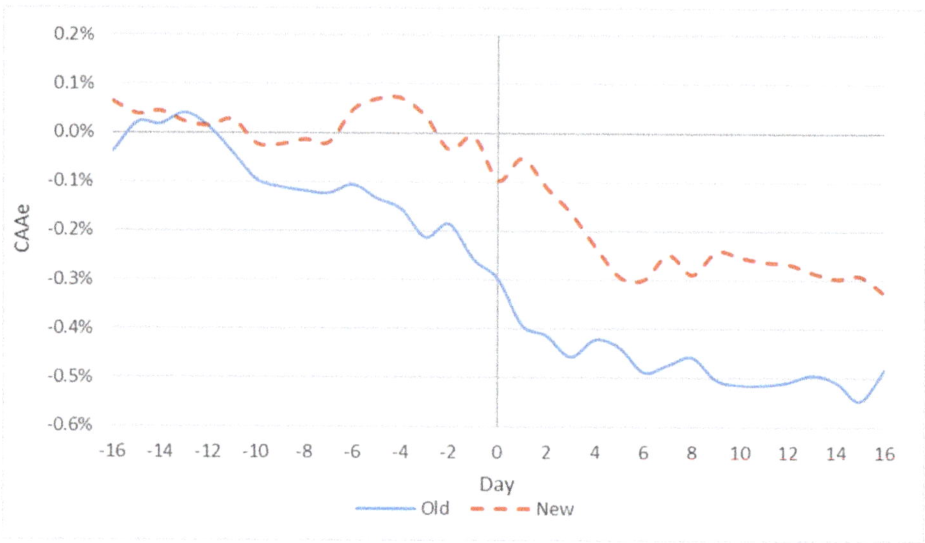

**Figure 6.** The CAAe behavior during the 33-day event window surrounding the event day for old and new announcements.

**Table 3.** The effect of gas discoveries on the exchange rate for old and new announcements.

|  | (−16,−4) | (−3,−1) | (0) | (+1,+3) | (+4,+6) | (+7,+16) |
|---|---|---|---|---|---|---|
| *Panel A: Old* | | | | | | |
| CAAe | −0.16% | −0.10% | −0.04% | −0.16% | −0.03% | 0.01% |
| Median CAe | −0.01% | −0.03% | −0.11% | −0.24% | −0.14% | −0.04% |
| Percent Positive | 62.33% | 54.79% | 54.11% | 55.48% | 51.37% | 61.64% |
| t-statistics | −0.75 | −1.03 | −0.40 | −1.58 | −0.31 | 0.03 |
| N | 146 | 146 | 146 | 146 | 146 | 146 |
| *Panel B: New* | | | | | | |
| CAAe | 0.07% | −0.08% | −0.09% *** | −0.07% | −0.14% ** | −0.03% |
| Median CAe | −0.02% | −0.06% | −0.12% | −0.08% | −0.21% | −0.14% |
| Percent Positive | 49.33% | 46.00% | 44.00% | 42.67% | 40.00% | 48.00% |
| t-statistics | 0.58 | −1.34 | −3.67 | −1.10 | −2.31 | −0.26 |
| N | 150 | 150 | 150 | 150 | 150 | 150 |

*** 99% significance level; ** 95% significance level.

The following table describes the effect of gas discoveries on the abnormal exchange rate for two announcement types: Panel A (Old) and Panel B (New).

In each panel, the cumulative average abnormal real exchange rate (CAAe), median cumulative abnormal real exchange rate (CAe), percentage of positive abnormal real exchange rate, t-statistics, and number of observations are reported for the six event windows.

Figure 6 and Table 3 show that there was no information leakage in any of the old and the new groups. Therefore, it may be concluded that there was no information leakage and that investors who were corporate insiders reacted similarly to new information and to old information before it was published for the general public. On the day of publication, old announcements had no influence on the exchange rate. However, in the group of new announcements, the AAe0 decreased significantly by 0.09% (t-statistics = −3.67). This indicates that the new announcements created an appreciation of the ILS. The Median CAe0 decreased in this group by 0.12%, and 44% of the announcements had a positive CAe. It may be concluded that since the connection of the Tamar gas field in 2013 for the supply of domestic NG to the Israeli economy, the potential for export became more realistic. Every announcement regarding NG strengthened the expectations for a future appreciation, and brought about a stronger reaction of the investors on the day of the publication. This influence also continued during the first six days following the announcement (+1,+6), so that the exchange rate decreased and the CAAe+1,+6 decreased by 0.20% (t-statistics = −2.41). During this period, investors that were exposed to the first announcement were able to use a profit strategy by shorting the exchange rate on the day of publication of the announcement and closing the position at the end of the trading day on the sixth day following publication. During this period, the Median CAe+1,+6 decreased by 0.29%, and 39.36% of the announcements had a positive CAe.

### 3.4. Single vs. Multiple Announcements

When the market has expectations for additional information, it is likely that investors' reactions will strengthen when given continuous information rather than a single information item. This section examines the difference in the behavior of investors toward announcements that appear only once, which are referred to as single announcements, versus those that appear a number of times, which are referred to as multiple announcements. The analysis was made on the first announcement in the series of multiple announcements. The group of single announcements includes 243 announcements, which were not republished during the period of three months from publication of the first announcement. In contrast, the group of multiple announcements includes 43 announcements, for which an additional announcement was issued within three months after the initial publication.

Table 4 and Figure 6 describe the CAAe−16,+16 during the 33 days surrounding the announcement, beginning from day (−16) before publication and continuing until day (16) following publication.

**Table 4.** The effect of gas discoveries on the real exchange rate for single and multiple announcements.

|  | (−16,−4) | (−3,−1) | (0) | (+1,+3) | (+4,+6) | (+7,+16) |
|---|---|---|---|---|---|---|
| *Panel A: Single* | | | | | | |
| CAAe | 0.00% | −0.08% | −0.05% | −0.06% | −0.08% | −0.11% |
| Median CAe | −0.11% | −0.04% | −0.10% | −0.24% | −0.22% | 0.29% |
| Percent Positive | 28.40% | 28.40% | 27.98% | 26.75% | 23.05% | 32.51% |
| t-statistics | 0.02 | −1.38 | −1.55 | −1.00 | −1.35 | −0.96 |
| N | 243 | 243 | 243 | 243 | 243 | 243 |
| *Panel B: Multiple* | | | | | | |
| CAAe | −0.30% | −0.13% | −0.12% | −0.34% ** | 0.26% | −0.15% |
| Median CAe | −0.44% | −0.02% | −0.12% | −0.16% | 0.12% | −0.43% |
| Percent Positive | 46.51% | 44.19% | 41.86% | 39.53% | 55.81% | 44.19% |
| t-statistics | −0.88 | −0.82 | −1.33 | −2.08 | 1.62 | −0.51 |
| N | 43 | 43 | 43 | 43 | 43 | 43 |

** 95% significance level.

The following table describes the effect of gas discoveries on the abnormal exchange rate for two announcement types: Panel A (Single) and Panel B (Multiple).

In each panel, the cumulative average abnormal exchange rate (CAAe), median cumulative abnormal exchange rate (CAe), percentage of positive abnormal exchange rate, t-statistics, and number of observations are reported for the six event windows.

Table 4 and Figure 7 show no influence of single announcements on the exchange rate, whereas for multiple announcements, information leakage began from the second day before the event and was accompanied by currency appreciation. The decrease continued on the day of publication of the announcement, as well as during the first three days following the event. It may be concluded that investors were able to identify initial single announcements and initial continuing announcements. Therefore, they were able to attain abnormal profits in the multiple announcements group. The profit strategy could be shorting the exchange rate either two days prior to the publication for investors with inside information, or on the day of the publication for the rest of the investors, and closing the position at the end of the third trading day following the event. A possible reason for this phenomenon is that the investors had expectations for future information that would be published in additional announcements.

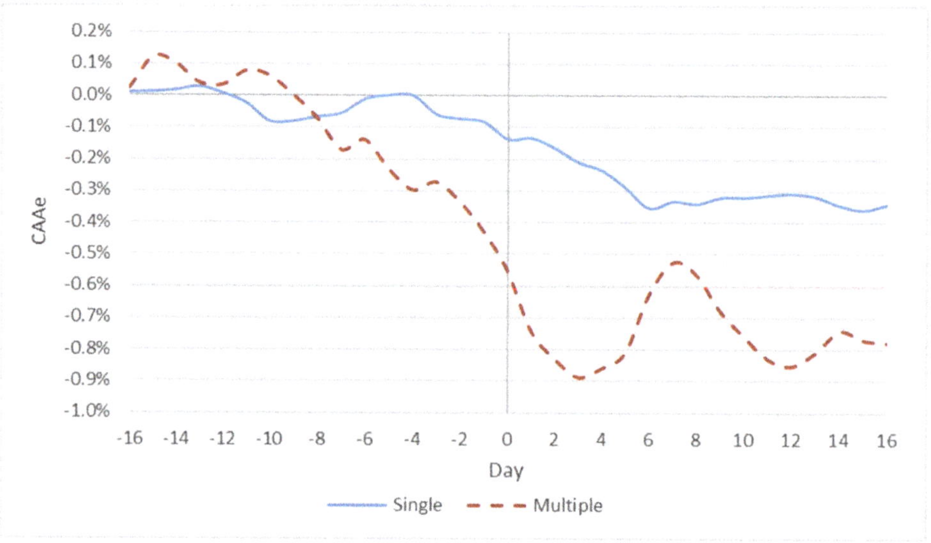


**Figure 7.** The CAAe behavior during the 33-day event window surrounding the event day for single and multiple announcements.

*3.5. GARCH Regression Results*

The validity of our results based is on GARCH modeling of exchange rate returns and the parametric test statistic. Table 5 describes the estimation of the parameters according to the expected normal return of exchange rate volatility (Equation (3)) during the period of the estimation window.

**Table 5.** Estimation of expected normal return for exchange rate and its volatility according to the GARCH(1,1) model during the period of the estimation window.

|  | $\beta_0$ | $\beta_1$ | $\gamma_1$ | Log Likelihood | F |
|---|---|---|---|---|---|
| GARCH | $1.345 \times 10^{-7}$ | 0.062 | 0.901 | 1851.774 | 6.35 |
| *p*-value | 0.099 | 0.050 | 0.000 |  |  |

The results in Table 5 show that the model is significant (F = 6.35), as well as the explanatory variables: change in the square exchange rate in day $t-1$ $\left(\varepsilon^2_{t-1}\right)$, and the variance in day $t-1$ $(\sigma^2_{t-1})$.

Using the estimations from GARCH(1,1) regression, MAR was examined to reveal the influence of NG announcements on the exchange rate for the entire sample of 296 pure announcements. Table 6 and describes the CAAe−16,+16 during 33 days surrounding the time of the announcement, beginning from day (−16) prior to publication, and until day (16) following publication.

**Table 6.** The effect of gas discoveries on the real exchange rate for the entire sample of 296 pure announcements with the GARCH model.

|  | (−16,−4) | (−3,−1) | (0) | (+1,+3) | (+4,+6) | (+7,+16) |
|---|---|---|---|---|---|---|
| CAAe | −0.12% | −0.06% | −0.08% ** | −0.15% ** | −0.03% | −0.13% |
| Median Cae | −0.01% | −0.06% | −0.11% | −0.10% | −0.17% | −0.03% |
| Percent Positive | 49.18% | 44.23% | 42.91% | 44.59% | 41.22% | 49.66% |
| t-statistics | −1.00 | −1.05 | −2.31 | −2.45 | −0.51 | −1.20 |
| N | 296 | 296 | 296 | 296 | 296 | 296 |

** 95% significance level.

The table describes the effect of gas discoveries on the abnormal real exchange rate for the entire sample with the GARCH model. In the table, the cumulative average abnormal real exchange rate (CAAe), median cumulative abnormal real exchange rate return (CAe), percentage of positive abnormal real exchange rate return, t-statistics, and number of observations are reported for the six event windows.

Table 6 reveals that during the period prior to publication of the announcement, the CAAe−16,−1 did not significantly decrease, indicating that in general during this period, there was no leakage of information regarding the content of the announcement. Alternatively, the influence of publication of the announcement occurred a long time before the actual publication of the real exchange rate. During this period, the Median CAe−16,−1 was almost unchanged, and 46.5% of the announcements had a positive CAe. On the day of the announcement, there was a statistically significant decrease of 0.08% in AAe0 (t-statistics = −2.31) in the exchange rate. This indicates that the announcements generally strengthened the ILS. The Median CAe0 decreased on this day by 0.11%, and 42.91% of the announcements had a positive CAe. Thus, it may be concluded that generally, the announcements provided new information to the investors on the day of publication and contributed to the appreciation of ILS.

During the first three days (+1,+3) following the announcement, the abnormal appreciation increased and the CAAe+1,+3 decreased by an additional 0.15% (t-statistics = −2.45). During this period, investors that were exposed to the first announcement were able to use a profit strategy by shorting the real exchange rate on the day of publication and closing the position at the end of the trading day on the third day following publication of the announcement. During this period, the Median CAe+1,+3 decreased by 0.10%, and 44.59% of the announcements had a positive CAe. To summarize, the GARCH regression results confirm the general outcome of event study evaluation and confirm the hypothesis that NG-related announcements strengthened ILS.

## 4. Discussion

The ongoing developments in the hydrocarbon discoveries in Israel's Exclusive Economic Zone (EEZ), as well as in the broader region, are taking front stage in the agendas of policy makers, politicians, and corporations. While there is a raging debate over all aspects of policy concerned with the present and future of these valuable resources, there seems to be a growing interest in understanding the magnitude of implications over Israel's economy, environment, and geopolitics. It is clear that NG will play a leading role in the foreseeable future of Israel's energy market. Moreover, its share of the country's economy is growing, influencing the industrial, transportation, and financial sectors among others, as well as macro factors of growth and employment rates.

The possible adverse effects of foreign exchange windfalls on the tradable sector have been a recurring theme in the literature on the Dutch disease and the resource curse [16] There are alternative windows through which researchers can get a view on the issues. Such effects can be associated with observable variations in the production structure or data from the balance of trade and payments. This approach is not yet possible to apply for investigating the NG impact on the Israeli economy.

Hence, we analyze the impact of recent developments in the NG sector on the Israeli economy through NG-related announcements and the exchange rate fluctuations.

In the investigated time period of the years 2008–2017, major NG reserves were discovered, and fiscal and export policies were adjusted. In addition, the energy flow has changed dramatically, crowding out imported fuels in favor of domestic NG and renewable energy in the supply mix. The share of NG as the core feedstock for electricity increased rapidly. The pipelines to connect the local industry to NG have been built. Contracts for Israeli NG export have been negotiated, and some have already been approved. The announcements reflected these transformations.

We found significant evidence indicating that investors already adjust the profit strategy according to the expectations of the large-scale export of NG. The appreciation of ILS on the day of announcements that related to the prospects of exports and development of the NG fields provides direct evidence for investors' strategy in view of the expected NG export, which will lead to foreign currency inflow and real exchange rate appreciation. Moreover, since the connection of the Tamar gas field in 2013, the announcements regarding NG strengthened the expectations for a future currency appreciation, and brought about a stronger reaction of foreign currency markets.

It may seem that the evidence for local currency appreciation that has occurred in relation to a potential increase in the export of NG supports non-governmental organizations (NGOs)' demand for a larger share of NG to be preserved for future generations. The argument is that fast NG depletion compromises the possibility for sustainable development. However, economic theory provides another path for sustainable economic development. The Hartwick rule for weak sustainability states that instead of natural resource preservation, the profit from the resource, if invested in the development of alternative energy sources, infrastructure, education, and research, can ensure sustainable development [39].In other words, rather than being used for consumption purposes in the public budget, the government take from NG should be transferred to what is known as Sovereign Wealth Fund (SWF). The SWF should be designed not only to implement the Hartwick rule, but also to insulate the public budget from the volatilities of NG prices, and protect the economy from Dutch disease. When invested in foreign financial markets, SWF offsets the foreign currency inflow caused by natural resource export and mitigates domestic currency appreciation.

Pursuant to the Sovereign Wealth Fund Law 5774-2014, a fund was established for the management of the proceeds that Israel receives from the impost on profits from NG. However, the government income from NG has not reached the fund yet, and the political pressure to use most of the revenue to cover public debt is high.

To summarize, the present study has shown that the classic symptom of the Dutch Disease—exchange rate appreciation—is already in place, even though the NG export has not yet reached its expected potential. In terms of policy implications, our research provides insights for policy makers to embark on strategic institutional reforms and develop regulations for efficient and sustainable natural gas extraction and utilization. Our findings indicate that the development of an optimal strategy for the government and the central bank is a key issue. The role of SWF in mitigating the foreign currency inflow is of high importance. Moreover, Israeli policy makers should not overlook lessons learned from the Dutch experience (and that other small developed countries) regarding the management of NG windfall. Appendix A: Adaptation of Events study methodology to estimation of abnormal return for exchange rate.

**Author Contributions:** Conceptualization, R.R.P.; Data curation, L.V.; Formal analysis, T.T., L.V. and R.R.P.; Investigation, L.V.; Methodology, T.T.; Validation, T.T.; Writing—original draft, R.R.P., T.T. and L.V.

**Funding:** This research received no external funding.

**Conflicts of Interest:** The authors declare no conflict of interest.

## Appendix A. Adaptation of Events Study Methodology to Estimation of Abnormal Return for Exchange Rate

First we calculate $\bar{e}_i$, the average of the daily change in the exchange rate in the estimation window for event $i$, $i \in (-300, -17)$ when $e_{i,t}$ is defined as the actual daily change in the ILS to USD exchange rate on day $t$ of event $i$.

$$\bar{e}_i = \frac{1}{284} \sum_{t=-300}^{-17} e_{i,t} \tag{A1}$$

Next we calculate the abnormal return for exchange rate $Ae_{i,t}$, that represents the difference between the actual exchange rate and the average daily exchange rate calculated in the estimation window on day $t$ of event $i$.

$$Ae_{i,t} = e_{i,t} - \bar{e}_i \tag{A2}$$

The average abnormal return for real exchange rate $AAe_t$ for the entire sample of events at time $t$, is calculated, where N represents the number of events examined at time $^t$.

$$AAe_t = \frac{1}{N} \sum_{i=1}^{N} Ae_{i,t} \tag{A3}$$

The cumulative average abnormal real exchange rate return $CAAe_{i,(-16,16)}$ in the event window $(-16,16)$ is calculated as:

$$CAAe_{i,(-16,16)} = \sum_{t=-16}^{16} AAe_{i,t} \tag{A4}$$

Most of the customary methods in the literature that use Event Study Methodology utilize t-tests in order to examine the significance of the cumulative abnormal return obtained in the previous section [22,40].

The classic test of Brown and Warner (1985) assumes the following null hypothesis:

$$H0: \ AAe_t \sim N\left(0, \sigma^2_i\right) \tag{A5}$$

where $AAe_{i,t}$ in our study is the abnormal real exchange rate return in the event window. The estimation of the variance on the event day $(\sigma^2_i)$ is based on the data deviations in the event window, when the classic model assumes that the variance in the event window is equivalent to the variance in the estimation window. In order to calculate the standard deviation of the sample:

$$\sigma(AAe_N) = \frac{1}{N} \sqrt{\left\{ \sum_{i=1}^{N} \left[ \frac{1}{K-2} \sum_{t=1}^{K} \left( Ae_{i,t} - \overline{Ae}_{i,t} \right)^2 \right] \right\}} \tag{A6}$$

where $K$ is defined as the number of observations during the estimation period, and $\overline{Ae}_{i,t}$ represents the average abnormal exchange rate return during the estimation period.

Therefore, a t-test can be formulated to analyze the average abnormal exchange rate return in the sample of N days as follows:

$$t_{CAAe_t} = \frac{CAAe_t}{\sigma_{AAe_t} \times \sqrt{N}} \tag{A7}$$

where $t_{CAAe_t}$ represents a statistical t-test of the average abnormal real exchange rate return during the continuous period. In addition, $CAAe_t$ represents the average abnormal real exchange rate return during the continuous period, $S_{AAe_t}$ represents the standard deviation of the average abnormal return in the estimation window, and N represents the number of days in the subperiod. According to the results obtained from the t-test for each of the subperiods in the different subgroups, the present research examined whether there is statistically significant abnormal return.

## References


1. Shaffer, B. Israel—New natural gas producer in the Mediterranean. *Energy Policy* **2011**, *39*, 5379–5387. [CrossRef]
2. Liebes, I. Economic Implications of Natural Gas Exports: A Case Study Analysis for Israel (Master's thesis). 2015. Available online: https://haifa.userservices.exlibrisgroup.com/view/delivery/972HAI_MAIN/1214394 5520002791?language=en (accessed on 1 June 2019).
3. Michaels, L.; Tal, A. Convergence and conflict with the 'National Interest': Why Israel abandoned its climate policy. *Energy Policy* **2015**, *87*, 480–485. [CrossRef]
4. Sheshinski, E.; Kandel, E.; Tzemach, S.; Mimran, Y.; Naserdishi, Y.; Nissan, U. Committee on Gas and Oil Royalties. Ministry of Finance: Jerusalem. 2011. Available online: http://mof.gov.il/Committees/Previously Committees/Pages/PhysicsPolicyCommittee.aspx (accessed on 11 November 2018).
5. IntelligenceUnit. Gas Framework Details Revealed. Retrieved January 2019. Available online: http://country.eiu.com/article.aspx?articleid=233308807&Country=Israel&topic=Economy&subtopic=For ecast&subsubtopic=Policy+trends&u=1&pid=903382274&oid=903382274&uid=1# (accessed on 6 July 2015).
6. Zemach, S.K.-Y. *The Inter-Ministerial Committee to Examine the Government's Policy Regarding Natural Gas in Israel*; Ministry of National Infrastructure: Jerusalem, Israel, 2012.
7. Adiri, U. *Conclusions of the Professional Team for the Periodic Examination of the Recommendations of the Committee on Examining the Government's Policy in the Natural Gas Market Adopted in Government Decision 442 Dated 23 June 2013*; Ministry of Energy: Jerusalem, Israel, 2018.
8. Sachs, J.D.; Warner, A.M. *Natural Resource Abundance and Economic Growth*; No. w5398; National Bureau of Economic Research: Cambridge, MA, USA, 1995; Available online: https://www.nber.org/papers/w5398 (accessed on 18 July 2019).
9. Ramírez-Cendrero, J.M.; Wirth, E. Is the Norwegian model exportable to combat Dutch disease? *Resour. Policy* **2016**, *48*, 85–96. [CrossRef]
10. Zweifel, P.; Praktiknjo, A.; Erdmann, G. *Energy Economics: Theory and Applications*; Texts in Business and Economics; Springer: Berlin, Germany, 2017.
11. Ebrahim-zadeh, C. Back to Basics Dutch Disease: Too much wealth. *Finance and Development*, March 2003; Volume 40.
12. Brahmbhatt, M.; Canuto, O.; Vostroknutova, E. Dealing with Dutch disease. In *Economic Premise, No 16, June*; The World Bank: Washington, DC, USA, 2010; Available online: http://documents.worldbank.org/curated/en /794871468161957086/Dealing-with-Dutch-disease (accessed on 18 July 2019).
13. Frankel, J.A. *The Natural Resource Curse: A Survey*; NBER Working Paper 15836; National Bureau of Economic Research, Inc.: Cambridge, MA, USA, 2010; Available online: https://www.nber.org/papers/w15836 (accessed on 18 July 2019).
14. Papyrakis, E.; Raveh, O. An Empirical Analysis of a Regional Dutch Disease: The Case of Canada. *Environ. Resour. Econ.* **2014**, *58*, 179–198. [CrossRef]
15. Van der Ploeg, F. Natural resources: Curse or blessing? *Econ. Lit.* **2011**, *49*, 366–420. [CrossRef]
16. Harding, T.; Venables, A.J. *Exports, Imports and Foreign Exchange Windfalls*; Oxcarre Research Paper; University of Oxford: Oxford, UK, 2010.
17. Friedman, A.; Galo, L. The Effective Exchange Rate in Israel. Retrieved 2019, from Bank of Israel. Available online: https://www.boi.org.il/en/NewsAndPublications/PressReleases/Documents/2015-10%20The%20Ef fective%20Exchange%20Rate%20in%20Israel.pdf (accessed on 18 July 2019).
18. BOI. Bank of Israel. Available online: https://www.boi.org.il/en/Markets/Pages/ForeignCurrency.asp (accessed on 1 June 2019).
19. Flug, K. *Remarks by the Governor of the Bank of Israel, Dr. Karnit Flug, at the Israel Economic Association's Annual Conference*; Israeli Economic Society: Jerusalem, Israel, 20 May 2014.
20. Ball, R.; Brown, P. An Empirical Evaluation of Accounting Income Numbers. *J. Account. Res.* **1968**, *6*, 159–178. [CrossRef]

21. Fama, E.; Fisher, C.; Jensen, M.; Roll, R. The Adjustment of Stock Prices to New Information. *Int. Econ. Rev.* **1969**, *10*, 1–21. [CrossRef]
22. Brown, S.J.; Warner, J.B. Using daily stock returns: The Case of Event Studies. *J. Financ. Econ.* **1985**, *14*, 3–31. [CrossRef]
23. Alkhatib, A.; Harasheh, M. Performance of Exchange Traded Funds during the Brexit referendum: An event study. *Int. J. Financ. Stud.* **2018**, *6*, 64. [CrossRef]
24. Jacobs, B.W.; Singhal, V.R. The effect of the Rana Plaza disaster on shareholder wealth of retailers: Implications for sourcing strategies and supply chain governance. *J. Opera. Manag.* **2017**, *49*, 52–66. [CrossRef]
25. Loipersberger, F. The effect of supranational banking supervision on the financial sector: Event study evidence from Europe. *J. Bank. Financ.* **2018**, *91*, 34–48. [CrossRef]
26. Lambertides, N. Sudden CEO vacancy and the long-run economic consequences. *Manag. Financ.* **2009**, *35*, 645–661. [CrossRef]
27. Shah, P.; Parvinder, A. M&A Announcements and Their Effect on Return to Shareholders: An Event Study. *Account. Financ. Res.* **2014**, *3*, 170–190.
28. Elad, F.I.; Bongbee, N.S. Event study on the Reaction of stock returns to Acquisition News. *Int. Financ. Bank.* **2017**, *4*, 33–43. [CrossRef]
29. Bollerslev, T. Generalized Autoregressive Conditional Heteroskedasticity. *J. Econ.* **1986**, *31*, 307–327. [CrossRef]
30. Hsieh, D.A. The Statistical Properties of Daily Foreign Exchange Rates: 1974–1983. *J. Int. Econ.* **1988**, *24*, 129–145. [CrossRef]
31. Milhoj, A. A Conditional Variance Model of Daily Deviations of an Exchange Rate. *J. Bus. Econ. Stat.* **1987**, *5*, 99–103.
32. Engle, R.F.; Bollerslev, T. Modeling the Persistence of Conditional Variances. *Econ. Rev.* **1986**, *5*, 1–50. [CrossRef]
33. Baillie, R.T.; Bollerslev, T. The Massage in Daily Exchange Rates: A Conditional-Variance Tale. *J. Bus. Econ. Stat.* **1989**, *7*, 297–305.
34. Chit, M.M.; Rizov, M.; Willenbockel, D. Exchange rate volatility and exports: New empirical evidence from the emerging East Asian economies. *World Econ.* **2010**, *33*, 239–263. [CrossRef]
35. Tulin, E.; Katz, M.L.; Sun, B. A simple test for distinguishing between internal reference price theories. *Quant. Mark. Econ.* **2010**, *8*, 303–332.
36. Epaphra, M. Modeling Exchange Rate Volatility: Application of the GARCH and EGARCH Models. *J. Math. Financ.* **2017**, *7*, 121–143. [CrossRef]
37. Latief, R.; Lin, L. The effect of exchange rate volatility on international trade and foreign direct investment (FDI) in developing countries along "one belt and one road". *Int. J. Financ. Stud.* **2018**, *6*, 86. [CrossRef]
38. Hansen, P.R.; Asger, L. A forecast comparison of volatility models: Does anything beat a GARCH(1,1)? *Appl. Econ.* **2005**, *20*, 873–889. [CrossRef]
39. Hartwick, J.M. International equity and the investing of rents from exhaustible resources. *Am. Econ. Rev.* **1977**, *67*, 972–974.
40. Barber, B.; Lyon, J.D. Detecting long-run abnormal stock returns: The empirical power and specification of test statistics. *J. Financ. Econ.* **1997**, *43*, 341–372. [CrossRef]

*Article*

# The Effects of Environmental Regulations on the Manufacturing Industry's Performance: A Comparison of Green and Non-Green Sectors in Korea

**Seulgi Yoo and Almas Heshmati ***

Department of Economics, Sogang University, Seoul 04107, Korea; sophyyoo@sogang.ac.kr
* Correspondence: heshmati@sogang.ac.kr; Tel.: +82-10-4513-1712

Received: 11 May 2019; Accepted: 12 June 2019; Published: 16 June 2019

**Abstract:** This study examines the effects of strengthened environmental regulations on employment and labor productivity in the Korean manufacturing industry using panel data from 2004 to 2015. It divides the industry into environmental (green and non-green) and carbon dioxide emitting (polluting and non-polluting) sectors to investigate the industrial sector's response heterogeneity to tightened regulations. We draw several conclusions on the basis of our empirical results. Firstly, environmental policies measured by enacting the LCGG (Low-carbon green growth) Act led to negative effects on labor productivity and employment in polluting industries. These negative effects show that the polluting industries take a higher cost burden because of the environmental policies as compared to the less-polluting industries; this finding is in line with previous studies in literature. Secondly, the green sector is experiencing higher labor productivity and employment as compared to the non-green sector after the tightened environmental regulations. Thirdly, the regulation-related negative effects anticipated in polluting industries are off-set if a firm is also included in the green sector which produces environment-related products. Hence, this result suggests that in terms of labor productivity and employment, it is possible that the manufacturing industry enables the achievement of sustainable development targets. While regulations negatively affect the performance of non-green firms by increasing the costs of highly contaminated ones, in the case of the green sector the regulations promote labor productivity and employment. This shows that a firm in the green sector which has high carbon dioxide emissions can adapt faster than its counterparts in a non-environmental sector in the polluting industry to the constraints imposed by strengthened environmental regulations. These empirical results imply that there will be labor reallocation from non-green to green sectors.

**Keywords:** environmental regulations; employment; manufacturing; act on low-carbon green growth; labor productivity; Korea

---

## 1. Introduction

A frequently asked question is, do environmental regulations reduce a firm's productivity and employment? While no strong conclusion has emerged, countries have strengthened their domestic and international environmental regulations. Although studies on the effects of environmental regulations on a firm's employment and productivity have increased [1], they mainly focus on the effects of polluting industries and do not consider the heterogeneity effects among different sectors. The debate on environmental regulations should be analyzed by distinguishing between green and non-green sectors. The two sectors are defined as industries producing goods and services related to the environment or resource recycling.

One segment that is the focus of policy is the green sector. Environmental regulations can create employment or so-called green jobs in this sector; this is important as these industries are regarded as the drivers of sustainable development.

This study examines how manufacturers respond to regulations measured by the Low-Carbon Green Growth (LCGG) strategy. Korea's environmental regulations were strengthened in the 2000s according to international standards. After 2010, the Government of Korea intended to simultaneously achieve both environmental and economic goals through the LCGG project [2]. Environmental regulations were strengthened after President Lee proclaimed 'Low-Carbon Green Growth,' as the government's goal for reducing greenhouse gas (GHG) emissions by 27–20 percent by 2020 relative to the 'business as usual' scenario of 2005. The Act on Low-Carbon Green Growth was enacted on 13 January 2010 to promote the development of the national economy by laying down the necessary foundation for low carbon, green growth and by utilizing green technologies and green industries as new engines of growth. This research is interested in the effects that these regulations had on economic outcomes defined as employment and labor productivity.

Greenstone [3] reviewed the effects of CAAA (Clean Air Act Amendment) on economic growth and industrial activities using US manufacturing plant-level data. US counties received non-attainment or attainment designations according to the air quality standards after CAAA was implemented. His study suggests that environmental regulations led to emitters in non-attainment counties who were subject to stricter regulatory oversight (treatment groups) losing output, jobs, and capital investments as compared to emitters in attainment counties. His results also suggest that there was a trade-off between environmental and economic outcomes. This is contrary to Porter's [4] argument that environmental regulations can promote investments and technology development. Morgenstern et al. [5] and Lanoie et al. [6] confirm the positive effects of regulations on employment and productivity respectively. The effects of environmental regulations on a firm's performance are still controversial because in addition to the enterprises' extensive margins of entry and exit the regulations also lead to labor reallocations between regulated and non-regulated firms asymmetrically depending on the type of regulation [7]. So, the total economic outcomes in the general equilibrium framework depend on the characteristics of the industries and the applied environmental policy.

In this study, we test three hypotheses. First, the Porter Hypothesis (PH) is tested in terms of labor productivity and employment in the polluting industry. The polluting industry's performances will decrease more than that of the less-polluting industry under environmental regulations in keeping with Greenstone [3] and Walker [8]. Second, dividing the firms by another criterion, namely whether they produce environmental goods, the green sector might benefit from the regulations. Lastly, this paper tests whether the negative effects of the regulations on a polluting establishment can be off-set if it is included in the green sector. We use Korea Statistics' manufacturing surveys from 2004 to 2015, which include the periods before and after LCGG's implementation. Manufacturing is identified as the major emitter of harmful substances in the production and processing of paper, rubber, chemicals, and petroleum refining. Hence, it responds sensitively and differently, depending on the emissions in the processes and the types of products produced. In addition, green manufacturers account for about 20 percent of this industry so it is useful to examine cross-sectional variations between these establishments as well.

Figure 1 shows the increased energy and environmental expenditure in the manufacturing industry from 2004 to 2016. The energy and environmental expenditure includes pollution abatement and control expenditure (PACE) and is a proxy for domestic environmental regulations since the expenditure is not directly related to the firms' profit maximization behavior but to environmental pressures by the authorities [9]. Expenditure increased sharply during the LCGG project period after 2010. This trend is useful for examining the effects of environmental regulations on economic growth over time using the difference-in-differences method.

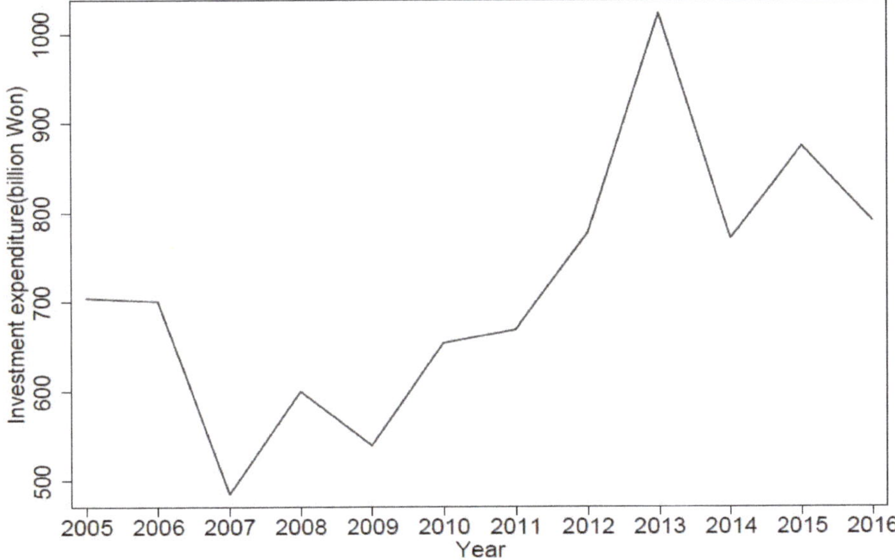

**Figure 1.** Energy and Environmental Expenditure of Korean Manufacturing (Source: Statistics Korea).

This study contributes to the environmental economic field in which studies that inspect the link between environmental policies and their effect on the economy are scarce outside the US. Environmental policies covered in literature using US data include Clean Water and Clean Air Acts [10,11], Clean Air Act Amendment [7], and Cross-State Air Pollution Rule [12]. Our empirical results suggest that employment and labor productivity were impacted by stricter environmental standards under LCGG in Korea and their effects were asymmetric depending on the industry's features. These findings have policy implications for 'the environment versus jobs' debate and the possibility of sustainable development in terms of productivity.

The rest of this study is organized as follows. Section 2 reviews previous studies on environmental regulations and firms' competitiveness. Section 3 documents the characteristics of industries according to $CO_2$ emissions and environmental classifications. Section 4 explains the theoretical framework of the relationship between regulations and a firm's performance and sets the empirical equation. Section 5 reports the results of the estimation analysis for employment and labor productivity. The conclusion is given in Section 6.

## 2. Literature Review

In the literature, the effect of regulations on a firm's economic outcomes has been examined in terms of productivity, investments, employment, and international trade [1]. Several studies also focus on the manufacturing industry since it is regarded as the main culprit responsible for emitting toxic substances and as such is a target of environmental regulations. After the famous Porter Hypothesis [4] researchers have tested whether this hypothesis can hold.

The following studies imply that the Porter Hypothesis holds at least weakly: Lanoie et al. [6] found the hypothesis to be consistent with Quebec manufacturing data. Their findings suggest that the direct effect of regulations on total factor productivity (TFP) growth was negative, but the lagged regulatory variable had a positive productivity effect. They also confirm that this effect was stronger when the industries were more exposed to international competition. Jaffe and Palmer [13] and Johnstone et al.'s [14] studies focus on investment activities in industries. Jaffe and Palmer [13] suggest that lagged stringency of environmental regulations measured by pollution control expenditure spurred R&D activities by using a three-digit industry level and a fixed effects model. Their results were the opposite when the model was estimated using the pooled ordinary least squares (POLS) method neglecting industry heterogeneity. Johnstone et al. [14] and Rubashkina et al. [15] found that innovations based on patents and R&D were not obstructed by environmental policies.

Yang et al. [16] support the Porter Hypothesis using data for Taiwanese manufacturing plants by using pollution abatement fees and R&D expenditure. In their study, capital expenditure was not significantly related to R&D in the case of pollution abatement. Molina-Azorín et al. [17] analyzed the relationship between a firm's performance and environmental practices in the Spanish hotel industry and found that a strong commitment to environmental practices was linked to the hotels' higher performance levels.

Greenstone [3] used US manufacturing data with fixed effects models like Jaffe and Palmer's [13] study, but his results are the opposite. Ozone regulations had the strongest contemporaneous negative effect and the overall effect of the regulations on the industry's TFP was also negative. The dynamic effects of regulations captured by the variations between attainment and non-attainment counties under CAAA were negative and the opposite of PH. According to Wagner et al. [18] the hypothesis did not hold in the European paper industry. Yuan and Xiang [19] maintain that the hypotheses for both weak and strong versions were not supported by data on Chinese manufacturing industries, but in the short term environmental regulations improved labor productivity and environmental and energy efficiency. Using German data, Rexhäuser and Rammer [20] show that PH held only in those innovations which increased a firm's energy efficiency. Thus, empirical research on environmental regulations' effects on a firm's competitiveness has reached no conclusive results [21].

The crowding-out effect in investments between clean and dirty industries is one of the reasons why it is hard to predict the net effect of regulations in an economy. Wang and Shen [22] examined the effects of environmental regulations in China, separating the industries on the basis of clean- and pollution-intensive production. They concluded that clean production industries had higher environmental productivity under regulations than dirty industries. Using the system generalized methods of moment (GMM) and threshold regression estimation methods, they also found that environmental regulations and environmental productivity had an inverted U-shaped relation. Gray and Shadbegian [10] and Kneller and Manderson [23] concentrated on the crowding-out effect between green and traditional investments (technology). According to them, environmental regulations promoted green investments while increased investments in green technology crowded-out existing investment activities. Gray and Shadbegian [10] explored investment decisions of US paper plants using a multinomial logit model. They found that pollution abatement costs and expenditure (PACE) used as a proxy of environmental investments crowded-out productive investments. Kneller and Manderson [23] also found this substitution in UK's manufacturing industry data using the two-step system, GMM. More pollution abatement pressures increased environmental research and development (R&D) and investments in environmental capital, but an increase in environmental R&D tended to crowd-out non-environmental R&D.

These asymmetric effects of environmental regulations occur not only at the industry or plant level but also at the common border between countries. Alpay et al. [24] used data from the food manufacturing industry in the US and Mexico during the North American Free Trade Agreement (NAFTA) to capture the effects of environmental regulations on each country's productivity under free trade. They argue that stricter environmental regulations in Mexico enhanced the industry's productivity growth thus corroborating PH, while US pollution regulations had an insignificant effect on the US food industry.

When it comes to employment, the impact of green innovations in literature is ambiguous. Kunapatarawong and Martínez-Ros [25] investigated the relationship between innovation activities and employment using firm-level Spanish panel data. They suggest that green innovations and employment were positively linked, especially in a dirty industry. Morgenstern et al. [5] found a weak positive relationship between stringent environmental regulations measured by PACE and jobs using a structural model in four polluting manufacturing industries. They used aggregated data from pulp and paper mills, plastic manufacturers, petroleum refiners, and iron and steel mills to estimate the structural model based on a translog cost functional form. Yamazaki [26] and Walker [8] focused on the re-allocative effects of regulations. Yamazaki [26] showed that in British Colombia employment fell in carbon-intensive and trade-sensitive industries, but jobs increased in clean service industries with the revenue-neutral carbon tax. Walker [8] found that workers reallocated from the regulated sector to other industries within the same labor market under CAAA. Aldy and Pizer [12] investigated power sector regulations and found that these affected jobs negatively, but they raised electricity rates and led to higher production costs in all US firms.

Previous research using Korean data includes that by Kang and Lee [27], Kim and Ha [9], Kang and Jo [28], Jung [29], and Lee and Choi [30]. According to Kang and Jo [28] the enforcement of the Indoor Air Quality Control Act enabled about 652 people to find jobs during 2015-19. The following papers show that the PH holds using proxies for environmental variables; Kang and Lee [27] concluded that PACE, a proxy for environmental regulations, increased R&D and this effect was more significant in the stringent regulation period (1992–2001) than during the less stringent regulation period in 1982-91. They also used industry-level manufacturing data and fixed-effects estimation. Kim and Ha's [9] research is similar to ours in terms of data. They used the ratio of energy and environmental expenditure to measure environmental pressures. The effects of environmental pressures on TFP were positive, especially in polluting industries through a lagged regulation proxy.

Studies assessing the effects of environmental regulations not the proxy variables are scarce in case of Korea, because domestic environmental regulations are not strict. The Low-Carbon Green Growth (LCGG) Act sets the fundamentals of environmental policies including the Emission Trading Scheme (ETS). Article 46 of the Act states that the introduction of a cap and trade system for emissions of GHG [2] and the pilot project for ETS was undertaken in January 2010. The carbon market will stimulate incentives for firms to reduce their emissions; as previously announced the Act was implemented in 2015. The GHG Inventory and Research Center was also established to construct GHG data and manage emissions according to Article 45 [2,31]. Lee and Choi [30] show that ETS implemented in 2015 promoted technical changes in the manufacturing industry, but this top-down approach by the government may have limits in enhancing environmental efficiency. Jung [29] suggests that LCGG was based on economic growth rather than ecological modernization and green jobs were only created in traditional environmental protection and pollution reduction areas. While this government-initiated project for achieving both economic growth and environmental conservation has been controversial [32], environmental and energy expenditure in the country reflects increased investments in the environmental parts (see Figure 1). We can directly analyze the effects of environmental regulations by dividing periods into before and after LCGG.

This study contributes to literature in two ways. First, it extends the field by shedding light on the relationship between regulations and firms' economic performance by distinguishing the sectors within the polluting industry. Second, it uses LCGG for measuring stringent environmental regulations which should be helpful in assessing policy. Therefore, the results suggest new academic research and important policy implications according to the types of sectors.

## 3. Data

For the empirical analysis, this study mainly uses data from the Survey on the Mining and Manufacturing Industry from 2004 to 2015 provided by the Korea National Statistical Office's (KOSTAT) Microdata Integrated Service (MDIS). The mining and manufacturing surveys cover establishment level data for firms with 10 or more employees. The surveys include the industry code at the five-digit level and financial information such as assets, capital stock and flow, output, value added, number of employees, wages, and the regional code.

Korea's Ministry of Environment classifies environmental industries at a five digit-level; this is in accordance with OECD's environment industry standards and we mainly call it the green sector in this paper. The classification of environmental industries in Korea is consistent with 'the Environmental Goods and Services Industry' as defined by the OECD. The definition of the environmental industry is: activities to design, manufacture, and install environmental facilities and measuring equipment for environmental conservation and management such as climate, air, water, environmental restoration, environmental safety, health, resource circulation, sustainable environmental resources or services for environmental techniques [33].

The environmental industry (or the green sector) is matched with the Korea Standard Industry Code (KSIC9); KOSTAT provides the matching code. The environmental industry is a comprehensive concept and is different from an eco-friendly or less-polluting industry. By definition, an industry belongs to the environmental sector if its production is environment related. This is why this paper uses two standards to divide establishments. For example, a plant that produces air purifiers may generate harmful substances but it is classified as an environmental industry because the air purifiers provide environmental services. Under this classification, retreatment of rubber tires (code 22112) and manufacture of other industrial glass—waste glass products (code 23129)—become green and polluting industries. Therefore, manufacturers who belong to the environmental industry are subject to regulations. However, as the degree of regulations increases, the demand for environmental goods and facilities also increases and this can lead to an improvement in a firm's performance.

To identify a polluting industry that would be sensitive to regulations, this paper uses $CO_2$ emissions data from the National GHG (Green House Gas) Emission Total Information System. The data is constructed at the 3-digit level and is only available for 2012–15. Because of this data limitation, we define polluting industries as those industries which are above the median of green gas emissions every year during the period. The criteria, 2nd and 3rd groups in terciles, are also used to see whether the main results are robust when the definition varies.

Summary statistics of the data are presented in Table 1. The results of the mean equality test (t-test) show that the mean variables between the green and non-green sectors in each industry differed statistically. The average size of green businesses such as output, capital, and age was larger than that of non-green businesses. In terms of the number of establishments, the businesses in the green sector accounted for about 21 percent of the manufacturing industry. Although the number of green establishments is relatively small, the average number of workers is similar or higher than in the non-green sector. The average wage of each group in the green sector is also higher. This higher salary reflects higher human capital and production technology levels.

This paper uses the capital variable as capital stock values at the beginning and end of the year. For calculating capital stock, we generally use the year-early data. However, there are some limitations with the 2010 and 2015 surveys since capital stock was omitted at the establishment level (financial and capital assets were surveyed at the firm level, so establishment level data was not reported). Hence, because of this missing data for 2010 and 2015 this paper uses capital stock at the end of 2009 and 2014 respectively. All nominal variables such as sales, value-added, and wages are converted to real values. In the case of sales and value added, the paper uses GDP deflators calculated by KOSTAT. We use the capital deflators and the consumer price index (CPI) obtained from the Bank of Korea to transform capital assets and wages to real values. The paper measures productivity using labor productivity, which is real value-added divided by the number of workers. Workers include full-time and irregular employees except dispatched employees. Age of the firm is measured in years from its date of establishment. The concentration index is measured as the Herfindahl Hirschman Index (HHI)—a square sum of shares of all establishments in the five-digit industries. The paper uses HHI for controlling the level of competition in the industry that affects firms' performance such as employment and labor productivity.

**Table 1.** Summary statistics of the data (by green sector and polluting industries).

| A. Less-polluting Industries | | | | | | | |
| --- | --- | --- | --- | --- | --- | --- | --- |
| | Green Sector | | | Non-Green Sector | | | t-test |
| Variable | Obs | Mean | Standard Deviation | Obs | Mean | Standard Deviation | Mean Difference |
| $CO_2$ | 40,676 | 3,400 | 7,121 | 155,374 | 3,782 | 13,655 | −382 *** (70.08) |
| Output | 112,930 | 13,918 | 97,069 | 459,676 | 13,249 | 250,934 | 668 (760.32) |
| Value added | 112,924 | 4,746 | 37,956 | 459,667 | 4,850 | 91,291 | −104 (277.37) |
| Total Employment | 112,930 | 35 | 100 | 459,676 | 35.8 | 196 | −0.80 (0.60) |
| Labor productivity (Output/Labor) | 112,930 | 267 | 611 | 459,676 | 214 | 1,188 | 56 *** (3.65) |
| Labor productivity (Vadd/Labor) | 112,924 | 95.4 | 169 | 459,667 | 80.4 | 315 | 15.1 *** (0.97) |
| Average Wage | 112,745 | 26.2 | 11.3 | 458,985 | 23.1 | 10.7 | 3.2 *** (0.36) |
| Capital | 107,459 | 4,659 | 34,217 | 439,110 | 4,263 | 63,609 | 396 ** (200.79) |
| Age | 112,930 | 10.7 | 9.0 | 459,676 | 10.3 | 9.1 | 0.41 *** (0.03) |
| HHI | 112,930 | 0.045 | 0.083 | 459,676 | 0.056 | 0.090 | −0.011 *** (0.0003) |
| B. Polluting Industries | | | | | | | |
| | Green Sector | | | Non-Green Sector | | | t-test |
| Variable | Obs | Mean | Standard Deviation | Obs | Mean | Standard Deviation | Mean Difference |
| $CO_2$ | 17,805 | 8,605 | 111,72 | 49,554 | 13,109 | 26,910 | −4504 *** (207.8) |
| Output | 42,089 | 60,918 | 606,650 | 116,041 | 38,030 | 608,978 | 22,888 *** (3461.6) |
| Value added | 42,089 | 18,050 | 198,400 | 116,041 | 12,752 | 226,866 | 5298 *** (1249.8) |
| Total Employment | 42,089 | 80 | 570 | 116,041 | 59 | 365 | 20.7 *** (2.44) |
| Labor productivity (Output/Labor) | 42,089 | 382 | 704 | 116,041 | 281 | 580 | 101 *** (3.50) |
| Labor productivity (Vadd/Labor) | 42,089 | 118 | 178 | 116,041 | 97 | 150 | 20.8 *** (0.90) |
| Average Wage | 41,994 | 29.2 | 12.5 | 115,925 | 27.2 | 10.6 | 1.9 *** (0.06) |
| Capital | 41,567 | 18,543 | 151,677 | 114,424 | 15,116 | 258,743 | 3426 ** (1346) |
| Age | 42,089 | 13.2 | 9.9 | 116,041 | 12.5 | 9.4 | 0.66 *** (0.05) |
| HHI | 42,089 | 0.038 | 0.076 | 116,041 | 0.041 | 0.062 | −0.003 *** (0.0004) |

Source: The Mining and Manufacturing Survey in Korean Statistics and National GHG Emission Total Information System. Notes: Employment is in units of people, the amount is one million won, and $CO_2$ is 1,000 tons. The asterisks *** and ** indicate significant at the 1% and 5% levels of significance respectively. Standard errors are in parenthesis and HHI is the Herfindahl Hirschman Index.

In particular, the production and employment in automobiles accounts for a large portion of the green sector such as manufacture of passenger motor vehicles (hydrogen, hybrid, photovoltaic, and natural gas automobiles) (code 30121), manufacture of other new parts and accessories for motor vehicles (exhaust gas reduction devices, DPF, diesel oxidation catalysts, and DOC devices) (code 30399), and manufacture of parts and accessories for motor engines (code 30310). This is consistent with the fact that domestic automobile manufacturing, which is highly dependent on exports increases investments in eco-friendly product development such as electric vehicles due to the strengthening of environmental regulations by importing countries. In the non-green sector, manufacture of semi-conductors and electronic components (codes 26110 and 26211), building ships and boats (codes 31114 and 31111), manufacture of plastic products for fabricating machines (code 22240), and manufacture of parts and accessories for motor vehicle bodies (code 30320) have soaring employment and output levels.

## 4. Methodology

### 4.1. Theoretical Framework

This section examines the mechanisms of how regulations affect a firm's performance and the related hypotheses. It is obvious that environmental policies increase firms' production costs directly. But this impact is asymmetric to economic units and depends on the differences in their production functions and the demand that they face. This means that besides divisions between regulation-targeting groups and others, systematic differences between the entities also lead to changes in relative costs. The Pollution Haven Hypothesis is also based on the asymmetric effects of environmental regulations between countries [24]. Figure 2 illustrates the asymmetric effects of environmental policies on a firm (or industry). Here $c_1$ and $c_2$ indicate each sector's compliance costs brought on by environmental regulations. Since the polluting industries have a higher cost burden of environmental regulations as compared to the less-polluting industries, they have higher compliance costs ($c_2 > c_1$).

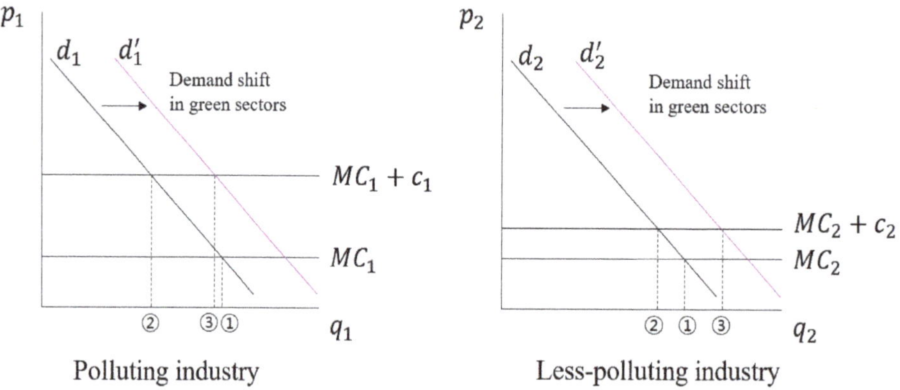


**Figure 2.** The effects of environmental regulations in industries differentiated by the level of pollution.

Although the contemporaneous effects of regulations on firms will be negative for their competitiveness (point ② in Figure 2), firms take optimal decisions to react to stricter environmental regulations. In this process, firms have the incentive to invest and develop new cost-efficient techniques which can reduce emissions at lower costs and subsequently off-set the direct negative effects of the policies (PH holds). Then increased pressure of the regulations in the past can have a positive impact on economic growth.

In research, the directions that indirect effects take in productivity, investments, and jobs have been controversial. However, if firms recover enough then their output will increase based on their improved productivity. In particular, strengthening regulations in the case of the green sector can lead to an increase in demand for environment-related goods and services over time by stimulating the need for green outputs in all the sectors. This is represented in Figure 2 by point ③. This upward demand shift (from $d_2$ to $d_2'$) should lead to an increase in output and employment.

If the industries can off-set their compliance costs (if PH holds), the regulation-induced losses in output and employment will dissolve. The green sector, in particular, could recover its losses faster than the non-green sector even in the polluting industries. It has an opportunity to expand its market under environmental regulations as compared to its non-green counterparts. Figure 2 explains this phenomenon with a demand shift excluding the price effect.

*4.2. Empirical Model*

This study categorizes the establishments into the environmental sector (green and non-green) and carbon dioxide emitting (polluting and non-polluting) industries for examining the effects of strengthened environmental regulations on employment and labor productivity. It allows estimation of the industrial sector's heterogeneity in response to tightened regulations in the Korean manufacturing industry. It estimates difference-in-differences regressions with three different specifications to shed light on the effects of tightened environmental regulations on the performance of the establishments.

The dependent variables ($\ln Y_{ijrt}$) are the logarithms of employment and labor productivity. Employment and labor productivity are often used for measuring a firm's performance. For calculating labor productivity, we use both output and value–added variables. The subscripts $i$, $j$, $r$, and $t$ denote establishment, industry, region, and year of observations respectively. The main estimated equation is specified as:

$$\ln Y_{ijrt} = \alpha + \beta\{\text{Post}_t \times \text{Poll}_j \times \text{Green}_{i\in S}\} + X\gamma + \mu_j + \theta_r + \lambda_t + \epsilon_{ijrt} \tag{1}$$

The purpose of the reduced form estimation is identifying the effects of environmental regulations in the context of LCGG using the DDD (difference-in-differences-in-differences) term. The variable $\text{Post}_t$ is an indicator of whether the year is after 2010 since the LCGG Act was enacted in January 2010. The Act on Low-Carbon Green Growth includes the following provisions: Reduction in the consumption of fossil fuels; reduction of GHG emissions by 27–30 percent by 2020 relative to the 'business as usual' of 2005; and increasing energy independence by using new and renewable energy sources. $\text{Poll}_j$ represents the polluting industry dummy variable and $\text{Green}_{i\in S}$ is another dummy variable which has a value of one if the establishment is included in the green sector or zero otherwise. So, the estimated coefficient of $\beta$ captures the asymmetric effects of environmental regulations between industries.

This study used three specifications. First, we did the DD (difference-in-differences) regression using interaction $\text{Post}_t \times \text{Poll}_j$ to confirm the effects of regulations on polluting industries; this has also been done by previous studies. Then, we compared the green and non-green sectors using $\text{Post}_t \times \text{Green}_{i\in S}$. And lastly, we applied Equation (1) to identify the regulations' effects on the dependent variables through an interaction among these three variables. The first and second terms of the DD term are all included and DD specifications are also saturated.

$X$ is a vector of other control variables such as $\text{HHI}_{jt}$, $\text{age}_{it}$, $\ln k_{it}$, and $\ln \text{wage}_{it}$. $\text{HHI}_{jt}$ is a concentration index measuring the degree of competition within the five-digit industries and $\text{age}_{it}$ is control for the size of a plant. $\ln k_{it}$ is the logarithm of capital intensity calculated by capital stock of labor and it is included in the labor productivity equation. We also use $\ln \text{wage}_{it}$ which is the logarithm of average wages of employees when estimating the employment equation. $\mu_j$, $\theta_r$, and $\lambda_t$ are industry, region, and time fixed effects. These fixed effects capture the unobservable industry, region, and time-specific characteristics. $\epsilon_{ijrt}$ is robust standard errors.

## 5. Results

This analysis focuses on examining how a plant responds to the adoption of new environmental regulations through the interaction term between the policy period's indicator and industry variables. Table 2 shows the results of labor productivity and employment estimations for polluting industries. Columns 1–3 give the results comparing six years before and after LCGG and Columns 4–six compare four-year effects. The DD terms in the table show significantly negative effects with the magnitude of the effects becoming larger when we consider a longer period. Labor productivity decreased about

5–6 percent and employment decreased 3 percent during the six years. These significantly negative effects were consistent when 2nd and 3rd groups in terciles were used instead of median to define the polluting industries. This implies that the increased costs induced by strengthened regulations led to a decrease in employment and productivity and this was not a temporary effect. This is consistent with previous studies [3,8,18].

**Table 2.** Estimation results based on labor productivity and employment as dependent variables for polluting industries.

| Dependent Variable | 2004–2009 vs. 2010–2015 | | | 2006–2009 vs. 2010–2013 | | |
|---|---|---|---|---|---|---|
| | ln(Vadd/L) | ln(Y/L) | ln(L) | ln(Vadd/L) | ln(Y/L) | ln(L) |
| Models | (1) | (2) | (3) | (4) | (5) | (6) |
| Post × Poll | −0.055 *** | −0.052 *** | −0.029 *** | −0.019 *** | −0.011 ** | −0.012 ** |
| | (0.003) | (0.004) | (0.004) | (0.004) | (0.005) | (0.005) |
| ln(K/L) | 0.181 *** | 0.249 *** | | 0.182 *** | 0.249 *** | |
| | (0.001) | (0.001) | | (0.001) | (0.001) | |
| ln(wage) | | | 0.357 *** | | | 0.360 *** |
| | | | (0.002) | | | (0.003) |
| Age | 0.004 *** | 0.004 *** | 0.018 *** | −0.023 | −0.02 | 0.045 |
| | (0.0001) | (0.0001) | (0.0001) | (0.027) | (0.032) | (0.030) |
| HHI | −0.024 | −0.019 | 0.001 | 0.004 *** | 0.005 *** | 0.019 *** |
| | (0.018) | (0.022) | (0.022) | (0.0001) | (0.0001) | (0.0001) |
| Constant | 3.433 *** | 4.206 *** | 2.226 *** | 3.515 *** | 4.319 *** | 2.185 *** |
| | (0.019) | (0.027) | (0.019) | (0.023) | (0.033) | (0.024) |
| Observations | 702,553 | 702,560 | 729,649 | 469,347 | 469,347 | 483,937 |
| R-squared | 0.324 | 0.392 | 0.264 | 0.315 | 0.389 | 0.267 |
| Industry FE | yes | yes | yes | yes | Yes | yes |
| Year FE | yes | yes | yes | yes | Yes | yes |
| Region FE | yes | yes | yes | yes | Yes | yes |

Notes: The robust standard errors are in parenthesis. Five-digit-level industry, region, and year fixed effects are used for all columns. Columns 1–3 give the results comparing the 2004–2009 and 2010–2015 periods and Columns 4–6 compare 4 years before and after LCGG. HHI is the Herfindahl Hirschman Index and K/L is capital intensity. Post indicates period after 2010 and Poll means polluting industry at the 2–3 digits level. The asterisks *** and ** indicate significant at the 1% and 5% levels of significance respectively.

In Table 3, we divide establishments into green and non-green sectors since we expect a positive demand shift to the green sector. This classification focuses on the product a plant supplies while polluting industries are defined as a sensitive group by the policy because of the emissions in their production processes. The estimated coefficients of interaction $Post_t \times Green_{i \in S}$ are positive and significant except for Column 6. This means that the green sector relatively experienced enhanced productivity and employment as compared to the non-green sector after LCGG.

Table 4 gives the results of the estimation of Equation (1). In terms of labor productivity, the DD terms show highly statistically significant effects. The green and non-green sectors in the polluting industries responded differently to the imposed regulations considering labor productivity. An establishment in a polluting industry had a decrease of about 6–7 percent in productivity after the regulations (Columns 1 and 2). However, these negative effects were off-set if the establishment was included in the green sector in the polluting industries. It also seems that there were a few productivity gains in this establishment. These positive effects on labor productivity in the green sector came from demand effects because we control the substitution between capital and labor using capital intensity. Hence, our results confirm that environment related industries are a key to understanding the consequences of environmental regulations. Although polluting industries face a higher burden because of the relatively higher compliance costs of the regulations, however, if they produce environment related goods or use environment related processes in their production, then these negative impacts could be canceled out.

**Table 3.** Estimation results based on labor productivity and employment as dependent variables for the green sector.

| Dependent Variable | 2004–2009 vs. 2010–2015 | | | 2006–2009 vs. 2010–2013 | | |
|---|---|---|---|---|---|---|
| | ln(Vadd/L) | ln(Y/L) | ln(L) | ln(Vadd/L) | ln(Y/L) | ln(L) |
| Models | (1) | (2) | (3) | (4) | (5) | (6) |
| Post × Green | 0.037 *** | 0.019 *** | 0.011 *** | 0.023 *** | 0.013 ** | 0.007 |
| | (0.004) | (0.004) | (0.004) | (0.005) | (0.005) | (0.005) |
| ln(K/L) | 0.183 *** | 0.252 *** | | 0.185 *** | 0.252 *** | |
| | (0.001) | (0.001) | | (0.001) | (0.001) | |
| ln(wage) | | | 0.378 *** | | | 0.381 *** |
| | | | (0.002) | | | (0.003) |
| Age | 0.004 *** | 0.005 *** | 0.019 *** | −0.025 | −0.02 | 0.046 |
| | (0.0001) | (0.0001) | (0.0001) | (0.027) | (0.032) | (0.030) |
| HHI | −0.025 | −0.017 | 0.01 | 0.004 *** | 0.005 *** | 0.019 *** |
| | (0.018) | (0.022) | (0.022) | (0.0001) | (0.0001) | (0.0001) |
| Constant | 3.421 *** | 4.186 *** | 2.135 *** | 3.499 *** | 4.297 *** | 2.096 *** |
| | (0.019) | (0.027) | (0.019) | (0.023) | (0.033) | (0.024) |
| Observations | 702,553 | 702,560 | 729,649 | 469,347 | 469,347 | 483,937 |
| R-squared | 0.322 | 0.39 | 0.252 | 0.312 | 0.386 | 0.255 |
| Industry FE | Yes | yes | yes | yes | Yes | yes |
| Year FE | Yes | yes | yes | yes | Yes | yes |
| Region FE | Yes | yes | yes | yes | Yes | yes |

Notes: The robust standard errors are in parenthesis. Five-digit level industry, region, and year fixed effects are used for all columns. Columns 1–3 give the results comparing the 2004–2009 and 2010–2015 periods and Columns 4–6 compare four years before and after LCGG. HHI is the Herfindahl Hirschman Index and K/L is capital intensity. Post indicates the period after 2010 and Green means environmental industry at the five-digit level. The asterisks *** and ** indicate significant at the 1% and 5% levels of significance respectively.

**Table 4.** Estimation results based on labor productivity and employment as dependent variables for the green sector in polluting industries.

| Dependent Variable | 2004–2009 vs. 2010–2015 | | | 2006–2009 vs. 2010–2013 | | |
|---|---|---|---|---|---|---|
| | ln(Vadd/L) | ln(Y/L) | ln(L) | ln(Vadd/L) | ln(Y/L) | ln(L) |
| Models | (1) | (2) | (3) | (4) | (5) | (6) |
| Post × Green × Poll | 0.040 *** | 0.069 *** | −0.008 | 0.025 ** | 0.044 *** | −0.004 |
| | (0.008) | (0.009) | (0.010) | (0.010) | (0.011) | (0.012) |
| Post × Poll | −0.068 *** | −0.071 *** | −0.028 *** | −0.027 *** | −0.023 *** | −0.012 ** |
| | (0.004) | (0.005) | (0.005) | (0.005) | (0.006) | (0.006) |
| Post × Green | 0.029 *** | 0.002 | 0.012 *** | 0.017 *** | −0.0001 | 0.006 |
| | (0.004) | (0.005) | (0.004) | (0.005) | (0.006) | (0.005) |
| ln(K/L) | 0.181 *** | 0.249 *** | | 0.182 *** | 0.249 *** | |
| | (0.001) | (0.001) | | (0.001) | (0.001) | |
| ln(wage) | | | 0.356 *** | | | 0.360 *** |
| | | | (0.002) | | | (0.003) |
| Age | 0.004 *** | 0.004 *** | 0.018 *** | −0.027 | −0.022 | 0.043 |
| | (0.0001) | (0.0001) | (0.0001) | (0.027) | (0.032) | (0.030) |
| HHI | −0.029 | −0.021 | −0.0005 | 0.004 *** | 0.005 *** | 0.019 *** |
| | (0.018) | (0.022) | (0.022) | (0.0001) | (0.0001) | (0.0001) |
| Constant | 3.435 *** | 4.206 *** | 2.227 *** | 3.517 *** | 4.319 *** | 2.186 *** |
| | (0.019) | (0.027) | (0.019) | (0.023) | (0.033) | (0.024) |
| Observations | 702,553 | 702,560 | 729,649 | 469,347 | 469,347 | 483,937 |
| R-squared | 0.324 | 0.393 | 0.264 | 0.315 | 0.389 | 0.267 |
| Industry FE | yes | yes | Yes | yes | Yes | Yes |
| Year FE | yes | yes | Yes | yes | Yes | Yes |
| Region FE | yes | yes | Yes | yes | Yes | Yes |

Notes: The robust standard errors are in parenthesis. Five-digit level industry, region, and year fixed effects are used for all columns. Columns 1–3 give the results comparing the 2004–2009 and 2010–2015 periods and Columns 4–6 compare 4 years before and after LCGG. HHI is the Herfindahl Hirschman Index and K/L is capital intensity. Post indicates the period after 2010 and Green means environmental industry at the five-digit level. Poll is dummy of polluting industry at the 2–3 digit level. The asterisks *** and ** indicate significant at the 1% and 5% levels of significance respectively.

In Columns 3 and 6, we use total employment as the dependent variable. The estimated coefficients of the DD terms are negative and not significant. This means that the green and non-green sectors in the polluting industries are not different considering the intensity of the regulations' effects. Plants in polluting industries reduce their employment but this reduction is not recovered within the industry and jobs increase in plants, including those in green and less polluting industries. This evidence implies a trade-off between green and non-green sectors within the brown industries only in labor productivity. And establishments in the green sector in less-polluting industries have a marginal increase in employment (Column 3). The main results are consistent when another criterion is used for a Poll indicator, but the off-set to a loss captured by DDD terms diminishes.

## 6. Conclusions

This study examined the effects of strengthened environmental regulations on employment and labor productivity by dividing the industries based on two criteria: environmental classification (green and non-green sectors) and carbon dioxide emissions (polluting and less-polluting industries). It used establishment-level panel data from 2004 to 2015 including the period in which the Government of Korea implemented the Low-Carbon Green Growth Act. Our empirical results lead to several conclusions. First, environmental regulations have negative effects on economic outcomes in polluting industries. Second, these effects are asymmetric between green and non-green sectors and this taxonomy is as important as whether a firm emits a larger quantity of $CO_2$ emissions which has been the main focus of previous studies. Finally, the positive effects in the green sector can off-set a part of the negative effects of the polluting industries. This suggests that plants producing environmental goods in polluting industries were not hit hard by the environmental regulations.

This study contributes to environmental policy related research as it not only considers the effects of the regulations on the manufacturing industry but also considers the asymmetric effects between the green and non-green sectors. In Korea, the environmental industry has attracted attention as the driving force for new growth in the low-carbon green growth policy [34]. As environmental regulations have been strengthened internationally, the need for eco-friendly processes and environmental goods has increased. Environmental regulations or pressures could stimulate job creation and labor productivity in the green sector or at least these will not hinder the performance of this group even though it emits substantial greenhouse gases. Nonetheless, the emitters in the non-green sector suffered because of the regulations and their productivity reduced. Therefore, restrictions and subsidies should be applied to firms considering their specific characteristics and GHG emissions.

This research has some limitations. First, it does not capture labor reallocations between industries or within local areas. Second, it uses 2–3 digit level data during 2012–15 to define polluting industries based on their $CO_2$ emissions. Accounting for other pollutants, such as $SO_2$ and use of an expanded dataset at the firm level will be informative. These limitations of the dataset may have hampered an accurate estimation of the effects. The empirical method in the paper was conducted without consideration of the firm dynamics accounting for the entry and exit of establishments. If future researches can capture input reallocation and decompose it between intensive and extensive margins, this enables distinguishing the differences in the regulations' effects on environmental and non-environmental sectors more clearly. In an attempt to reduce uncertainty, given data availability, future research can conduct more systematic sensitivity analysis of the results by accounting for the weaknesses listed above.

**Author Contributions:** Conceptualization, S.Y. and A.H.; methodology, S.Y. and A.H.; STATA software, S.Y.; validation, S.Y. and A.H.; formal analysis, S.Y. and A.H.; investigation, S.Y.; resources, S.Y.; data curation, S.Y.; writing—original draft preparation, S.Y.; writing—review and editing, A.H.; visualization, S.Y.; supervision, A.H.; project administration, A.H.; no funding acquired.

**Funding:** This research received no external funding.

**Acknowledgments:** The authors are grateful to the Department of Economics, Sogang University for facilitating access to industry data. We are grateful to the three anonymous referees and an editor of the journal for their comments and suggestions on an earlier version of this paper.

**Conflicts of Interest:** The authors declare no conflict of interest.

## References


1. Dechezleprêtre, A.; Sato, M. The impacts of environmental regulations on competitiveness. *Rev. Environ. Econ. Policy* **2017**, *11*, 183–206. [CrossRef]
2. Ministry of Government Legislation. Framework Act on Low Carbon, Green Growth. Available online: http://www.moleg.go.kr/english/korLawEng?pstSeq=54792 (accessed on 20 May 2019).
3. Greenstone, M. The impacts of environmental regulations on industrial activity: Evidence from the 1970 and 1977 clean air act amendments and the census of manufactures. *J. Polit. Econ.* **2002**, *110*, 1175–1219. [CrossRef]
4. Porter, M.E.; van der Linde, C. Green and competitive: Ending the stalemate. *Harv. Bus. Rev.* **1995**, *73*, 120–134.
5. Morgenstern, R.D.; Pizer, W.A.; Shih, J.S. Jobs versus the environment: An industry-level perspective. *J. Environ. Econ. Manag.* **2002**, *43*, 412–436. [CrossRef]
6. Lanoie, P.; Patry, M.; Lajeunesse, R. Environmental regulation and productivity: Testing the porter hypothesis. *J. Prod. Anal.* **2008**, *30*, 121–128. [CrossRef]
7. Coglianese, C.; Carrigan, C. The Jobs and Regulation Debate. In *Does Regulation Kill Jobs?* University of Pennsylvania Press: Philadelphia, PA, USA, 2014; pp. 1–30.
8. Walker, W.R. The transitional costs of sectoral reallocation: Evidence from the clean air act and the workforce. *Q. J. Econ.* **2013**, *128*, 1787–1835. [CrossRef]
9. Kim, J.H.; Ha, B.C. The Effects of Environmental Regulations on the Industrial Productivity: An Empirical Analysis of the Korean Case. *J. Ind. Econ. Bus.* **2012**, *25*, 1711–1727. Available online: https://www.kci.go.kr/kciportal/ci/sereArticleSearch/ciSereArtiView.kci?sereArticleSearchBean.artiId=ART001655613 (accessed on 31 May 2019).
10. Gray, W.B.; Shadbegian, R.J. Environmental regulation, investment timing, and technology choice. *J. Ind. Econ.* **1998**, *46*, 235–256. [CrossRef]
11. Popp, D. Pollution control innovations and the Clean Air Act of 1990. *J. Policy Anal. Manag.* **2003**, *22*, 641–660. [CrossRef]
12. Aldy, J.E.; Pizer, W.A. The competitiveness impacts of climate change mitigation policies. *J. Assoc. Environ. Resour. Econ.* **2015**, *2*, 565–595. [CrossRef]
13. Jaffe, A.B.; Palmer, K. Environmental regulation and innovation: A panel data study. *Rev. Econ. Stat.* **1997**, *79*, 610–619. [CrossRef]
14. Johnstone, N.; Haščič, I.; Popp, D. Renewable energy policies and technological innovation: Evidence based on patent counts. *Environ. Resour. Econ.* **2010**, *45*, 133–155. [CrossRef]
15. Rubashkina, Y.; Galeotti, M.; Verdolini, E. Environmental regulation and competitiveness: Empirical evidence on the Porter Hypothesis from European manufacturing sectors. *Energy Policy* **2015**, *83*, 288–300. [CrossRef]
16. Yang, C.H.; Tseng, Y.H.; Chen, C.P. Environmental regulations, induced R&D, and productivity: Evidence from Taiwan's manufacturing industries. *Resour. Energy Econ.* **2012**, *34*, 514–532.
17. Molina-Azorín, J.F.; Claver-Cortés, E.; Pereira-Moliner, J.; Tarí, J.J. Environmental practices and firm performance: An empirical analysis in the Spanish hotel industry. *J. Clean. Prod.* **2009**, *17*, 516–524. [CrossRef]
18. Wagner, M.; Van Phu, N.; Azomahou, T.; Wehrmeyer, W. The relationship between the environmental and economic performance of firms: An empirical analysis of the European paper industry. *Corp. Soc. Responsib. Environ. Manag.* **2002**, *9*, 133–146. [CrossRef]
19. Yuan, B.; Xiang, Q. Environmental regulation, industrial innovation and green development of Chinese manufacturing: Based on an extended CDM model. *J. Clean. Prod.* **2018**, *176*, 895–908. [CrossRef]
20. Rexhäuser, S.; Rammer, C. Environmental innovations and firm profitability: Unmasking the Porter hypothesis. *Environ. Resour. Econ.* **2014**, *57*, 145–167. [CrossRef]

21. KozIuk, T.; Zipperer, V. Environmental Policies and Productivity Growth: A Critical Review of Empirical Findings. *OECD J. Econ. Stud.* **2015**, *2014*, 155–185.
22. Wang, Y.; Shen, N. Environmental regulation and environmental productivity: The case of China. *Renew. Sustain. Energy Rev.* **2016**, *62*, 758–766. [CrossRef]
23. Kneller, R.; Manderson, E. Environmental regulations and innovation activity in UK manufacturing industries. *Resour. Energy Econ.* **2012**, *34*, 211–235. [CrossRef]
24. Alpay, E.; Kerkvliet, J.; Buccola, S. Productivity growth and environmental regulation in Mexican and US food manufacturing. *Am. J. Agric. Econ.* **2002**, *84*, 887–901. [CrossRef]
25. Kunapatarawong, R.; Martínez-Ros, E. Towards green growth: How does green innovation affect employment? *Res. Policy.* **2016**, *45*, 1218–1232. [CrossRef]
26. Yamazaki, A. Jobs and climate policy: Evidence from British Columbia's revenue-neutral carbon tax. *J. Environ. Econ. Manag.* **2017**, *83*, 197–216. [CrossRef]
27. Kang, M.O.; Lee, S.Y. A panel data analysis of the effects of environmental regulation on competitiveness of the Korean manufacturing sector. *J. Environ. Policy Adm.* **2006**, *14*, 169–193. Available online: https://www.kci.go.kr/kciportal/ci/sereArticleSearch/ciSereArtiView.kci?sereArticleSearchBean.artiId=ART001011530 (accessed on 31 May 2019).
28. Kang, M.O.; Jo, Y.H. Effects of environmental policies on the creation of jobs in Korea. Korean Environment Institute. 2015. Available online: http://kiss.kstudy.com/thesis/thesis-view.asp?key=3425980 (accessed on 31 May 2019).
29. Jung, Y.M. Is South Korea's Green Job Policy Sustainable? *Sustainability* **2015**, *7*, 8748–8767. [CrossRef]
30. Lee, H.S.; Choi, Y. Environmental Performance Evaluation of the Korean Manufacturing Industry Based on Sequential DEA. *Sustainability* **2019**, *11*, 874. [CrossRef]
31. Greenhouse Gas Inventory and Research Center. 2018 National Inventory Report (NIR). 2018. Available online: http://www.gir.go.kr/home/board/read.do?pagerOffset=0&maxPageItems=10&maxIndexPages=10&searchKey=&searchValue=&menuId=36&boardId=43&boardMasterId=2&boardCategoryId= (accessed on 20 May 2019).
32. Yun, S.J. The Ideological Basis and the Reality of "Low Carbon Green Growth". *Korean Assoc. Environ. Sociol.* **2009**, *13*, 219–266. Available online: https://www.kci.go.kr/kciportal/ci/sereArticleSearch/ciSereArtiView.kci?sereArticleSearchBean.artiId=ART001361435 (accessed on 31 May 2019).
33. Statistics Korea. Report on the Environment Industry Survey. 2015. Available online: http://www.me.go.kr/home/web/policy_data/read.do;jsessionid=SX4NjLTdCU8q5dq7UX3d-3-b.mehome1?pagerOffset=150&maxPageItems=10&maxIndexPages=10&searchKey=&searchValue=&menuId=10259&orgCd=&condition.orderSeqId=6643&condition.rnSeq=417&condition.deleteYn=N&seq=7148 (accessed on 20 May 2019).
34. Jones, R.S.; Yoo, B. *Korea's Green Growth Strategy: Mitigating Climate Change and Developing New Growth Engines*; OECD Economic Department Working Papers; OECD Publishing: Paris, France, 2011.

MDPI

St. Alban-Anlage 66

4052 Basel

Switzerland

Tel. +41 61 683 77 34

Fax +41 61 302 89 18

www.mdpi.com

*Energies* Editorial Office

E-mail: energies@mdpi.com

www.mdpi.com/journal/energies